A Step-by-Step Approach to Using SAS® for

Factor Analysis and Structural Equation Modeling

Second Edition

Norm O'Rourke and Larry Hatcher

support.sas.com/bookstore

The correct bibliographic citation for this manual is as follows: O'Rouke, Norm, Ph.D. and Hatcher, Larry, Ph.D. 2013. *A Step-by-Step Approach to Using SAS® for Factor Analysis and Structural Equation Modeling, Second Edition*. Cary, NC: SAS Institute Inc.

A Step-by-Step Approach to Using SAS® for Factor Analysis and Structural Equation Modeling, Second Edition

ISBN 978-1-59994-230-8

SAS Institute Inc., SAS Campus Drive, Cary, North Carolina 27513-2414.

February 2014

I dedicate this book to my parents, who worked hard and sacrificed so that I would have the opportunities that they never had.

L.H.

Contents

About This Book

Purpose

This book provides a comprehensive introduction to many of the statistical procedures most common in social science research today. We describe these statistical procedures in detail and list the mathematical assumptions underpinning these statistical procedures. Moreover, we progress step-to-step through detailed examples, provide the code and output, and interpret the results. We also provide examples that show how to summarize and describe study findings for written research reports.

Is This Book for You?

This book is intended for senior undergraduate and graduate statistics courses—for those users with and without prior SAS exposure—and for those users with and without prior statistics knowledge. The core content is described in detail in the book's chapters; yet for those users with no prior SAS knowledge, we provide several appendices that describe the basics of working with SAS (e.g., working with data files, raw data, correlation, and covariance matrices).

Prerequisites

There are few prerequisites for this book. Appendices at the end of this book provide the novice SAS user with foundational information that is required to begin working with SAS. Even without extensive prior experience, users of this book can learn the basics of factor analyses, path analyses, and structural equation modeling (SEM).

What's New in This Edition

In this second edition, we include an extended discussion of statistical power analyses and sample size requirements for path analyses, confirmatory factor analyses (CFA), and SEM. More precisely, we provide an easy-to-use table to help users determine sample size requirements for path analyses. With latent variable models (e.g., CFA and SEM), we provide SAS code to estimate statistical power. We also provide SAS code to calculate sample size requirements when planning your research to ensure that you will have sufficient statistical power when later conducting these analyses.

Additionally, we describe contemporary goodness-of-fit statistics (and threshold values) to examine when reporting CFA and SEM results, describe how and when to revise hypothesized models, and identify procedures to follow when selecting which goodness-of-fit indices to report.

About the Examples

Software Used to Develop the Book's Content

The examples in this book were computed using SAS 9.3. We walk the user through examples using PROC FACTOR, PROC CORR, and PROC CALIS.

The data and programs used in this book are available from the authors' pages at http://support.sas.com/orourke and http://support.sas.com/hatcher.

Example Code and Data

You can access the example code and data for this book by linking to its authors' pages at http://support.sas.com/orourke and http://support.sas.com/hatcher. Look for the cover thumbnail of this book, and select Example Code and Data to display the SAS programs that are included for this book.

For an alphabetical listing of all books for which example code and data are available, see http://support.sas.com/bookcode. Select a title to display the book's example code.

If you are unable to access the code through the Web site, send an e-mail to saspress@sas.com.

Additional Resources

SAS offers you a rich variety of resources to help build your SAS skills and explore and apply the full power of SAS software. Whether you are in a professional or academic setting, we have learning products that can help you maximize your investment in SAS.

Bookstore	http://support.sas.com/bookstore/
Training	http://support.sas.com/training/
Certification	http://support.sas.com/certify/
SAS Global Academic Program	http://support.sas.com/learn/ap/
SAS OnDemand	http://support.sas.com/learn/ondemand/

Or

Knowledge Base	http://support.sas.com/resources/
Support	http://support.sas.com/techsup/
Training and Bookstore	http://support.sas.com/learn/
Community	http://support.sas.com/community/

Keep in Touch

We look forward to hearing from you. We invite questions, comments, and concerns. If you want to contact us about a specific book, please include the book title in your correspondence.

To Contact the Author through SAS Press

By e-mail: saspress@sas.com

Via the Web: http://support.sas.com/author_feedback

SAS Books

For a complete list of books available through SAS, visit http://support.sas.com/bookstore.

Phone: 1-800-727-3228

Fax: 1-919-677-8166

E-mail: sasbook@sas.com

SAS Book Report

Receive up-to-date information about all new SAS publications via e-mail by subscribing to the SAS Book Report monthly eNewsletter. Visit http://support.sas.com/sbr.

About These Authors

Norm O'Rourke, Ph.D., R.Psych., is a clinical psychologist and associate professor with the Interdisciplinary Research in the Mathematical and Computational Sciences (IRMACS) Centre at Simon Fraser University in Burnaby (BC), Canada. He sits on the executive board of the American Psychological Association's Society for Clinical Geropsychology and the National Mental Health Commission of Canada. To date, he has published two governmental reports and seventy peer-reviewed publications in leading gerontology, measurement, and mental health academic journals. As co-applicant, Dr. O'Rourke has been part of teams awarded $4M in research funding, and $1.3M as principal applicant in governmental and foundation funding as team leader.

Larry Hatcher, Ph.D., is a professor of psychology at Saginaw Valley State University in Saginaw, Michigan, where he teaches classes in general psychology, industrial psychology, elementary statistics, advanced statistics, and computer applications in data analysis. The author of several books dealing with statistics and data analysis, Hatcher has taught at the college level since 1984 after earning his doctorate in industrial and organizational psychology from Bowling Green State University in 1983.

Learn more about these authors by visiting their author pages, where you can download free book excerpts, access example code and data, read the latest reviews, get updates, and more:

http://support.sas.com/orourke

http://support.sas.com/hatcher

Acknowledgments from the First Edition

I learned about structural equation modeling while on a sabbatical at Bowling Green State University during the 1990–1991 academic year. My thanks to Joe Cranny, who was chair of the Psychology Department at BGSU at the time, and who helped make the sabbatical possible.

My department chair, Mel Goldstein, encouraged me to complete this book and made many accommodations in my teaching schedule so that I would have time to do so. My friend and department colleague, Heidar Modaresi, encouraged me to begin this project and offered useful comments on how to proceed. My secretary, Cathy Carter, eased my workload by performing many helpful tasks. My friend Nancy Stepanski edited an early draft of Chapter 4, and provided many constructive comments that helped shape the final book. My thanks to all. Special thanks to my wife, Ellen, who, as usual, offered encouragement and support every step of the way.

Many people at SAS Institute were very helpful in reviewing and editing chapters, and in answering hundreds of questions. These include David Baggett, Jennifer Ginn, Jeff Lopes, Blanche Phillips, Jim Ashton, Cathy Maahs-Fladung, and David Teal. All of these were consistently positive, patient, and constructive, and I appreciate their contributions.

L.H.

Chapter 1: Principal Component Analysis

Introduction: The Basics of Principal Component Analysis

Principal component analysis is used when you have obtained measures for a number of observed variables and wish to arrive at a smaller number of variables (called "principal components") that will account for, or capture, most of the variance in the observed variables. The principal components may then be used as predictors or criterion variables in subsequent analyses.

A Variable Reduction Procedure

Principal component analysis is a variable reduction procedure. It is useful when you have obtained data for a number of variables (possibly a large number of variables) and believe that there is redundancy among those variables. In this case, redundancy means that some of the variables are correlated with each other, often because they are measuring the same construct. Because of this redundancy, you believe that it should be possible to reduce the observed variables into a smaller number of principal components that will account for most of the variance in the observed variables.

Because it is a variable reduction procedure, principal component analysis is similar in many respects to exploratory factor analysis. In fact, the steps followed when conducting a principal component analysis are virtually identical to those followed when conducting an exploratory factor analysis. There are significant conceptual differences between the two, however, so it is important that you do not mistakenly claim that you are performing factor analysis when you are actually performing principal component analysis. The differences between these two procedures are described in greater detail in a later subsection titled "Principal Component Analysis Is *Not* Factor Analysis."

An Illustration of Variable Redundancy

We now present a fictitious example to illustrate the concept of variable redundancy. Imagine that you have developed a seven-item measure to gauge job satisfaction. The fictitious instrument is reproduced here:

Please respond to the following statements by placing your response to the left of each statement. In making your ratings, use a number from 1 to 7 in which 1 = "Strongly Disagree" and 7 = "Strongly Agree."

_____ 1. My supervisor(s) treats me with consideration.
_____ 2. My supervisor(s) consults me concerning important decisions that affect my work.
_____ 3. My supervisor(s) gives me recognition when I do a good job.
_____ 4. My supervisor(s) gives me the support I need to do my job well.
_____ 5. My pay is fair.
_____ 6. My pay is appropriate, given the amount of responsibility that comes with my job.
_____ 7. My pay is comparable to that of other employees whose jobs are similar to mine.

Perhaps you began your investigation with the intention of administering this questionnaire to 200 employees using their responses to the seven items as seven separate variables in subsequent analyses.

There are a number of problems with conducting the study in this manner, however. One of the more important problems involves the concept of redundancy as previously mentioned. Examine the content of the seven items in the questionnaire. Notice that items 1 to 4 each deal with employees' satisfaction with their supervisors. In this way, items 1 to 4 are somewhat redundant or overlapping in terms of what they are measuring. Similarly, notice that items 5 to 7 each seem to deal with the same topic: employees' satisfaction with their pay.

Statistical findings may further support the likelihood of item redundancy. Assume that you administer the questionnaire to 200 employees and compute all possible correlations between responses to the seven items. Fictitious correlation coefficients are presented in Table 1.1:

Table 1.1: Correlations among Seven Job Satisfaction Items

	Correlations						
Variable	**1**	**2**	**3**	**4**	**5**	**6**	**7**
1	1.00						
2	.75	1.00					
3	.83	.82	1.00				
4	.68	.92	.88	1.00			
5	.03	.01	.04	.01	1.00		
6	.05	.02	.05	.07	.89	1.00	
7	.02	.06	.00	.03	.92	.76	1.00

NOTE: $N = 200$.

When correlations among several variables are computed, they are typically summarized in the form of a **correlation matrix** such as the one presented in Table 1.1; this provides an opportunity to review how a correlation matrix is interpreted. (See Appendix A.5 for more information about correlation coefficients.)

The rows and columns of Table 1.1 correspond to the seven variables included in the analysis. Row 1 (and column 1) represents variable 1, row 2 (and column 2) represents variable 2, and so forth. Where a given row and column intersect, you will find the correlation coefficient between the two corresponding variables. For example, where the row for variable 2 intersects with the column for variable 1, you find a coefficient of .75; this means that the correlation between variables 1 and 2 is .75.

The correlation coefficients presented in Table 1.1 show that the seven items seem to *hang together* in two distinct groups. First, notice that items 1 to 4 show relatively strong correlations with each another. This could be because items 1 to 4 are measuring the same construct. In the same way, items 5 to 7 correlate strongly with one another, a possible indication that they also measure a single construct. Even more interesting, notice that items 1 to 4 are very weakly correlated with items 5 to 7. This is what you would expect to see if items 1 to 4 and items 5 to 7 were measuring two different constructs.

Given this apparent redundancy, it is possible that the seven questionnaire items are not really measuring seven different constructs. More likely, items 1 to 4 are measuring a single construct that could reasonably be labeled "satisfaction with supervision," whereas items 5 to 7 are measuring a different construct that could be labeled "satisfaction with pay."

If responses to the seven items actually display redundancy as suggested by the pattern of correlations in Table 1.1, it would be advantageous to reduce the number of variables in this dataset, so that (in a sense) items 1 to 4 are collapsed into a single new variable that reflects employees' satisfaction with supervision and items 5 to 7 are collapsed into a single new variable that reflects satisfaction with pay. You could then use these two new variables (rather than the seven original variables) as predictor variables in multiple regression, for instance, or another type of statistical analysis.

In essence, this is what is accomplished by principal component analysis: it allows you to reduce a set of observed variables into a smaller set of variables called principal components. The resulting principal components may then be used in subsequent analyses.

What Is a Principal Component?

How Principal Components Are Computed

A **principal component** can be defined as a linear combination of optimally weighted observed variables. In order to understand the meaning of this definition, it is necessary to first describe how participants' scores on a principal component are computed.

In the course of performing a principal component analysis, it is possible to calculate a score for each participant for each principal component. In the preceding study, for example, each participant would have scores on two components: one score on the "satisfaction with supervision" component; and one score on the "satisfaction with pay" component. Participants' actual scores on the seven questionnaire items would be optimally weighted and then summed to compute their scores for a given component.

Below is the general formula to compute scores on the first component extracted (created) in a principal component analysis:

$$C_1 = b_{11}(X_1) + b_{12}(X_2) + \ldots b_{1p}(X_p)$$

where

C_1 = the participant's score on principal component 1 (the first component extracted)

b_{1p} = the coefficient (or weight) for observed variable p, as used in creating principal component 1

X_p = the participant's score on observed variable p

For example, assume that component 1 in the present study was "satisfaction with supervision." You could determine each participant's score on principal component 1 by using the following fictitious formula:

$$C_1 = .44\ (X_1) + .40\ (X_2) + .47\ (X_3) + .32\ (X_4)$$

$$+ .02\ (X_5) + .01\ (X_6) + .03\ (X_7)$$

In this case, the observed variables (the "X" variables) are participant responses to the seven job satisfaction questions: X_1 represents question 1; X_2 represents question 2; and so forth. Notice that different coefficients or weights were assigned to each of the questions when computing scores on component 1: questions 1 to 4 were assigned relatively large weights that range from .32 to .47, whereas questions 5 to 7 were assigned very small weights ranging from .01 to .03. This makes sense, because component 1 is the satisfaction with supervision component and satisfaction with supervision was measured by questions 1 to 4. It is therefore appropriate that items 1 to 4 would be given a good deal of weight in computing participant scores on this component, while items 5 to 7 would be given comparatively little weight.

Because component 2 measures a different construct, a different equation with different weights would be used to compute scores for this component (i.e., "satisfaction with pay"). Below is a fictitious illustration of this formula:

$$C_2 = .01\ (X_1) + .04\ (X_2) + .02\ (X_3) + .02\ (X_4)$$

$$+ .48\ (X_5) + .31\ (X_6) + .39\ (X_7)$$

The preceding example shows that, when computing scores for the second component, considerable weight would be given to items 5 to 7, whereas comparatively little would be given to items 1 to 4. As a result, component 2 should account for much of the variability in the three satisfaction with pay items (i.e., it should be strongly correlated with those three items).

But how are these weights for the preceding equations determined? PROC FACTOR in SAS generates these weights by using what is called an **eigenequation**. The weights produced by these eigenequations are optimal weights in the sense that, for a given set of data, no other set of weights could produce a set of components that are more effective in accounting for variance among observed variables. These weights are created to satisfy what is known as *the principle of least squares*. Later in this chapter we will show how PROC FACTOR can be used to extract (create) principal components.

It is now possible to understand the definition provided at the beginning of this section more fully. A principal component was defined as a linear combination of optimally weighted observed variables. The words "linear combination" mean that scores on a component are created by adding together scores for the observed variables being analyzed. "Optimally weighted" means that the observed variables are weighted in such a way that the resulting components account for a maximal amount of observed variance in the dataset.

Number of Components Extracted

The preceding section may have created the impression that, if a principal component analysis were performed on data from our fictitious seven-item job satisfaction questionnaire, only two components would be created. Such an impression would not be entirely correct.

In reality, the number of components extracted in a principal component analysis is equal to the number of observed variables being analyzed. This means that an analysis of responses to the seven-item questionnaire would actually result in seven components, not two.

In most instances, however, only the first few components account for meaningful amounts of variance; only these first few components are retained, interpreted, and used in subsequent analyses. For example, in your analysis of the seven-item job satisfaction questionnaire, it is likely that only the first two components would account for, or capture, meaningful amounts of variance. Therefore, only these would be retained for interpretation. You could assume that the remaining five components capture only trivial amounts of variance. These latter components would therefore not be retained, interpreted, or further analyzed.

Characteristics of Principal Components

The first component extracted in a principal component analysis accounts for a maximal amount of total variance among the observed variables. Under typical conditions, this means that the first component will be correlated with at least some (often many) of the observed variables.

The second component extracted will have two important characteristics. First, this component will account for a maximal amount of variance in the dataset that was not accounted for or captured by the first component. Under typical conditions, this again means that the second component will be correlated with some of the observed variables that did not display strong correlations with component 1.

The second characteristic of the second component is that it will be uncorrelated with the first component. Literally, if you were to compute the correlation between components 1 and 2, that coefficient would be zero. (For the exception, see the following section regarding oblique solutions.)

The remaining components that are extracted exhibit the same two characteristics: each accounts for a maximal amount of variance in the observed variables that was not accounted for by the preceding components; and each is uncorrelated with all of the preceding components. Principal component analysis proceeds in this manner with each new component accounting for progressively smaller amounts of variance. This is why only the first few components are retained and interpreted. When the analysis is complete, the resulting components will exhibit varying degrees of correlation with the observed variables, but will be uncorrelated with each another.

What is meant by "total variance" in the dataset? To understand the meaning of "total variance" as it is used in a principal component analysis, remember that the observed variables are standardized in the course of the analysis. This means that each variable is transformed so that it has a mean of zero and a standard deviation of one (and hence a variance of one). The "total variance" in the dataset is simply the sum of variances for these observed variables. Because they have been standardized to have a standard deviation of one, each observed variable contributes one unit of variance to the total variance in the dataset. Because of this, total variance in principal component analysis will always be equal to the number of observed variables analyzed. For example, if seven variables are being analyzed, the total variance will equal seven. The components that are extracted in the analysis will partition this variance. Perhaps the first component will account for 3.2 units of total variance; perhaps the second component will account for 2.1 units. The analysis continues in this way until all variance in the dataset has been accounted for or explained.

Orthogonal versus Oblique Solutions

This chapter will discuss only principal component analyses that result in orthogonal solutions. An **orthogonal solution** is one in which the components are uncorrelated ("orthogonal" means uncorrelated).

It is possible to perform a principal component analysis that results in correlated components. Such a solution is referred to as an **oblique solution**. In some situations, oblique solutions are preferred to orthogonal solutions because they produce cleaner, more easily interpreted results.

However, oblique solutions are often complicated to interpret. For this reason, this chapter will focus only on the interpretation of orthogonal solutions. The concepts discussed will provide a good foundation for the somewhat more complex concepts discussed later in this text.

Principal Component Analysis Is Not Factor Analysis

Principal component analysis is commonly confused with factor analysis. This is understandable because there are many important similarities between the two. Both are methods that can be used to identify groups of observed variables that tend to hang together empirically. Both procedures can also be performed with PROC FACTOR, and they generally provide similar results.

Nonetheless, there are some important conceptual differences between principal component analysis and factor analysis that should be understood at the outset. Perhaps the most important difference deals with the **assumption of an underlying causal structure**. Factor analysis assumes that covariation among the observed variables is due to the presence of one or more latent variables that exert directional influence on these observed variables. An example of such a structure is presented in Figure 1.1.

Figure 1.1: Example of the Underlying Causal Structure That Is Assumed in Factor Analysis

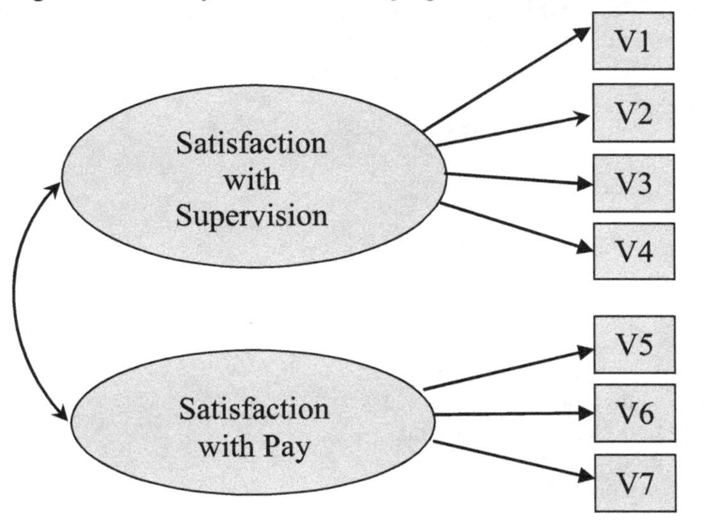

The ovals in Figure 1.1 represent the latent (unmeasured) factors of "satisfaction with supervision" and "satisfaction with pay." These factors are latent in the sense that it is assumed employees hold these beliefs but that these beliefs cannot be measured directly; however, they do influence employees' responses to the items that constitute the job satisfaction questionnaire described earlier. (These seven items are represented as the squares labeled V1 to V7 in the figure.) It can be seen that the "supervision" factor exerts influence on items V1 to V4 (the supervision questions), whereas the "pay" factor exerts influence on items V5 to V7 (the pay items).

Researchers use factor analysis when they believe that one or more unobserved or latent factors exert directional influence on participants' responses. Exploratory factor analysis helps the researcher identify the number and nature of such latent factors. These procedures are described in the next chapter.

In contrast, principal component analysis makes no assumptions about underlying causal structures; it is simply a variable reduction procedure that (typically) results in a relatively small number of components accounting for, or capturing, most variance in a set of observed variables (i.e., groupings of observed variables versus latent constructs).

Another important distinction between the two is that principal component analysis assumes no measurement error whereas factor analysis separates true variance from measurement error. Acknowledging measurement error is particularly germane to social science research because our instruments are invariably incomplete measures of underlying constructs. Principal component analysis is sometimes used in instrument construction studies to overestimate precision of measurement (i.e., overestimate the effectiveness of the scale).

In summary, both factor analysis and principal component analysis are important in social science research, but their conceptual foundations are quite distinct.

Example: Analysis of the Prosocial Orientation Inventory

Assume that you have developed an instrument called the Prosocial Orientation Inventory (POI) that assesses the extent to which a person has engaged in helping behaviors over the preceding six months. This fictitious instrument contains six items and is presented here:

Instructions: Below are a number of activities in which people sometimes engage. For each item, please indicate how frequently you have engaged in this activity over the past six months. Provide your response by circling the appropriate number to the left of each item using the response key below:

7 = Very Frequently
6 = Frequently
5 = Somewhat Frequently
4 = Occasionally
3 = Seldom
2 = Almost Never
1 = Never

1 2 3 4 5 6 7 1. I went out of my way to do a favor for a coworker.
1 2 3 4 5 6 7 2. I went out of my way to do a favor for a relative.
1 2 3 4 5 6 7 3. I went out of my way to do a favor for a friend.
1 2 3 4 5 6 7 4. I gave money to a religious charity.
1 2 3 4 5 6 7 5. I gave money to a charity not affiliated with a religion.
1 2 3 4 5 6 7 6. I gave money to a panhandler.

When this instrument was developed, the intent was to administer it to a sample of participants and use their responses to the six items as separate predictor variables. As previously stated, however, you learned that this is a problematic practice and have decided, instead, to perform a principal component analysis on responses to see if a smaller number of components can successfully account for most variance in the dataset. If this is the case, you will use the resulting components as predictor variables in subsequent analyses.

At this point, it may be instructive to examine the content of the six items that constitute the POI to make an informed guess as to what is likely to result from the principal component analysis. Imagine that when you first constructed the instrument, you assumed that the six items were assessing six different types of prosocial behavior. Inspection of items 1 to 3, however, shows that these three items share something in common: they all deal with "going out of one's way to do a favor for someone else." It would not be surprising then to learn that these three items will hang together empirically in the principal component analysis to be performed. In the same way, a review of items 4 to 6 shows that each of these items involves the activity of "giving money to those in need." Again, it is possible that these three items will also group together in the course of the analysis.

In summary, the nature of items suggests that it may be possible to account for variance in the POI with just two components: a "helping others" component and a "financial giving" component. At this point, this is only speculation, of course; only a formal analysis can determine the number and nature of components measured by the inventory of items. (Remember that the preceding instrument is fictitious and used for purposes of illustration only and should not be regarded as an example of a good measure of prosocial orientation. Among other problems, this questionnaire deals with only two forms of helping behavior.)

Preparing a Multiple-Item Instrument

The preceding section illustrates an important point about how *not* to prepare a multiple-item scale to measure a construct. Generally speaking, it is poor practice to throw together a questionnaire, administer it to a sample, and then perform a principal component analysis (or factor analysis) to determine what the questionnaire is measuring.

Better results are much more likely when you make *a priori* decisions about what you want the questionnaire to measure, and then take steps to ensure that it does. For example, you would have been more likely to obtain optimal results if you:

- began with a thorough review of theory and research on prosocial behavior
- used that review to determine how many types of prosocial behavior may exist
- wrote multiple questionnaire items to assess each type of prosocial behavior

Using this approach, you could have made statements such as "There are three types of prosocial behavior: acquaintance helping; stranger helping; and financial giving." You could have then prepared a number of items to assess each of these three types, administered the questionnaire to a large sample, and performed a principal component analysis to see if three components did, in fact, emerge.

Number of Items per Component

When a variable (such as a questionnaire item) is given a weight in computing a principal component, we say that the variable **loads** on that component. For example, if the item "Went out of my way to do a favor for a coworker" is given a lot of weight on the "helping others" component, we say that this item "loads" on that component.

It is highly desirable to have a minimum of three (and preferably more) variables loading on each retained component when the principal component analysis is complete (see Clark and Watson 1995). Because some items may be dropped during the course of the analysis (for reasons to be discussed later), it is generally good practice to write at least five items for each construct that you wish to measure. This increases your chances that at least three items per component will survive the analysis. Note that we have violated this recommendation by writing only three items for each of the two *a priori* components constituting the POI.

Keep in mind that the recommendation of three items per scale should be viewed as an absolute minimum and certainly not as an optimal number. In practice, test and attitude scale developers normally desire that their scales contain many more than just three items to measure a given construct. It is not unusual to see individual scales that include 10, 20, or even more items to assess a single construct (e.g., Chou and O'Rourke 2012; O'Rourke and Cappeliez 2002). Up to a point, the greater the number of scale items, the more reliable it will be (Henson 2001). The recommendation of three items per scale should therefore be viewed as a rock-bottom lower bound, appropriate only if practical concerns prevent you from including more items (e.g., total questionnaire length). For more information on scale construction, see DeVellis (2012) and, Saris and Gallhofer (2007).

Minimal Sample Size Requirements

Principal component analysis is a large-sample procedure. To obtain reliable results, the minimal number of participants providing usable data for the analysis should be the larger of 100 participants or 5 times the number of variables being analyzed (Streiner 1994).

To illustrate, assume that you wish to perform an analysis on responses to a 50-item questionnaire. (Remember that when responses to a questionnaire are analyzed, the number of variables is equal to the number of items on that questionnaire.) Five times the number of items on the questionnaire equals 250. Therefore, your final sample should provide usable (complete) data from at least 250 participants. Note, however, that any participant who fails to answer just one item will not provide usable data for the principal component analysis and will therefore be excluded from the final sample. A certain number of participants can always be expected to leave at least one question blank. To ensure that the final sample includes at least 250 usable responses, you would be wise to administer the questionnaire to perhaps 300 to 350 participants (see Little and Rubin 1987). A preferable alternative is to use an imputation procedure that assigns values for skipped items (van Buuren 2012). A number of such procedures are available in SAS but are not covered in this text.

These rules regarding the number of participants per variable again constitute a lower bound, and some have argued that they should be applied only under two optimal conditions for principal component analysis: 1) when many variables are expected to load on each component, and 2) when variable communalities are high. Under less optimal conditions, even larger samples may be required.

What is a communality? A **communality** refers to the percent of variance in an observed variable that is accounted for by the retained components (or factors). A given variable will display a large communality if it loads heavily on at least one of the study's retained components. Although communalities are computed in both procedures, the *concept* of variable communality is more relevant to factor analysis than principal component analysis.

SAS Program and Output

You may perform principal component analysis using the PRINCOMP, CALIS, or FACTOR procedures. This chapter will show how to perform the analysis using PROC FACTOR since this is a somewhat more flexible SAS procedure. (It is also possible to perform an exploratory factor analysis with PROC FACTOR or PROC CALIS.) Because the analysis is to be performed using PROC FACTOR, the output will at times make reference to factors rather than to principal components (e.g., component 1 will be referred to as FACTOR1 in the output). It is important to remember, however, that you are performing principal component analysis, not factor analysis.

This section will provide instructions on writing the SAS program and an overview of the SAS output. A subsequent section will provide a more detailed treatment of the steps followed in the analysis as well as the decisions to be made at each step.

Writing the SAS Program

The DATA Step

To perform a principal component analysis, data may be entered as raw data, a correlation matrix, a covariance matrix, or some other format. (See Appendix A.2 for further description of these data input options.) In this chapter's first example, raw data will be analyzed.

Assume that you administered the POI to 50 participants, and entered their responses according to the following guide:

Line	Column	Variable Name	Explanation
1	1–6	V1–V6	Participants' responses to survey questions 1 through 6. Responses were provided along a 7-point scale.

Here are the statements to enter these responses as raw data. The first three observations and the last three observations are reproduced here; for the entire dataset, see Appendix B.

```
data D1;
      input V1-V6 ;

datalines;
556754
567343
777222
.
.
.
767151
455323
455544
;
run;
```

The dataset in Appendix B includes only 50 cases so that it will be relatively easy to enter the data and replicate the analyses presented here. It should be restated, however, that 50 observations is an unacceptably small sample for principal component analysis. Earlier it was noted that a sample should provide usable data from the larger of either 100 cases or 5 times the number of observed variables. A small sample is being analyzed here for illustrative purposes only.

The PROC FACTOR Statement

The general form for the SAS program to perform a principal component analysis is presented here:

```
proc factor   data=dataset-name
              simple
              method=prin
              priors=one
              mineigen=p
              rotate=varimax
              round
              flag=desired-size-of-"significant"-factor-loadings ;
   var  variables-to-be-analyzed ;
run;
```

Options Used with PROC FACTOR

The PROC FACTOR statement begins the FACTOR procedure and a number of options may be requested in this statement before it ends with a semicolon. Some options that are especially useful in social science research are:

FLAG

causes the output to flag (with an asterisk) factor loadings with absolute values greater than some specified size. For example, if you specify

```
flag=.35
```

an asterisk will appear next to any loading whose absolute value exceeds .35. This option can make it much easier to interpret a factor pattern. Negative values are not allowed in the FLAG option, and the FLAG option can be used in conjunction with the ROUND option.

METHOD=factor-extraction-method

specifies the method to be used in extracting the factors or components. The current program specifies

```
method=prin
```

to request that the principal axis (principal factors) method be used for the initial extraction. This is the appropriate method for a principal component analysis.

MINEIGEN=p

specifies the critical eigenvalue a component must display if that component is to be retained (here, p = the critical eigenvalue). For example, the current program specifies

```
mineigen=1
```

This statement will cause PROC FACTOR to retain and rotate any component whose eigenvalue is 1.00 or larger. Negative values are not allowed.

NFACT=n

allows you to specify the number of components to be retained and rotated where n = the number of components.

OUT=name-of-new-dataset
creates a new dataset that includes all of the variables in the existing dataset, along with factor scores for the components retained in the present analysis. Component 1 is given the variable name FACTOR1, component 2 is given the name FACTOR2, and so forth. It must be used in conjunction with the NFACT option, and the analysis must be based on raw data.

PRIORS=prior-communality-estimates
specifies prior communality estimates. Users should always specify PRIORS=one to perform a principal component analysis.

ROTATE=rotation-method
specifies the rotation method to be used. The preceding program requests a varimax rotation that provides orthogonal (uncorrelated) components. Oblique rotations may also be requested (correlated components).

ROUND
factor loadings and correlation coefficients in the matrices printed by PROC FACTOR are normally carried out to several decimal places. Requesting the ROUND option, however, causes all coefficients to be limited to two decimal places, rounded to the nearest integer, and multiplied by 100 (thus eliminating the decimal point). This generally makes it easier to read the coefficients.

PLOTS=SCREE
creates a plot that graphically displays the size of the eigenvalues associated with each component. This can be used to perform a scree test to visually determine how many components should be retained.

SIMPLE
requests simple descriptive statistics: the number of usable cases on which the analysis was performed and the means and standard deviations of the observed variables.

The VAR Statement

The variables to be analyzed are listed on the VAR statement with each variable separated by at least one space. Remember that the VAR statement is a *separate* statement and not an option within the FACTOR statement, so don't forget to end the FACTOR statement with a semicolon before beginning the VAR statement.

Example of an Actual Program

The following is an actual program, including the DATA step, that could be used to analyze some fictitious data. Only a few sample lines of data appear here; the entire dataset can be found in Appendix B.

```
data D1;
     input   #1    @1   (V1-V6)    (1.);

datalines;
556754
567343
777222
.
.
.
767151
455323
455544
;
run;

proc factor    data=D1
               simple
               method=prin
               priors=one
               mineigen=1
               plots=scree
```

```
                rotate=varimax
                round
                flag=.40   ;
      var V1 V2 V3 V4 V5 V6;
run;
```

Results from the Output

The preceding program would produce three pages of output. Here is a list of some of the most important information provided by the output and the page on which it appears:

- page 1 includes simple statistics (mean values and standard deviations)
- page 2 includes scree plot of eigenvalues and cumulative variance explained
- page 3 includes the final communality estimates

The output created by the preceding program is presented here as Output 1.1.

Output 1.1: Results of the Initial Principal Component Analysis of the Prosocial Orientation Inventory (POI) Data (Page 1)

The FACTOR Procedure

Input Data Type	Raw Data
Number of Records Read	50
Number of Records Used	50
N for Significance Tests	50

Means and Standard Deviations from 50 Observations		
Variable	Mean	Std Dev
V1	5.1800000	1.3951812
V2	5.4000000	1.1065667
V3	5.5200000	1.2162170
V4	3.6400000	1.7929567
V5	4.2200000	1.6695349
V6	3.1000000	1.5551101

Output 1.1 (Page 2)

The FACTOR Procedure
Initial Factor Method: Principal Components

Prior Communality Estimates: ONE

Eigenvalues of the Correlation Matrix: Total = 6 Average = 1				
	Eigenvalue	Difference	Proportion	Cumulative
1	2.26643553	0.29182092	0.3777	0.3777
2	1.97461461	1.17731470	0.3291	0.7068
3	0.79729990	0.35811605	0.1329	0.8397
4	0.43918386	0.14791916	0.0732	0.9129
5	0.29126470	0.06006329	0.0485	0.9615
6	0.23120141		0.0385	1.0000

2 factors will be retained by the MINEIGEN criterion.

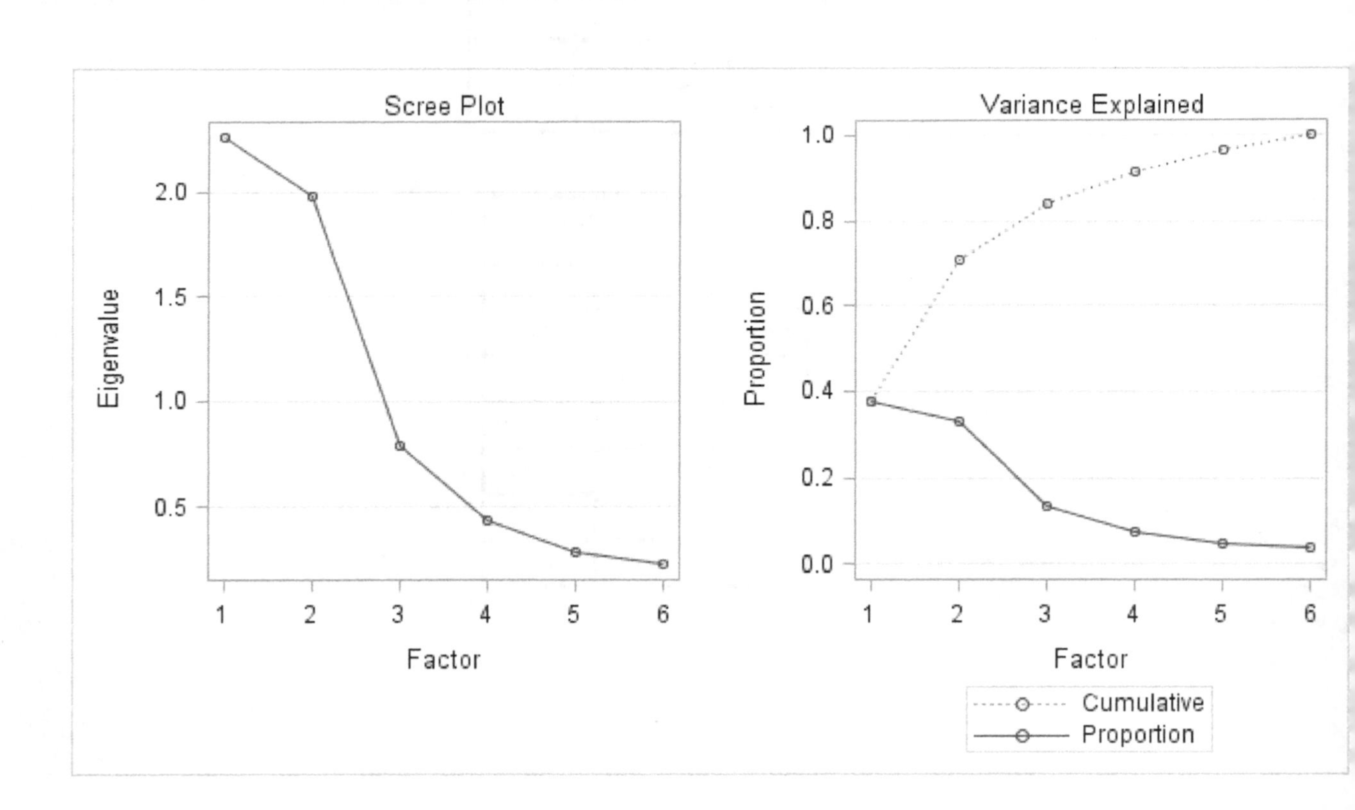

Factor Pattern				
	Factor1		Factor2	
V1	58	*	70	*
V2	48	*	53	*
V3	60	*	62	*
V4	64	*	-64	*
V5	68	*	-45	*
V6	68	*	-46	*
Printed values are multiplied by 100 and rounded to the nearest integer. Values greater than 0.4 are flagged by an '*'.				

Variance Explained by Each Factor	
Factor1	Factor2
2.2664355	1.9746146

Final Communality Estimates: Total = 4.241050					
V1	V2	V3	V4	V5	V6
0.82341782	0.50852894	0.74399020	0.82257428	0.66596347	0.67657543

Output 1.1 (Page 3)

The FACTOR Procedure
Rotation Method: Varimax

Orthogonal Transformation Matrix		
	1	2
1	0.76914	0.63908
2	-0.63908	0.76914

Rotated Factor Pattern				
	Factor1		Factor2	
V1	0		91	*
V2	3		71	*
V3	7		86	*
V4	90	*	-9	
V5	81	*	9	
V6	82	*	8	
Printed values are multiplied by 100 and rounded to the nearest integer. Values greater than 0.4 are flagged by an '*'.				

Variance Explained by Each Factor	
Factor1	Factor2
2.1472475	2.0938026

Final Communality Estimates: Total = 4.241050					
V1	V2	V3	V4	V5	V6
0.82341782	0.50852894	0.74399020	0.82257428	0.66596347	0.67657543

Page 1 from Output 1.1 provides simple statistics for the observed variables included in the analysis. Once the SAS log has been checked to verify that no errors were made in the analysis, these simple statistics should be reviewed to determine how many usable observations were included in the analysis, and to verify that the means and standard deviations are in the expected range. On page 1, it says "Means and Standard Deviations from 50 Observations," meaning that data from 50 participants were included in the analysis.

Steps in Conducting Principal Component Analysis

Principal component analysis is normally conducted in a sequence of steps, with somewhat subjective decisions being made at various points. Because this chapter is intended as an introduction to the topic, this text will not provide a comprehensive discussion of all of the options available at each step; instead, specific recommendations will be made, consistent with common practice in applied research. For a detailed treatment of principal component analysis and factor analysis, see Stevens (2002).

Step 1: Initial Extraction of the Components

In principal component analysis, the number of components extracted is equal to the number of variables being analyzed. Because six variables are analyzed in the present study, six components are extracted. The first can be expected to account for a fairly large amount of the total variance. Each subsequent component will account for progressively smaller amounts of variance. Although a large number of components may be extracted in this way, only the first few components will be sufficiently important to be retained for interpretation.

Page 2 from Output 1.1 provides the eigenvalue table from the analysis. (This table appears just below the heading "Eigenvalues of the Correlation Matrix: Total = 6 Average = 1".) An **eigenvalue** represents the amount of variance captured by a given component. In the column heading "Eigenvalue," the eigenvalue for each component is presented. Each row in the matrix presents information for each of the six components. Row 1 provides information about the first component extracted, row 2 provides information about the second component extracted, and so forth.

Where the column heading "Eigenvalue" intersects with rows 1 and 2, it can be seen that the eigenvalue for component 1 is approximately 2.27, while the eigenvalue for component 2 is 1.97. This pattern is consistent with our earlier statement that the first components tend to account for relatively large amounts of variance, whereas the later components account for comparatively smaller amounts.

Step 2: Determining the Number of "Meaningful" Components to Retain

Earlier it was stated that the number of components extracted is equal to the number of variables analyzed. This requires that you decide just how many of these components are truly meaningful and worthy of being retained for rotation and interpretation. In general, you expect that only the first few components will account for meaningful amounts of variance and that the later components will tend to account for only trivial variance. The next step, therefore, is to determine how many meaningful components should be retained to interpret. This section will describe four criteria that may be used in making this decision: the eigenvalue-one criterion, the scree test, the proportion of variance accounted for, and the interpretability criterion.

The Eigenvalue-One Criterion

In principal component analysis, one of the most commonly used criterion for solving the number-of-components problem is the eigenvalue-one criterion, also known as the Kaiser-Guttman criterion (Kaiser 1960). With this method, you retain and interpret all components with eigenvalues greater than 1.00.

The rationale for this criterion is straightforward: each observed variable contributes one unit of variance to the total variance in the dataset. Any component with an eigenvalue greater than 1.00 accounts for a greater amount of variance than had been contributed by one variable. Such a component therefore accounts for a meaningful amount of variance and (in theory) is worthy of retention.

On the other hand, a component with an eigenvalue less than 1.00 accounts for less variance than contributed by one variable. The purpose of principal component analysis is to reduce a number of observed variables into a relatively smaller number of components. This cannot be effectively achieved if you retain components that account for less variance than had been contributed by individual variables. For this reason, components with eigenvalues less than 1.00 are viewed as trivial and are not retained.

The eigenvalue-one criterion has a number of positive features that contribute to its utility. Perhaps the most important reason for its use is its simplicity. It does not require subjective decisions; you merely retain components with eigenvalues greater than 1.00.

Yet this criterion often results in retaining too many components, particularly when a small to moderate number of variables are analyzed and the variable communalities are high. Stevens (2002) reviews studies that have investigated the accuracy of the eigenvalue-one criterion and recommends its use when fewer than 30 variables are being analyzed and communalities are greater than .70, or when the analysis is based on more than 250 observations and the mean communality is greater than .59.

There are, however, various problems associated with the eigenvalue-one criterion. As suggested in the preceding paragraph, it can lead to retaining too many components under circumstances that are often encountered in research (e.g., when many variables are analyzed, when communalities are small). Also, the reflexive application of this criterion can lead to retaining a certain number of components when the actual difference in the eigenvalues of successive components is trivial. For example, if component 2 has an eigenvalue of 1.01 and component 3 has an eigenvalue of 0.99, then component 2 will be retained but component 3 will not. This may mistakenly lead you to believe that the third component was meaningless when, in fact, it accounted for almost the same amount of variance as the second component. In short, the eigenvalue-one criterion can be helpful when used judiciously, yet the reflexive application of this approach can lead to serious errors of interpretation. Almost always, the eigenvalue-one criterion should be considered in conjunction with other criteria (e.g., scree test, the proportion of variance accounted for, and the interpretability criterion) when deciding how many components to retain and interpret.

With SAS, the eigenvalue-one criterion can be applied by including the MINEIGEN=1 option in the PROC FACTOR statement and not including the NFACT option. The use of the MINEIGEN=1 will cause PROC FACTOR to retain any component with an eigenvalue greater than 1.00.

The eigenvalue table from the current analysis appears on page 2 of Output 1.1. The eigenvalues for components 1, 2, and 3 are 2.27, 1.97, and 0.80, respectively. Only components 1 and 2 have eigenvalues greater than 1.00, so the eigenvalue-one criterion would lead you to retain and interpret only these two components.

Fortunately, the application of the criterion is fairly unambiguous in this case. The last component retained (2) has an eigenvalue of 1.97, which is substantially greater than 1.00, and the next component (3) has an eigenvalue of 0.80, which is clearly lower than 1.00. In this instance, you are not faced with the difficult decision of whether to retain a component with an eigenvalue approaching 1.00 (e.g., an eigenvalue of .99). In situations such as we describe here, the eigenvalue-one criterion may be used with greater confidence.

The Scree Test

With the scree test (Cattell 1966), you plot the eigenvalues associated with each component and look for a definitive "break" between the components with relatively large eigenvalues and those with relatively small eigenvalues. The components that appear *before* the break are assumed to be meaningful and are retained for rotation, whereas those appearing *after* the break are assumed to be unimportant and are not retained. Sometimes a scree plot will display several large breaks. When this is the case, you should look for the last big break before the eigenvalues begin to level off. Only the components that appear before this last large break should be retained.

Specifying the PLOTS=SCREE option in the PROC FACTOR statement tells SAS to print an eigenvalue plot as part of the output. This appears as page 2 of Output 1.1.

You can see that the component numbers are listed on the horizontal axis, while eigenvalues are listed on the vertical axis. With this plot, notice there is a relatively small break between components 1 and 2, and a relatively large break following component 2. The breaks between components 3, 4, 5, and 6 are all relatively small. It is often helpful to draw long lines with extended tails connecting successive pairs of eigenvalues so that these breaks are more apparent (e.g., measure degrees separating lines with a protractor).

Because the large break in this plot appears between components 2 and 3, the scree test would lead you to retain only components 1 and 2. The components appearing after the break (3 to 6) would be regarded as trivial.

The scree test can be expected to provide reasonably accurate results, provided that the sample is large (over 200) and most of the variable communalities are large (Stevens 2002). This criterion too has its weaknesses, however, most notably the ambiguity of scree plots under common research conditions. Very often, it is difficult to determine precisely where in the scree plot a break exists, or even if a break exists at all. In contrast to the eigenvalue-one criterion, the scree test is often more subjective.

The break in the scree plot on page 3 of Output 1.1 is unusually obvious. In contrast, consider the plot that appears in Figure 1.2.

Figure 1.2: A Scree Plot with No Obvious Break

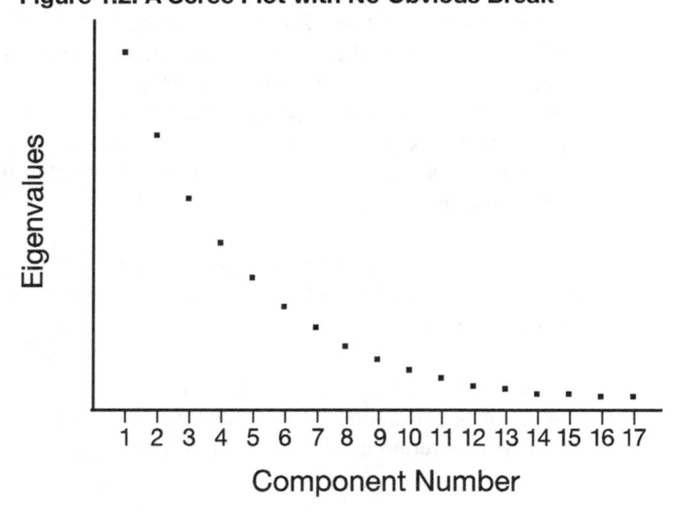

Figure 1.2 presents a fictitious scree plot from a principal component analysis of 17 variables. Notice that there is no obvious break in the plot that separates the meaningful components from the trivial components. Most researchers would agree that components 1 and 2 are probably meaningful whereas components 13 to 17 are probably trivial; but it is difficult to decide exactly where you should draw the line. This example underscores the qualitative nature of judgments based solely on the scree test.

Scree plots such as the one presented in Figure 1.2 are common in social science research. When encountered, the use of the scree test must be supplemented with additional criteria such as the "variance accounted for" criterion and the interpretability criterion, to be described later.

Why do they call it a "scree" test? The word "scree" refers to the loose rubble that lies at the base of a cliff or glacier. When performing a scree test, you normally hope that the scree plot will take the form of a cliff. At the top will be the eigenvalues for the few meaningful components, followed by a definitive break (the edge of the cliff). At the bottom of the cliff will lay the scree (i.e., eigenvalues for the trivial components).

Proportion of Variance Accounted For

A third criterion to address the number of factors problem involves retaining a component if it accounts for more than a specified proportion (or percentage) of variance in the dataset. For example, you may decide to retain any component that accounts for at least 5% or 10% of the total variance. This proportion can be calculated with a simple formula:

$$\text{Proportion} = \frac{\text{Eigenvalue for the component of interest}}{\text{Total eigenvalues of the correlation matrix}}$$

In principal component analysis, the "total eigenvalues of the correlation matrix" is equal to the total number of variables being analyzed (because each variable contributes one unit of variance to the analysis).

Fortunately, it is not necessary to actually compute these percentages by hand since they are provided in the results of PROC FACTOR. The proportion of variance captured by each component is printed in the eigenvalue table (page 2) and appears below the "Proportion" heading.

The eigenvalue table for the current analysis appears on page 2 of Output 1.1. From the "Proportion" column, you can see that the first component alone accounts for 38% of the total variance, the second component alone accounts for 33%, the third component accounts for 13%, and the fourth component accounts for 7%. Assume that you have decided to retain any component that accounts for at least 10% of the total variance in the dataset. With the present results, this criterion leads you to retain components 1, 2, and 3. (Notice that use of this criterion would result in retaining more components than would be retained using the two preceding criteria.)

An alternative criterion is to retain enough components so that the *cumulative* percent of variance is equal to some minimal value. For example, recall that components 1, 2, 3, and 4 accounted for approximately 38%, 33%, 13%, and 7% of the total variance, respectively. Adding these percentages together results in a sum of 91%. This means that the *cumulative* percent of variance accounted for by components 1, 2, 3, and 4 is 91%. When researchers use the "cumulative percent of variance accounted for" criterion for solving the number-of-components problem, they usually retain enough components so that the cumulative percent of variance is at least 70% (and sometimes 80%).

With respect to the results of PROC FACTOR, the cumulative percent of variance accounted for is presented in the eigenvalue table (from page 2), below the "Cumulative" heading. For the present analysis, this information appears in the eigenvalue table on page 2 of Output 1.1. Notice the values that appear below the heading "Cumulative." Each value indicates the percent of variance accounted for by the present component as well as all preceding components. For example, the value for component 2 is approximately .71 (intersection of the column labeled "Cumulative" and the second row). This value of .71 indicates that approximately 71% of the total variance is accounted for by components 1 and 2. The corresponding entry for component 3 is approximately .84, indicating that 84% of the variance is accounted for by components 1, 2, and 3. If you were to use 70% as the "critical value" for determining the number of components to retain, you would retain only components 1 and 2 in the present analysis.

The primary advantage of the proportion of variance criterion is that it leads you to retain a group of components that combined account for a relatively large proportion of variance in the dataset. Nonetheless, the critical values discussed earlier (10% for individual components and 70% to 80% for the combined components) are quite arbitrary. Because of this and related problems, this approach has been criticized for its subjectivity.

The Interpretability Criterion

Perhaps the most important criterion for solving the number-of-components problem is the **interpretability criterion**: interpreting the substantive meaning of the retained components and verifying that this interpretation makes sense in terms of what is known about the constructs under investigation. The following list provides four rules to follow when applying this criterion. A later section (titled "Step 4: Interpreting the Rotated Solution") shows how to actually interpret the results of a principal component analysis. The following rules will be more meaningful after you have completed that section.

1. **Are there at least three variables (items) with significant loadings on each retained component?** A solution is less satisfactory if a given component is measured by fewer than three variables.
2. **Do the variables that load on a given component share the same conceptual meaning?** For example, if three questions on a survey all load on component 1, do all three of these questions appear to be measuring the same construct?
3. **Do the variables that load on different components seem to be measuring different constructs?** For example, if three questions load on component 1 and three other questions load on component 2, do the first three questions seem to be measuring a construct that is conceptually distinct from the construct measured by the other three questions?
4. **Does the rotated factor pattern demonstrate "simple structure"? Simple structure** means that the pattern possesses two characteristics: (a) most of the variables have relatively high factor loadings on only one component and near zero loadings on the other components; and (b) most components have relatively high loadings for some variables and near-zero loadings for the remaining variables. This concept of simple structure will be explained in more detail in "Step 4: Interpreting the Rotated Solution."

Recommendations

Given the preceding options, what procedures should you actually follow in solving the number-of-components problem? We recommend combining all four in a structured sequence. First, use the MINEIGEN=1 option to implement the eigenvalue-one criterion. Review this solution for interpretability but use caution if the break between the components with eigenvalues above 1.00 and those below 1.00 is not clear-cut (e.g., if component 1 has an eigenvalue of 1.01 and component 2 has an eigenvalue of 0.99).

Next, perform a scree test and look for obvious breaks in the eigenvalues. Because there will often be more than one break in the scree plot, it may be necessary to examine two or more possible solutions.

Next, review the amount of common variance accounted for by each individual component. You probably should not rigidly use some specific but arbitrary cutoff point such as 5% or 10%. Still, if you are retaining components that account for as little as 2% or 4% of the variance, it may be wise to take a second look at the solution and verify that these latter components are truly of substantive importance. In the same way, it is best if the combined components account for at least 70% of the cumulative variance. If less than 70% is captured, it may be prudent to consider alternate solutions that include a larger number of components.

Finally, apply the interpretability criteria to each solution. If more than one solution can be justified on the basis of the preceding criteria, which of these solutions is the most interpretable? By seeking a solution that is both interpretable and satisfies one or more of the other three criteria, you maximize chances of retaining the optimal number of components.

Step 3: Rotation to a Final Solution

Factor Patterns and Factor Loadings

After extracting the initial components, PROC FACTOR will create an unrotated **factor pattern matrix**. The rows of this matrix represent the variables being analyzed, and the columns represent the retained components. (Note that even though we are performing principal component analysis, components are labeled as FACTOR1, FACTOR2, and so forth in the output.)

The entries in the matrix are factor loadings. A **factor loading** (or, more correctly, a *component loading*) is a general term for a coefficient that appears in a factor pattern matrix or a factor structure matrix. In an analysis that results in oblique (correlated) components, the definition of a factor loading is different depending on whether it is in a factor *pattern* matrix or in a factor *structure* matrix. The situation is simpler, however, in an analysis that results in orthogonal components (as in the present chapter). In an orthogonal analysis, factor loadings are equivalent to bivariate correlations between the observed variables and the components.

For example, the factor pattern matrix from the current analysis appears on page 2 of Output 1.1. Where the rows for observed variables intersect with the column for FACTOR1, you can see that the correlation between V1 and the first component is .58, the correlation between V2 and the first component is .48, and so forth.

Rotations

Ideally, you would like to review the correlations between the variables and the components, and use this information to *interpret* the components. In other words, you want to determine what construct seems to be measured by component 1, what construct seems to be measured by component 2, and so forth. Unfortunately, when more than one component has been retained in an analysis, the interpretation of an unrotated factor pattern is generally quite difficult. To facilitate interpretation, you will normally perform an operation called a "rotation." A **rotation** is a linear transformation that is performed on the factor solution for the purpose of making the solution easier to interpret.

PROC FACTOR allows you to request several different types of rotations. The preceding program that analyzed data from the POI study included the statement

```
rotate=varimax
```

A **varimax rotation** is an orthogonal rotation, meaning that it results in uncorrelated components. Compared to some other types of rotations, a varimax rotation tends to maximize the variance of a column of the factor pattern matrix (as opposed to a row of the matrix). This rotation is probably the most commonly used orthogonal rotation in the social sciences (e.g., Chou and O'Rourke 2012). The results of the varimax rotation for the current analysis appear on page 5 of Output 1.1.

Step 4: Interpreting the Rotated Solution

Interpreting a rotated solution means determining just what is measured by each of the retained components. Briefly, this involves identifying the variables with high loadings on a given component and determining what these variables share in common. Usually, a brief name is assigned to each retained component to describe what it appears to measure.

The first decision to be made at this stage is how large a factor loading must be to be considered "large." Stevens (2002) discusses some of the issues relevant to this decision and even provides guidelines for testing the statistical significance of factor loadings. Given that this is an introductory treatment of principal component analysis, simply consider a loading to be "large" if its absolute value exceeds .40.

The rotated factor pattern for the POI study appears on page 3 of Output 1.1. The following provides a structured approach for interpreting this factor pattern.

5. **Read across the row for the first variable.** All "meaningful loadings" (i.e., loadings greater than .40) have been flagged with an asterisk ("*"). This was accomplished by including the FLAG=.40 option in the preceding program. If a given variable has a meaningful loading on more than one component, cross out that variable and ignore it in your interpretation. In many situations, researchers drop variables that load on more than one component because the variables are not pure measures of any one construct. (These are sometimes referred to as *complex items.*) In the present case, this means looking at the row heading "V1" and reading to the right to see if it loads on more than one component. In this case it does not, so you may retain this variable.
6. **Repeat this process for the remaining variables, crossing out any variable that loads on more than one component.** In this analysis, none of the variables have high loadings on more than one component, so none will have to be deleted. In other words, there are no complex items.
7. **Review all of the surviving variables with high loadings on component 1 to determine the nature of this component.** From the rotated factor pattern, you can see that only items 4, 5, and 6 load on component 1 (note the asterisks). It is now necessary to turn to the questionnaire itself and review the content in order to decide what a given component should be named. What do questions 4, 5, and 6 have in common? What common construct do they appear to be measuring? For illustration, the questions being analyzed in the present case are reproduced here. Remember that question 4 was represented as V4 in the SAS program, question 5 was V5, and so forth. Read questions 4, 5, and 6 to see what they have in common.

1 2 3 4 5 6 7	1. Went out of my way to do a favor for a coworker.
1 2 3 4 5 6 7	2. Went out of my way to do a favor for a relative.
1 2 3 4 5 6 7	3. Went out of my way to do a favor for a friend.
1 2 3 4 5 6 7	4. Gave money to a religious charity.
1 2 3 4 5 6 7	5. Gave money to a charity not affiliated with a religion.
1 2 3 4 5 6 7	6. Gave money to a panhandler.

Questions 4, 5, and 6 all seem to deal with giving money to persons in need. It is therefore reasonable to label component 1 the "financial giving" component.

8. **Repeat this process to name the remaining retained components.** In the present case, there is only one remaining component to name: component 2. This component has high loadings for questions 1, 2, and 3. In reviewing these items, it is apparent that each seems to deal with helping friends, relatives, or other acquaintances. It is therefore appropriate to name this the "helping others" component.
9. **Determine whether this final solution satisfies the interpretability criteria.** An earlier section indicated that the overall results of a principal component analysis are satisfactory only if they meet a number of interpretability criteria. The adequacy of the rotated factor pattern presented on page 3 of Output 1.1 is assessed in terms of the following criteria:
   a. **Are there at least three variables (items) with significant loadings on each retained component?** In the present example, three variables loaded on component 1 and three also loaded on component 2, so this criterion was met.
   b. **Do the variables that load on a given component share similar conceptual meaning?** All three variables loading on component 1 measure giving to those in need, while all three loading on component 2 measure prosocial acts performed for others. Therefore, this criterion is met.
   c. **Do the variables that load on different components seem to be measuring different constructs?** The items loading on component 1 measure respondents' financial contributions, while the items loading on component 2 measure helpfulness toward others. Because these seem to be conceptually distinct constructs, this criterion appears to be met as well.
   d. **Does the rotated factor pattern demonstrate "simple structure"?** Earlier, it was noted that a rotated factor pattern demonstrates simple structure when it has two characteristics. First, most of the variables should have high loadings on one component and near-zero loadings on other components. It can be seen that the pattern obtained here meets that requirement: items 1 to 3 have high loadings on component 2 and near-zero loadings on component 1. Similarly, items 4 to 6 have high loadings on component 1 and near-zero loadings on component 2. The second

characteristic of simple structure is that each component should have high loadings for some variables and near-zero loadings for the others. The pattern obtained here also meets this requirement: component 1 has high loadings for items 4 to 6 and near-zero loadings for the other items whereas component 2 has high loadings for items 1 to 3 and near-zero loadings on the remaining items. In short, the rotated component pattern obtained in this analysis does appear to demonstrate simple structure.

Step 5: Creating Factor Scores or Factor-Based Scores

Once the analysis is complete, it is often desirable to assign scores to participants to indicate where they stand on the retained components. For example, the two components retained in the present study were interpreted as "financial giving" and "helping others." You may now want to assign one score to each participant to indicate that participant's standing on the "financial giving" component and a second score to indicate that participant's standing on the "helping others" component. Once assigned, these component scores could be used either as predictor variables or as criterion variables in subsequent analyses.

Before discussing the options for assigning these scores, it is important to first draw a distinction between factor scores and factor-based scores. In principal component analysis, a **factor score** (or **component score**) is a linear composite of the optimally weighted observed variables. If requested, PROC FACTOR will compute each participant's factor scores for the two components by:

- determining the optimal weights
- multiplying participant responses to questionnaire items by these weights
- summing the products

The resulting sum will be a given participant's score on the component of interest. Remember that a separate equation with different weights is computed for each retained component.

A **factor-based score**, on the other hand, is merely a linear composite of the variables that demonstrate meaningful loadings for the component in question. In the preceding analysis, for example, items 4, 5, and 6 demonstrated meaningful loadings for the "financial giving" component. Therefore, you could calculate the factor-based score on this component for a given participant by simply adding together her responses to items 4, 5, and 6. Notice that, with a factor-based score, the observed variables are not multiplied by optimal weights before they are summed.

Computing Factor Scores

Factor scores are requested by including the NFACT and OUT options in the PROC FACTOR statement. Here is the general form for a SAS program that uses the NFACT and OUT option to compute factor scores:

```
proc factor   data=dataset-name
              simple
              method=prin
              priors=one
              nfact=number-of-components-to-retain
              rotate=varimax
              round
              flag=desired-size-of-"significant"-factor-loadings
              out=name-of-new-SAS-dataset   ;
   var  variables-to-be-analyzed ;
run;
```

Here are the actual program statements (minus the DATA step) that could be used to perform a principal component analysis and compute factor scores for the POI study:

```
proc factor   data=D1
        simple
        method=prin
        priors=one
        nfact=2
        rotate=varimax
        round
        flag=.40
❶       out=D2   ;
    var V1 V2 V3 V4 V5 V6;
run;
```

Notice how this program differs from the original program presented earlier in the chapter (in the section titled "SAS Program and Output"). The MINEIGEN=1 option has been removed and replaced with the NFACT=2 option. The OUT=D2 option has also been added.

Line ❶ of the preceding program asks that an output dataset be created and given the name D2. This name is arbitrary; any name consistent with SAS requirements would be acceptable. The new dataset named D2 will contain all variables contained in the previous dataset (D1), as well as new variables named FACTOR1 and FACTOR2. FACTOR1 will contain factor scores for the first retained component, and FACTOR2 will contain scores for the second. The number of new "FACTOR" variables created will be equal to the number of components retained by the NFACT statement.

The OUT option may be used to create component scores only if the analysis has been performed on raw data as opposed to a correlation or covariance matrix. The use of the NFACT statement is also required.

Having created the new variables named FACTOR1 and FACTOR2, you may be interested to see how they relate to the study's original observed variables. This can be done by appending PROC CORR statements to the SAS program, following the last of the PROC FACTOR statements. The full program minus the DATA step is presented here:

```
proc factor  data=D1
        simple
        method=prin
        priors=one
        nfact=2
        rotate=varimax
        round
        flag=.40
❶ out=D2   ;
   var V1 V2 V3 V4 V5 V6;
  run;

❷ proc corr   data=D2;
   var FACTOR1 FACTOR2;
   with V1 V2 V3 V4 V5 V6 FACTOR1 FACTOR2;
  run;
```

Notice that the PROC CORR statement on line ❷ specifies DATA=D2. This dataset (D2) is the name of the output dataset created on line ❶ the PROC FACTOR statement. The PROC CORR statement requests that the factor score variables (FACTOR1 and FACTOR2) be correlated with participants' responses to questionnaire items 1 to 6 (V1 to V6).

The preceding program produces five pages of output. Pages 1 and 2 (not shown) provide simple statistics, the eigenvalue table, and the unrotated factor pattern. Page 3 provides the rotated factor pattern and final communality estimates (same as before). Page 4 provides the standardized scoring coefficients used in creating factor scores. Finally, page 5 provides the correlations requested by the corr procedure. Pages 3, 4, and 5 of the output created by the preceding program are presented here as Output 1.2.

Output 1.2: Output Pages 3, 4, and 5 from the Analysis of POI Data from Which Factor Scores Were Created (Page 3)

The FACTOR Procedure
Rotation Method: Varimax

Orthogonal Transformation Matrix		
	1	2
1	0.76914	0.63908
2	-0.63908	0.76914

Rotated Factor Pattern				
	Factor1		Factor2	
V1	0		91	*
V2	3		71	*
V3	7		86	*
V4	90	*	-9	
V5	81	*	9	
V6	82	*	8	
Printed values are multiplied by 100 and rounded to the nearest integer. Values greater than 0.4 are flagged by an '*'.				

Variance Explained by Each Factor	
Factor1	Factor2
2.1472475	2.0938026

Final Communality Estimates: Total = 4.241050					
V1	V2	V3	V4	V5	V6
0.82341782	0.50852894	0.74399020	0.82257428	0.66596347	0.67657543

Output 1.2 (Page 4)

The FACTOR Procedure
Rotation Method: Varimax
Scoring Coefficients Estimated by Regression

Squared Multiple Correlations of the Variables with Each Factor	
Factor1	Factor2
1.0000000	1.0000000

Standardized Scoring Coefficients		
	Factor1	Factor2
V1	-0.03109	0.43551
V2	-0.00726	0.34071
V3	0.00388	0.41044
V4	0.42515	-0.07087
V5	0.37618	0.01947
V6	0.38020	0.01361

Output 1.2 (Page 5)

The CORR Procedure

8 With Variables:	V1 V2 V3 V4 V5 V6 Factor1 Factor2
2 Variables:	Factor1 Factor2

Simple Statistics						
Variable	N	Mean	Std Dev	Sum	Minimum	Maximum
V1	50	5.18000	1.39518	259.00000	1.00000	7.00000
V2	50	5.40000	1.10657	270.00000	3.00000	7.00000
V3	50	5.52000	1.21622	276.00000	2.00000	7.00000
V4	50	3.64000	1.79296	182.00000	1.00000	7.00000
V5	50	4.22000	1.66953	211.00000	1.00000	7.00000
V6	50	3.10000	1.55511	155.00000	1.00000	7.00000

Simple Statistics						
Variable	**N**	**Mean**	**Std Dev**	**Sum**	**Minimum**	**Maximum**
Factor1	50	0	1.00000	0	-1.87908	2.35913
Factor2	50	0	1.00000	0	-2.95892	1.58951

Pearson Correlation Coefficients, N = 50 Prob > |r| under H0: Rho=0		
	Factor1	**Factor2**
V1	-0.00429 0.9764	0.90741 <.0001
V2	0.03328 0.8185	0.71234 <.0001
V3	0.06720 0.6429	0.85993 <.0001
V4	0.90274 <.0001	-0.08740 0.5462
V5	0.81055 <.0001	0.09474 0.5128
V6	0.81834 <.0001	0.08303 0.5665
Factor1	1.00000	0.00000 1.0000
Factor2	0.00000 1.0000	1.00000

The simple statistics for PROC CORR appear on page 5 in Output 1.2. Notice that the simple statistics for the observed variables (V1 to V6) are identical to those that appeared at the beginning of the factor output discussed earlier (at the top of Output 1.1, page 1). In contrast, note the simple statistics for FACTOR1 and FACTOR2 (the factor score variables for components 1 and 2, respectively). Both have means of 0 and standard deviations of 1; these variables were constructed to be standardized variables.

The correlations between FACTOR1 and FACTOR2 and the original observed variables appear on page 5. You can see that the correlations between FACTOR1 and V1 to V6 on page 4 of Output 1.2 are identical to the factor loadings of V1 to V6 on FACTOR1 on page 3 of Output 1.1, under "Rotated Factor Pattern." This makes sense, as the elements of a factor pattern (in an orthogonal solution) are simply correlations between the observed variables and the components themselves. Similarly, you can see that the correlations between FACTOR2 and V1 to V6 from page 5 of Output 1.2 are also identical to the corresponding factor loadings from page 5 of Output 1.1.

Of particular interest is the correlation between FACTOR1 and FACTOR2, as computed by PROC CORR. This appears on page 5 of Output 1.2, where the row for FACTOR2 intersects with the column for FACTOR1. Notice that the observed correlation between these two components is zero. This is as expected; the rotation method used in the principal component analysis was the varimax method which produces orthogonal, or uncorrelated, components.

Computing Factor-Based Scores

A second (and less sophisticated) approach to scoring involves the creation of new variables that contain factor-based scores rather than true factor scores. A variable that contains factor-based scores is sometimes referred to as a **factor-based scale**.

Although factor-based scores can be created in a number of ways, the following method has the advantage of being relatively straightforward:

1. To calculate factor-based scores for component 1, first determine which questionnaire items had high loadings on that component.
2. For a given participant, add together that participant's responses to these items. The result is that participant's score on the factor-based scale for component 1.
3. Repeat these steps to calculate each participant's score on the remaining retained components.

Although this may sound like a cumbersome task, it is actually quite simple with the use of data manipulation statements contained in a SAS program. For example, assume that you have performed the principal component analysis on your questionnaire responses and have obtained the findings reported in this chapter. Specifically, you found that survey items 4, 5, and 6 loaded on component 1 (the "financial giving" component), whereas items 1, 2, and 3 loaded on component 2 (the "helping others" component).

You would now like to create two new SAS variables. The first variable, called GIVING, will include each participant's factor-based score for financial giving. The second variable, called HELPING, will include each participant's factor-based score for helping others. Once these variables are created, they can be used as criterion or predictor variables in subsequent analyses. To keep things simple, assume that you are simply interested in determining whether there is a significant correlation between GIVING and HELPING.

At this time, it may be useful to review Appendix A.3, "Working with Variables and Observations in SAS Datasets," particularly the section on creating new variables from existing variables. This review should make it easier to understand the data manipulation statements used here.

Assume that earlier statements in the SAS program have already entered responses to the six questionnaire items. These variables are included in a dataset called D1. The following are the subsequent lines that will then create a new dataset called D2. This dataset will include all of the variables in D1 as well as the newly created factor-based scales called GIVING and HELPING.

```
❶  data D2;
❷     set D1;

❸  GIVING    = (V4 + V5 + V6);
   HELPING   = (V1 + V2 + V3);

❹  proc corr   data=D2;
❺     var GIVING  HELPING;
❻  run;
```

Lines ❶ and ❷ request that a new dataset be created called D2, and that it be set up as a duplicate of existing dataset D1. On line ❸, the new variable called GIVING is created. For each participant, the responses to items 4, 5, and 6 are added together. The result is each participant's score on the factor-based scale for the first component. These scores are summed and labeled GIVING. The component-based scale for the "helping others" component is created on line ❹, and these scores are stored as the variable called HELPING. Lines ❺ to ❻ request the correlations between GIVING and HELPING be computed. GIVING and HELPING can now be used as predictor or criterion variables in subsequent analyses. To save space, the results of this program will not be presented here. However, note that this output would probably display a nonzero correlation between GIVING and HELPING. This may come as a surprise because earlier it was shown that the factor scores contained in FACTOR1 and FACTOR2 (counterparts to GIVING and HELPING) were uncorrelated.

The reason for this apparent contradiction is simple: FACTOR1 and FACTOR2 are true principal components, and true principal components (created in an orthogonal solution) are always created with optimally weighted equations so that they will be mutually uncorrelated.

In contrast, GIVING and HELPING are not true principal components that consist of true factor scores; they are merely variables *based* on the results of a principal component analysis. Optimal weights (that would ensure orthogonality) were not used in the creation of GIVING and HELPING. This is why factor-based scales generally demonstrate nonzero correlations while true principal components (from an orthogonal solution) will not.

Recoding Reversed Items Prior to Analysis

It is almost always best to recode any reversed or negatively keyed items before conducting any of the analyses described here. In particular, it is essential that reversed items be recoded prior to the program statements that produce factor-based scales. For example, the three questionnaire items that assess financial giving appear again here:

1 2 3 4 5 6 7	4. Gave money to a religious charity.
1 2 3 4 5 6 7	5. Gave money to a charity not affiliated with a religion.
1 2 3 4 5 6 7	6. Gave money to a panhandler.

None of these items are reversed. With each item, a response of "7" indicates a high level of financial giving. In the following, however, item 4 is a reversed item; a response of "7" indicating a low level of giving:

1 2 3 4 5 6 7	4. Chose not to give money to a religious charity.
1 2 3 4 5 6 7	5. Gave money to a charity not affiliated with a religion.
1 2 3 4 5 6 7	6. Gave money to a panhandler.

If you were to perform a principal component analysis on responses to these items, the factor loading for item 4 would most likely have a sign that is the opposite of the sign of the loadings for items 5 and 6 (e.g., if items 5 and 6 had positive loadings, then item 4 would have a negative loading). This would complicate the creation of a component-based scale: with items 5 and 6, higher scores indicate greater giving whereas with item 4, lower scores indicate greater giving. You would not want to sum these three items as they are presently coded. First, it will be necessary to reverse item 4. Notice how this is done in the following program (assume that the data have already been input in a SAS dataset named D1):

```
data D2;
   set D1;

❶  V4 = 8 - V4;

   GIVING   = (V4 + V5 + V6);
   HELPING  = (V1 + V2 + V3);

proc corr   DATA=D2;
   var GIVING    HELPING;
run;
```

Line ❶ of the preceding program created a new, recoded version of variable V4. Values on this new version of V4 are equal to the quantity 8 minus the value of the old version of V4. For participants whose score on the old version of V4 was 1, their value on the new version of V4 is 7 (because 8 – 1 = 7) whereas for those whose score is 7, their value on the new version of V4 is 1 (because 8 – 7 = 1). Again, see Appendix A.3 for further description of this procedure.

The general form of the formula used to recode reversed items is

```
variable-name = constant - variable-name ;
```

In this formula, the "constant" is the following quantity:

the number of points on the response scale used with the questionnaire item plus 1

Therefore, if you are using the 4-point response format, the constant is 5. If using a 9-point scale, the constant is 10.

If you have prior knowledge about which items are going to appear as reversed (with reversed component loadings) in your results, it is best to place these recoding statements early in your SAS program, before the PROC FACTOR statements. This will make interpretation of the components more straightforward because it will eliminate significant loadings with opposite signs from appearing on the same component. In any case, it is essential that the statements used to recode reversed items appear before the statements that create any factor-based scales.

Step 6: Summarizing the Results in a Table

For reports that summarize the results of your analysis, it is generally desirable to prepare a table that presents the rotated factor pattern. When analyzed variables contain responses to questionnaire items, it can be helpful to reproduce the questionnaire items within this table. This is presented in Table 1.2:

Table 1.2: Rotated Factor Pattern and Final Communality Estimates from Principal Component Analysis of Prosocial Orientation Inventory

Component			
1	**2**	**h^2**	**Items**
.00	.91	.82	Went out of my way to do a favor for a coworker.
.03	.71	.51	Went out of my way to do a favor for a relative.
.07	.86	.74	Went out of my way to do a favor for a friend.
.90	-.09	.82	Gave money to a religious charity.
.81	.09	.67	Gave money to a charity not associated with a religion.
.82	.08	.68	Gave money to a panhandler.

Note: $N = 50$. Communality estimates appear in column headed h^2.

The final communality estimates from the analysis are presented under the heading "**h^2**" in the table. These estimates appear in the SAS output following "Variance Explained by Each Factor" (page 3 of Output 1.2).

Very often, the items that constitute the questionnaire are lengthy, or the number of retained components is large, so that it is not possible to present the factor pattern, the communalities, and the items themselves in the same table. In such situations, it may be preferable to present the factor pattern and communalities in one table and the items in a second. Shared item numbers (or single words or defining phrases) may then be used to associate each item with its corresponding factor loadings and communality.

Step 7: Preparing a Formal Description of the Results for a Paper

The preceding analysis could be summarized in the following way:

Principal component analysis was performed on responses to the 6-item questionnaire using ones as prior communality estimates. The principal axis method was used to extract the components, and this was followed by a varimax (orthogonal) rotation.

Only the first two components had eigenvalues greater than 1.00; results of a scree test also suggested that only the first two were meaningful. Therefore, only the first two components were retained for rotation. Combined, components 1 and 2 accounted for 71% of the total variance (38% plus 33%, respectively).

Questionnaire items and corresponding factor loadings are presented in Table 1.2. When interpreting the rotated factor pattern, an item was said to load on a given component if the factor loading was .40 or greater for that component and less than .40 for the other. Using these criteria, three items were found to load on the first component, which was subsequently labeled "financial giving." Three items also loaded on the second component labeled "helping others."

An Example with Three Retained Components

The Questionnaire

The next example involves fictitious research that examines Rusbult's (1980) investment model (see Le and Agnew 2003 for a review). This model identifies variables believed to affect a person's commitment to a romantic relationship. In this context, **commitment** refers to the person's intention to maintain the relationship and stay with a current romantic partner.

One version of the investment model predicts that commitment will be affected by three antecedent variables: satisfaction, investment size, and alternative value. **Satisfaction** refers to a person's affective (emotional) response to the relationship. Among other things, people report high levels of satisfaction when their current relationship comes close to their perceived ideal relationship. **Investment size** refers to the amount of time, energy, and personal resources that an individual has put into the relationship. For example, people report high investments when they have spent a lot of time with their current partner and have developed mutual friends that may be lost if the relationship were to end. Finally, **alternative value** refers to the attractiveness of alternatives to one's current partner. A person would score high on alternative value if, for example, it would be appealing to date someone else or perhaps just be alone for a while.

Assume that you wish to conduct research on the investment model and are in the process of preparing a 12-item questionnaire to assess levels of satisfaction, investment size, and alternative value in a group of participants involved in romantic relationships. Part of the instrument used to assess these constructs is presented here:

Indicate the extent to which you agree or disagree with each of the following statements by specifying the appropriate response in the space to the left of the statement. Please use the following response format to make these ratings:

7 = Strongly Agree
6 = Agree
5 = Slightly Agree
4 = Neither Agree Nor Disagree
3 = Slightly Disagree
2 = Disagree
1 = Strongly Disagree

_____ 1. I am satisfied with my current relationship.
_____ 2. My current relationship comes close to my ideal relationship.
_____ 3. I am more satisfied with my relationship than the average person.
_____ 4. I feel good about my current relationship.
_____ 5. I have invested a great deal of time in my current relationship.
_____ 6. I have invested a great deal of energy in my current relationship.
_____ 7. I have invested a lot of my personal resources (e.g., money) in developing my current relationship.
_____ 8. My partner and I have established mutual friends that I might lose if we were to break up.
_____ 9. There are plenty of other attractive people for me to date if I were to break up with my current partner.
_____ 10. It would be appealing to break up with my current partner and date someone else.
_____ 11. It would be appealing to break up with my partner to be alone for a while.
_____ 12. It would be appealing to break up with my partner and "play the field."

In the preceding questionnaire, items 1 to 4 were written to assess satisfaction, items 5 to 8 were written to assess investment size, and items 9 to 12 were written to assess alternative value. Assume that you administer this questionnaire to 300 participants and now want to perform a principal component analysis on their responses.

Writing the Program

Earlier, it was noted that it is possible to perform a principal component analysis on a correlation matrix (or covariance matrix) as well as on raw data. This section shows how the former is done. The following program includes the correlation matrix that provides all possible correlation coefficients between responses to the 12 questionnaire items and performs a principal component analysis on these fictitious data:

```
data D1(type=corr)  ;
     input    _type_   $
              _name_   $
              V1-V12   ;
  datalines;
  n     .    300  300  300  300  300  300  300  300  300  300  300  300
  std   .   2.48 2.39 2.58 3.12 2.80 3.14 2.92 2.50 2.10 2.14 1.83 2.26
  corr V1   1.00  .    .    .    .    .    .    .    .    .    .    .
  corr V2    .69 1.00  .    .    .    .    .    .    .    .    .    .
  corr V3    .60  .79 1.00  .    .    .    .    .    .    .    .    .
  corr V4    .62  .47  .48 1.00  .    .    .    .    .    .    .    .
  corr V5    .03  .04  .16  .09 1.00  .    .    .    .    .    .    .
  corr V6    .05 -.04  .08  .05  .91 1.00  .    .    .    .    .    .
  corr V7    .14  .05  .06  .12  .82  .89 1.00  .    .    .    .    .
  corr V8    .23  .13  .16  .21  .70  .72  .82 1.00  .    .    .    .
  corr V9   -.17 -.07 -.04 -.05 -.33 -.26 -.38 -.45 1.00  .    .    .
  corr V10  -.10 -.08  .07  .15 -.16 -.20 -.27 -.34  .45 1.00  .    .
  corr V11  -.24 -.19 -.26 -.28 -.43 -.37 -.53 -.57  .60  .22 1.00  .
  corr V12  -.11 -.07  .07  .08 -.10 -.13 -.23 -.31  .44  .60  .26 1.00
  ;
run ;
    proc factor   data=D1
                  method=prin
                  priors=one
                  mineigen=1
                  plots=scree
                  rotate=varimax
                  round
                  flag=.40;
      var  V1-V12;
run;
```

The PROC FACTOR statement in the preceding program follows the general form recommended for the previous data analyses. Notice that the MINEIGEN=1 statement requests that all components with eigenvalues greater than 1.00 be retained and the PLOTS=SCREE option requests a scree plot of eigenvalues. These options are particularly helpful for the initial analysis of data as they can help determine the correct number of components to retain. If the scree test (or the other criteria) suggests retaining some number of components other than what would be retained using the MINEIGEN=1 option, that option may be dropped and replaced with the NFACT option.

Results of the Initial Analysis

The preceding program produced three pages of output, with the following information appearing on each page:

- page 1 reports the data input procedure and sample size
- page 2 includes the eigenvalue table and scree plot of eigenvalues
- page 3 includes the rotated factor pattern and final communality estimates

The eigenvalue table from this analysis appears on page 1 of Output 1.3. The eigenvalues themselves appear in the left-hand column under the heading "Eigenvalue." From these values, you can see that components 1, 2, and 3 have eigenvalues of 4.47, 2.73, and 1.70, respectively. Furthermore, you can see that only these first three components have eigenvalues greater than 1.00. This means that three components will be retained by the MINEIGEN criterion. Notice that the first nonretained component (component 4) has an eigenvalue of approximately 0.85 which, of course, is well below 1.00. This is encouraging, as you have more confidence in the eigenvalue-one criterion when the solution does not contain "near-miss" eigenvalues (e.g., .98 or .99).

Output 1.3: Results of the Initial Principal Component Analysis of the Investment Model Data (page 1)

The FACTOR Procedure

Input Data Type	Correlations
N Set/Assumed in Data Set	300
N for Significance Tests	300

Output 1.3 (page 2)

The FACTOR Procedure
Initial Factor Method: Principal Components

Prior Communality Estimates: ONE

Eigenvalues of the Correlation Matrix: Total = 12 Average = 1				
	Eigenvalue	Difference	Proportion	Cumulative
1	4.47058134	1.73995858	0.3725	0.3725
2	2.73062277	1.02888853	0.2276	0.6001
3	1.70173424	0.85548155	0.1418	0.7419
4	0.84625269	0.22563029	0.0705	0.8124
5	0.62062240	0.20959929	0.0517	0.8642
6	0.41102311	0.06600575	0.0343	0.8984
7	0.34501736	0.04211948	0.0288	0.9272
8	0.30289788	0.07008042	0.0252	0.9524
9	0.23281745	0.04595812	0.0194	0.9718
10	0.18685934	0.08061799	0.0156	0.9874
11	0.10624135	0.06091129	0.0089	0.9962
12	0.04533006		0.0038	1.0000

3 factors will be retained by the MINEIGEN criterion.

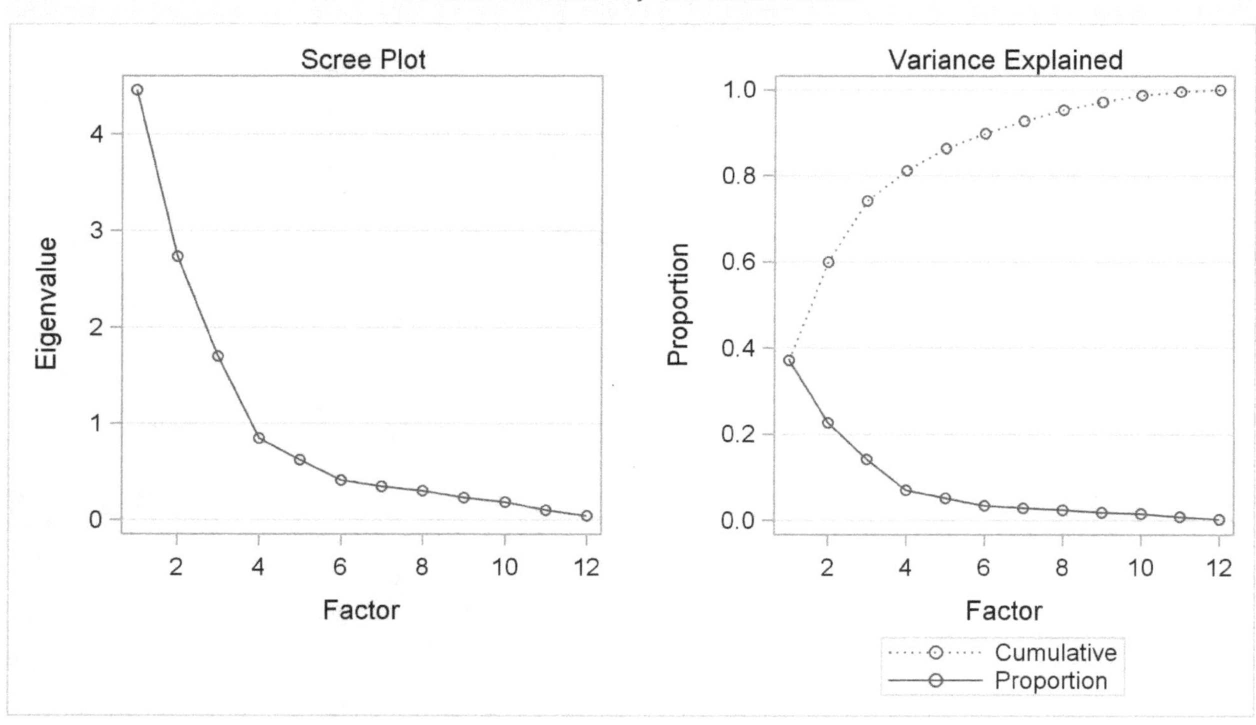

Factor Pattern						
	Factor1		Factor2		Factor3	
V1	39		76	*	-14	
V2	31		82	*	-12	
V3	34		79	*	9	
V4	31		69	*	15	
V5	80	*	-26		41	*
V6	79	*	-32		41	*
V7	87	*	-27		26	
V8	88	*	-14		9	
V9	-61	*	14		47	*
V10	-43	*	23		68	*
V11	-72	*	-6		12	
V12	-40		19		72	*
Printed values are multiplied by 100 and rounded to the nearest integer. Values greater than 0.4 are flagged by an '*'.						

Variance Explained by Each Factor		
Factor1	Factor2	Factor3
4.4705813	2.7306228	1.7017342

Output 1.3 (page 3)

The FACTOR Procedure
Rotation Method: Varimax

Orthogonal Transformation Matrix			
	1	2	3
1	0.83136	0.34431	-0.43623
2	-0.29481	0.93864	0.17902
3	0.47110	-0.02022	0.88185

Rotated Factor Pattern						
	Factor1		Factor2		Factor3	
V1	3		85	*	-16	
V2	-4		88	*	-10	
V3	9		86	*	8	
V4	13		75	*	12	
V5	93	*	2		-3	

Rotated Factor Pattern						
	Factor1		Factor2		Factor3	
V6	95	*	-4		-4	
V7	93	*	4		-19	
V8	81	*	17		-33	
V9	-32		-9		71	*
V10	-11		6		82	*
V11	-52	*	-30		41	*
V12	-5		3		84	*
Printed values are multiplied by 100 and rounded to the nearest integer. Values greater than 0.4 are flagged by an '*'.						

Variance Explained by Each Factor		
Factor1	Factor2	Factor3
3.7048597	2.9364774	2.2616012

The eigenvalue table in Output 1.3 also shows that the first three components combined account for slightly more than 74% of the total variance. (This variance value can be observed at the intersection of the column labeled "Cumulative" and row "3".) The "percentage of variance accounted for" criterion suggests that it may be appropriate to retain three components.

The scree plot from this solution appears on page 2 of Output 1.3. This scree plot shows that there are several large breaks in the data following components 1, 2, and 3, and then the line begins to flatten beginning with component 4. The last large break appears after component 3, suggesting that only components 1 to 3 account for meaningful variance. This suggests that only these first three components should be retained and interpreted. Notice how it is almost possible to draw a straight line through components 4 to 12. The components that lie along a semi-straight line such as this are typically assumed to be measuring only trivial variance (i.e., components 4 to 12 constitute the "scree" of your scree plot).

So far, the results from the eigenvalue-one criterion, the variance accounted for criterion, and the scree plot are in agreement, suggesting that a three-component solution may be most appropriate. It is now time to review the rotated factor pattern to see if such a solution is interpretable. This matrix is presented on page 3 of Output 1.3.

Following the guidelines provided earlier, you begin by looking for factorially complex items (i.e., items with meaningful loadings on more than one component). A review shows that item 11 (variable V11) is a complex item, loading on both components 1 and 3. Item 11 should therefore be discarded. Except for this item, the solution is otherwise fairly straightforward.

To interpret component 1, you read down the column for FACTOR1 and see that items 5 to 8 load significantly on this component. These items are:

_____ 5. I have invested a great deal of time in my current relationship.
_____ 6. I have invested a great deal of energy in my current relationship.
_____ 7. I have invested a lot of my personal resources (e.g., money) in developing my current relationship.
_____ 8. My partner and I have established mutual friends that I might lose if we were to break up.

All of these items deal with the investments that participants have made in their relationships, so it makes sense to label this the "investment size" component.

The rotated factor pattern shows that items 1 to 4 have meaningful loadings on component 2. These items are:

_____ 1. I am satisfied with my current relationship.
_____ 2. My current relationship comes close to my ideal relationship.
_____ 3. I am more satisfied with my relationship than the average person.
_____ 4. I feel good about my current relationship.

Given the content of the preceding items, it seems reasonable to label component 2 the "satisfaction" component.

Finally, items 9, 10, and 12 have meaningful loadings on component 3. (Again, remember that item 11 has been discarded.) These items are:

_____ 9. There are plenty of other attractive people around for me to date if I were to break up with my current partner.
_____ 10. It would be appealing to break up with my current partner and date someone else.
_____ 12. It would be appealing to break up with my partner and "play the field."

These items all seem to deal with the attractiveness of alternatives to one's current relationship, so it makes sense to label this the "alternative value" component.

You may now step back and determine whether this solution satisfies the interpretability criteria presented earlier.

1. Are there at least three variables with meaningful loadings on each retained component?
2. Do the variables that load on a given component share the same conceptual meaning?
3. Do the variables that load on different components seem to be measuring different constructs?
4. Does the rotated factor pattern demonstrate "simple structure"?

In general, the answer to each of these questions is "yes," indicating that the current solution is, in most respects, satisfactory. There is, however, a problem with item 11, which loads on both components 1 and 3. This problem prevents the current solution from demonstrating a perfectly "simple structure" (criterion 4 from above). To eliminate this problem, it may be desirable to repeat the analysis, this time analyzing all of the items *except* for item 11. This will be done in the second analysis of the investment model data described below.

Results of the Second Analysis

To repeat the current analysis with item 11 deleted, it is necessary only to modify the VAR statement of the preceding program. This may be done by changing the VAR statement so that it appears as follows:

```
var V1-V10 V12;
```

All other aspects of the program will remain as they were previously. The eigenvalue table, scree plot, the unrotated factor pattern, the rotated factor pattern, and final communality estimates obtained from this revised program appear in Output 1.4:

Output 1.4: Results of the Second Analysis of the Investment Model Data (Page 1)

The FACTOR Procedure

Input Data Type	Correlations
N Set/Assumed in Data Set	300
N for Significance Tests	300

Output 1.4 (page 2)

The FACTOR Procedure
Initial Factor Method: Principal Components

Prior Communality Estimates: ONE

Eigenvalues of the Correlation Matrix: Total = 11 Average = 1				
	Eigenvalue	Difference	Proportion	Cumulative
1	4.02408599	1.29704748	0.3658	0.3658
2	2.72703851	1.03724743	0.2479	0.6137
3	1.68979108	1.00603918	0.1536	0.7674
4	0.68375190	0.12740106	0.0622	0.8295
5	0.55635084	0.16009525	0.0506	0.8801
6	0.39625559	0.08887964	0.0360	0.9161
7	0.30737595	0.04059618	0.0279	0.9441
8	0.26677977	0.07984443	0.0243	0.9683
9	0.18693534	0.07388104	0.0170	0.9853
10	0.11305430	0.06447359	0.0103	0.9956
11	0.04858072		0.0044	1.0000

3 factors will be retained by the MINEIGEN criterion.

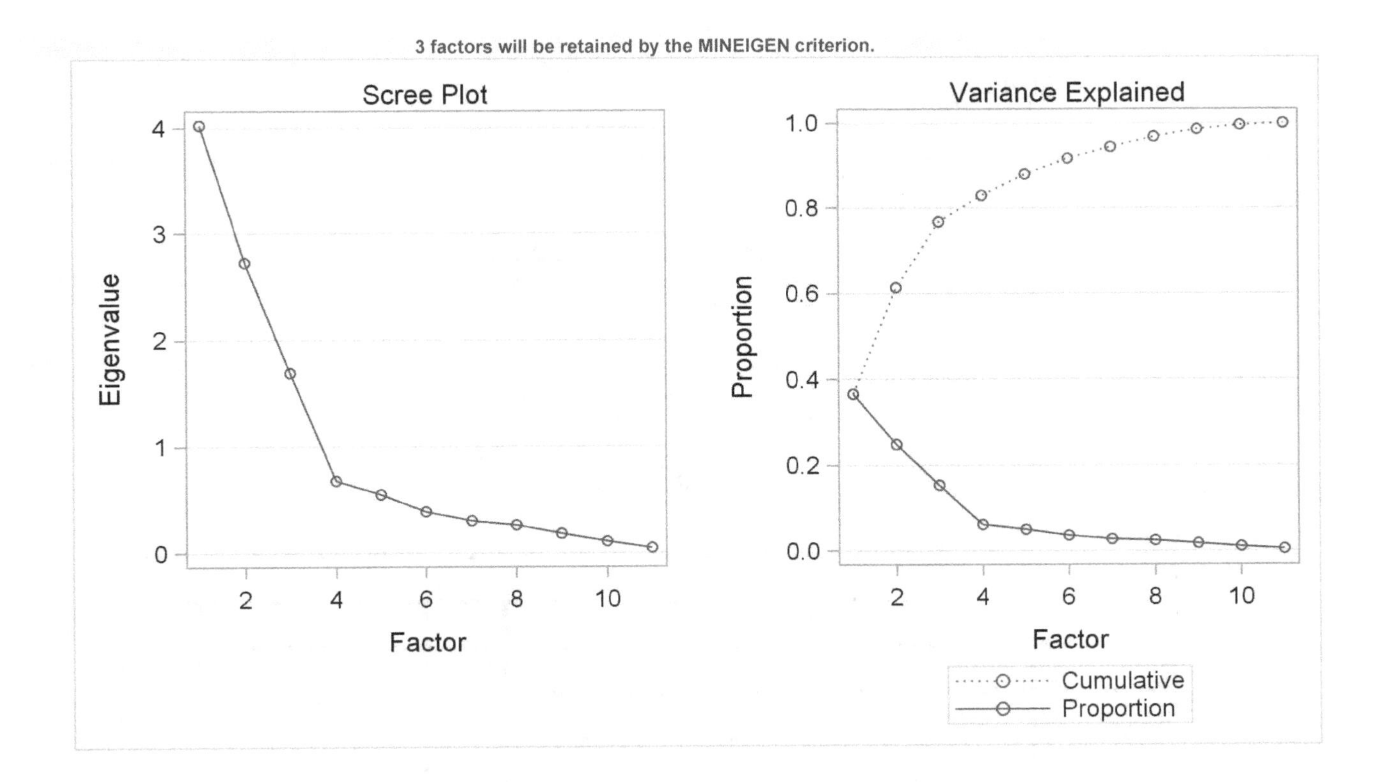

Factor Pattern						
	Factor1		Factor2		Factor3	
V1	38		77	*	-17	
V2	30		83	*	-15	
V3	32		80	*	8	
V4	29		70	*	15	
V5	83	*	-23		38	
V6	83	*	-30		38	
V7	89	*	-24		24	
V8	88	*	-12		7	
V9	-56	*	13		47	*
V10	-44	*	22		70	*
V12	-40		18		74	*

Printed values are multiplied by 100 and rounded to the nearest integer. Values greater than 0.4 are flagged by an '*'.

Variance Explained by Each Factor		
Factor1	Factor2	Factor3
4.0240860	2.7270385	1.6897911

Output 1.4 (Page 3)

The FACTOR Procedure
Rotation Method: Varimax

Orthogonal Transformation Matrix			
	1	2	3
1	0.84713	0.32918	-0.41716
2	-0.27774	0.94354	0.18052
3	0.45303	-0.03706	0.89073

Rotated Factor Pattern						
	Factor1		Factor2		Factor3	
V1	3		86	*	-17	
V2	-4		89	*	-11	
V3	8		86	*	8	
V4	12		75	*	14	
V5	94	*	4		-4	
V6	96	*	-2		-6	
V7	93	*	5		-20	
V8	81	*	18		-33	
V9	-30		-8		68	*
V10	-12		4		85	*
V12	-5		1		86	*
Printed values are multiplied by 100 and rounded to the nearest integer. Values greater than 0.4 are flagged by an '*'.						

Variance Explained by Each Factor		
Factor1	Factor2	Factor3
3.4449528	2.8661574	2.1298054

The results obtained when item 11 is deleted from the analysis are very similar to those obtained when it was included. The eigenvalue table of Output 1.4 shows that the eigenvalue-one criterion would again result in retaining three components. The first three components account for close to 77% of the total variance, which means that three components would also be retained if you used the variance-accounted-for criterion. Also, the scree plot from page 2 of Output 1.4 is cleaner than observed with the initial analysis; the break between components 3 and 4 is now more distinct and the eigenvalues again level off after this break. This means that three components would also likely be retained if the scree test were used to solve the number-of-components problem.

The biggest change can be seen in the rotated factor pattern that appears on page 4 of Output 1.4. The solution is now cleaner in the sense that no item loads on more than one component (i.e., no complex items). The current results now demonstrate a somewhat simpler structure than the initial analysis of the investment model data.

Conclusion

Principal component analysis is an effective procedure for reducing a number of observed variables into a smaller number that account for most of the variance in a dataset. This technique is particularly useful when you need a data reduction procedure that makes no assumptions concerning an underlying causal structure responsible for covariation in the data.

Appendix: Assumptions Underlying Principal Component Analysis

Because a principal component analysis is performed on a matrix of Pearson correlation coefficients, the data should satisfy the assumptions for this statistic. These assumptions are described in Appendix A.5, "Preparing Scattergrams and Computing Correlations," and are briefly reviewed here:

- **Interval- or ratio-level measurement.** All variables should be assessed on an interval or ratio level of measurement.
- **Random sampling.** Each participant will contribute one score on each observed variable. These sets of scores should represent a random sample drawn from the population of interest.
- **Linearity.** The relationship between all observed variables should be linear.
- **Bivariate normal distribution.** Each pair of observed variables should display a bivariate normal distribution (e.g., they should form an elliptical scattergram when plotted).

References

Cattell, R. B. (1966). The scree test for the number of factors. *Multivariate Behavioral Research, 1,* 245–276.

Chou, P. H. B., and O'Rourke, N. (2012). Development and initial validation of the Therapeutic Misunderstanding Scale for use with clinical trial research participants. *Aging and Mental Health, 16,* 45–15.

Clark, L. A., and Watson, D. (1995). Constructing validity: Basic issues in objective scale development. *Psychological Assessment, 7,* 309–319.

DeVellis, R. F. (2012). *Scale development theory and applications* (3rd Ed.). Thousand Oaks, CA: Sage.

Henson, R. K. (2001). Understanding internal consistency estimates: A conceptual primer on coefficient alpha. *Measurement and Evaluation in Counseling and Development, 34,* 177–189.

Kaiser, H. F. (1960). The application of electronic computers to factor analysis. *Educational and Psychological Measurement, 20,* 141–151.

Le, B., and Agnew, C. R. (2003). Commitment and its theorized determinants: A meta-analysis of the investment model. *Personal Relationships, 10,* 37–57.

Little, R. J. A., and Rubin, D. B. (1987). *Statistical analyses with missing data.* New York: Wiley.

O'Rourke, N., and Cappeliez, P. (2002). Development and validation of a couples measure of biased responding: The Marital Aggrandizement Scale. *Journal of Personality Assessment, 78,* 301–320.

Rusbult, C. E. (1980). Commitment and satisfaction in romantic associations: A test of the investment model. *Journal of Experimental Social Psychology, 16,* 172–186.

Saris, W. E., and Gallhofer, I. N. (2007). *Design, evaluation, and analysis of questionnaires for survey research.* Hoboken, NJ: Wiley InterScience.

Stevens, J. (2002). *Applied multivariate statistics for the social sciences* (4th Ed.). Mahwah, NJ: Lawrence Erlbaum.

Streiner, D. L. (1994). Figuring out factors: The use and misuse of factor analysis. *Canadian Journal of Psychiatry, 39,* 135–140.

van Buuren, S. (2012). *Flexible imputation of missing data.* Boca Raton, FL. Chapman and Hall.

Chapter 2: Exploratory Factor Analysis

Introduction: When Is Exploratory Factor Analysis Appropriate?

Exploratory factor analysis is used when you have obtained responses to several of measures and wish to identify the number and nature of the underlying factors that are responsible for covariation in the data. In other words, exploratory factor analysis is appropriate when you wish to identify the **factor structure** underlying a set of data.

For example, imagine that you are a political scientist who has developed a 50-item questionnaire to assess political attitudes. You administer the questionnaire to 500 people, and perform a factor analysis on their responses. The results of the analysis suggest that although the questionnaire contained 50 items, it really just measures two underlying factors, or constructs. You decided to label the first construct the **social conservatism** factor. Individuals who scored high on this construct tended to agree with statements such as "People should be married before living together," and "Children should respect their elders." You chose to label the second construct **economic conservatism**. Individuals who scored high on this factor tended to agree with statements such as "The size of the federal government should be reduced," and "Our taxes should be lowered."

In short, by performing a factor analysis on responses to this questionnaire, you were able to determine the number of constructs measured by this questionnaire (two) as well as the nature of those constructs. The results of the analysis showed which questionnaire items were measuring the social conservatism factor, and which were measuring economic conservatism.

The use of factor analysis assumes that each of the observed variables being analyzed is measured on an interval or ratio scale. Some additional assumptions underlying the use of factor analysis are listed in an appendix at the end of this chapter.

NOTE: You will see a good deal of similarity between the issues discussed in this chapter and those discussed in the preceding chapter on principal component analysis. This is because there are many similarities in terms of how principal component analysis and exploratory factor analysis are conducted even though there are conceptual differences between the two. Some of these differences and similarities are discussed in a later section titled "Exploratory Factor Analysis versus Principal Component Analysis."

It is likely that some users will read this chapter without first reviewing the previous chapter on principal component analysis; this makes it necessary to present much of the material that was already covered in the principal component chapter. Readers who have already covered the principal component chapter should be able to skim this material more quickly.

Introduction to the Common Factor Model

Example: Investment Model Questionnaire

Exploratory factor analysis will be demonstrated by performing a factor analysis on fictitious data from a questionnaire designed to measure constructs from Rusbult's investment model (1980). The investment model was introduced in the preceding chapter; you will remember that this model describes certain constructs that affect an individual's **commitment** to a romantic relationship (i.e., one's intention to maintain the relationship). Two of the constructs that are believed to influence commitment are alternative value and investment size. **Alternative value** refers to the attractiveness of alternatives to one's current romantic partner. For example, a woman would score high on alternative value if it would be appealing for her to leave her current partner for a different partner, or simply to leave her current partner and be unattached. **Investment size** refers to the time or personal resources that a person has put into a relationship with a current partner. For example, a woman would score high on investment size if she has invested a lot of time and effort in developing her current relationship, or if she and her partner have many mutual friendships that may be lost if the relationship were to end.

Imagine that you have developed a short questionnaire to assess alternative value and investment size. The questionnaire is to be completed by persons who are currently involved in romantic relationships. With this questionnaire, items 1 to 3 were designed to assess investment size, and items 4 to 6 were designed to assess alternative value. Part of the questionnaire is reproduced below:

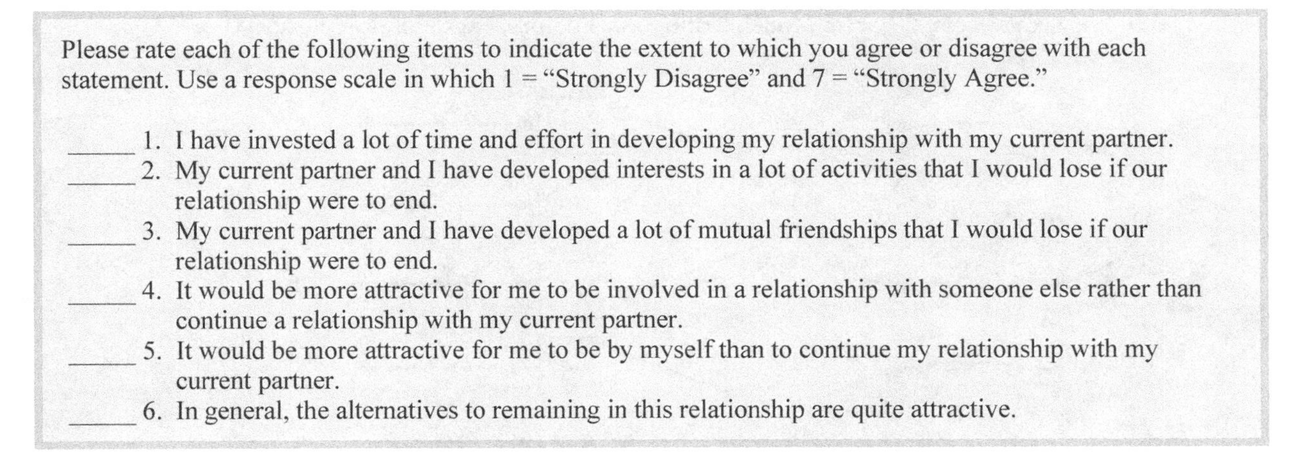
Please rate each of the following items to indicate the extent to which you agree or disagree with each statement. Use a response scale in which 1 = "Strongly Disagree" and 7 = "Strongly Agree."

_____ 1. I have invested a lot of time and effort in developing my relationship with my current partner.
_____ 2. My current partner and I have developed interests in a lot of activities that I would lose if our relationship were to end.
_____ 3. My current partner and I have developed a lot of mutual friendships that I would lose if our relationship were to end.
_____ 4. It would be more attractive for me to be involved in a relationship with someone else rather than continue a relationship with my current partner.
_____ 5. It would be more attractive for me to be by myself than to continue my relationship with my current partner.
_____ 6. In general, the alternatives to remaining in this relationship are quite attractive.

Assume that this questionnaire was administered to 200 participants, and their responses were entered so that responses to question 1 were coded as variable V1, responses to question 2 were coded as variable V2, and so forth. The correlations between the six variables are presented in Table 2.1.

Table 2.1: Correlations Coefficients between Questions Assessing Investment Size and Alternative Value

	Intercorrelations					
Question	**V1**	**V2**	**V3**	**V4**	**V5**	**V6**
V1	1.00					
V2	.81	1.00				
V3	.79	.92	1.00			
V4	-.03	-.07	-.01	1.00		
V5	-.06	-.01	-.11	.78	1.00	
V6	-.10	-.08	-.04	.79	.85	1.00

NOTE: N=200.

The preceding matrix of correlation coefficients consists of six rows (running horizontally) and six columns (running vertically). Where the row for one variable intersects with the column for a second variable, you will find the correlation coefficient for that pair of variables. For example, where the row for V2 intersects with the column for V1, you can see that the correlation between these items is .81.

Notice the pattern of intercorrelations. Questions 1, 2, and 3 are strongly correlated with one another, but these variables are essentially uncorrelated with questions 4, 5, and 6; similarly, question 4, 5, and 6 are strongly correlated with one another, but are essentially uncorrelated with questions 1, 2, and 3. Reviewing the complete matrix reveals that there are two sets of variables that seem to "hang together:" Variables 1, 2, and 3 form one group, and variables 4, 5, and 6 form the second group. But why do responses group together in this manner?

The Common Factor Model: Basic Concepts

One possible explanation for this pattern of intercorrelations may be found in Figure 2.1. In this figure, responses to questions 1 through 6 are represented as the six squares labeled V1 through V6. This model suggests that variables V1, V2, and V3 are correlated with one another because they are all influenced by the same underlying factor. A **factor** is an unobserved variable (or latent variable). Being "latent" means that you cannot measure a factor directly like you would measure an observed variable such as height or weight. A factor is a hypothetical construct: You believe it exists and that it influences certain manifest (or observed) variables that can be measured directly. In the present study, the manifest or observed variables are participant responses to items 1 through 6.

Figure 2.1: Six Variable, 2-Factor Model, Orthogonal Factors, Factorial Complexity=1

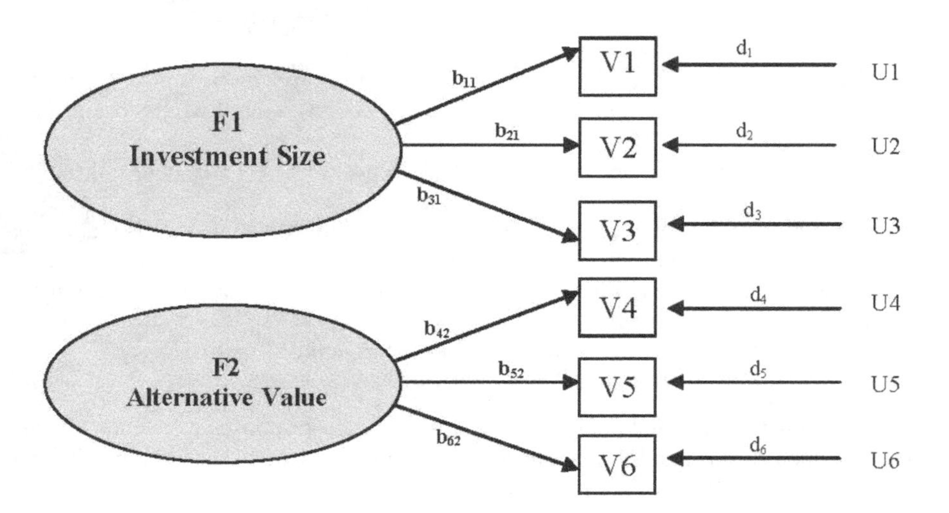

When representing models as figures, it is conventional to represent observed variables as squares or rectangles, and to represent latent factors as circles or ovals. You can therefore see that two factors appear in Figure 2.1. The first is labeled "F1: Investment Size," and the second is labeled "F2: Alternative Value."

We now return to the original question: Why do variables V1, V2, and V3 correlate so strongly with one another? According to the model presented in Figure 2.1, these variables are intercorrelated because they are all measuring aspects of the same latent factor: the underlying "investment size" construct. This model proposes that, within participants' belief systems, there is a construct that you might call "investment size." Furthermore, this construct influences the way that participants respond to questions 1, 2, and 3 (notice the arrows going from the oval factor to the squares). Even though you cannot directly measure someone's standing on the factor (i.e., it is a hypothetical construct), you can infer that it exists by:

- noting that questions 1, 2, and 3 correlate highly with one another
- reviewing the content of questionnaire items 1, 2, and 3 (i.e., noting what these questions actually say)noting that all three questions seem to be measuring the same basic construct that could reasonably be named "investment size"

(Please don't misunderstand, the preceding is not a description of how to perform factor analysis; it is just an example to help convey the conceptual meaning of our hypothetical model.)

Common Factors

The investment size factor (F1) presented in Figure 2.1 is known as a common factor. A **common factor** is a one that influences more than one observed variable. In this case, you can see that variables V1, V2, and V3 are all influenced by the investment size factor. It is called a *common* factor because more than one variable shares it in common. Because of this terminology, the type of analyses discussed in this chapter is sometimes referred to as **common factor analysis**.

In the lower half of Figure 2.1, you can see that there is a second common factor (F2) representing the "alternative value" hypothetical construct. This factor affects responses to items 4, 5, and 6 (notice the directional arrows). In short, variables V4, V5, and V6 are intercorrelated because they have this alternative value factor in common. In contrast, variables V4, V5, and V6 are not influenced by the investment size factor (notice that there are no arrows going from F1 to these variables); similarly, V1, V2, and V3 are not influenced by the alternative value factor, F2. This should help clarify why variables V1, V2, and V3 tend to be uncorrelated with variables V4, V5, and V6.

Orthogonal versus Oblique Models

A few more points must be made in order to understand the factor model presented in Figure 2.1 more fully. Notice that there is no arrow connecting F1 and F2. If it were hypothesized that the factors were correlated with one another, there would be a curved double-headed or bidirectional arrow connecting the two ovals. A double-headed arrow indicates that two constructs are correlated with no cause-and-effect relationship specified. The absence of a double-headed arrow in Figure 2.1 means that the researcher expects these factors are uncorrelated, or **orthogonal**. If a double-headed arrow did connect them, we would say that the factors are correlated, or **oblique**. Oblique factor models will be discussed later in this chapter.

In some factor models, a single-headed arrow connects two latent factors, indicating that one factor is expected to have a directional effect on the other. Such models are normally not examined with exploratory factor analysis, however, and will not be discussed in this chapter. For information on models that predict relationships between latent factors, see Chapter 5 "Developing Measurement Models with Confirmatory Factor Analysis" and Chapter 6: "Structural Equation Modeling."

Unique Factors

Notice that the two common factors are not the only ones that influence the observed variables. For example, you can see that there are actually two factors that influence variable V1: (a) the common factor, F1; and (b) a second factor, "U1." Here, U1 is a **unique factor**: One that influences only one observed variable. A unique factor represents all of the independent factors that are unique to that single variable including the error component that is unique to that variable. In the figure, the unique factor U1 affects only V1, U2 affects only V2, and so forth.

Factor Loadings

In Figure 2.1, each of the arrows going from a common factor to an observed variable is identified with a specific coefficient such as b_{11}, b_{21}, or b_{42}. The convention used in labeling these coefficients is quite simple: The first number in the subscript represents the number of the variable that the arrow points toward, and the second number in the subscript represents the number of the factor where the arrow originates. In this way, the coefficient "b_{21}" represents the arrow that goes to variable 2 from Factor 1; the coefficient "b_{52}" represents the arrow that goes to variable 5 from Factor 2; and so forth.

These coefficients represent **factor loadings**. But what exactly is a factor loading? Technically, it is a coefficient that appears in either a factor pattern matrix or a factor structure matrix. (These matrices are included in the output of an oblique factor analysis.) When one conducts an oblique factor analysis, the loadings in the pattern matrix will have a definition that is different from the definition given to loadings in the structure matrix. We will discuss these definitions later in the chapter. To keep things simple, however, we will skip the oblique analysis for the moment, and instead describe what the loadings represent when one performs an analysis in which the factors are orthogonal (uncorrelated). Factor loadings have a more simple interpretation in an orthogonal solution.

When examining orthogonal factors, the b coefficients may be understood in a number of different ways. For example, they may be viewed as:

- **Standardized regression coefficients**. The factor loadings obtained in an analysis with orthogonal factors may be thought of as standardized regression weights. If all variables (including the factors) are standardized to have unit variance (i.e., variance = 1.00), the b coefficients are analogous to the standardized regression coefficients (or regression weights) obtained in regression analysis. In other words, the b weights may be thought of as optimal linear weights by which the F factors are multiplied in calculating participant scores on the V variables (i.e., the weights used in predicting the variables from the factors).
- **Correlation coefficients**. Factor loadings also represent the product-moment correlation coefficients between an observed variable and its underlying factor. For example, if $b_{52} = .85$, this would indicate that the correlation between V5 and F2 is .85. This may surprise you if you are familiar with multiple regression, because most textbooks on multiple regression point out that standardized multiple

- regression coefficients and correlation coefficients are different things. However, standardized regression coefficients are equivalent to correlation coefficients when predictor variables are completely uncorrelated with each other. And that is the case in factor analysis with orthogonal factors: The factors serve as *predictor* variables in predicting the observed variables. Because the factors are uncorrelated, the factor loadings may be interpreted as both standardized regression weights and as correlation coefficients.
- **Path coefficients**. Finally, b coefficients are also analogous to the path coefficients obtained in path analysis. That is, they may be seen as standardized linear weights that represent the size of the effect that an underlying factor has in predicting variability in the observed variable. (Path analysis is covered in Chapter 4 of this text.)

Factor loadings are important because they help you interpret the factors that are responsible for covariation in the data. This means that, after the factors are rotated, you can review the nature of the variables that have significant loadings for a given factor (i.e., the variables that are most strongly related to the factor). The nature of these variables will help you understand the nature of that factor.

Factorial Complexity

Factorial complexity is a characteristic of an observed variable. The factorial complexity of a variable refers to the number of common factors that have a significant loading for that variable. For example, in Figure 2.1 you can see that the factorial complexity of V1 is one: V1 displays a significant loading for F1, but not for F2. The factorial complexity of V4 is also one: It displays a significant loading for F2 but not for F1.

Although the Figure 2.1 factor model is fairly simple, Figure 2.2 depicts a more complex example. As with the previous model, two common factors are again responsible for covariation in the dataset. However, you can see that both common factors in Figure 2.2 have significant loadings on all six observed variables. In the same way, you can see that each variable is influenced by both common factors. Because each variable in the figure has significant loadings for two common factors, each variable has a factorial complexity of two.

Figure 2.2: Six Variable, 2-Factor Model, Orthogonal Factors, Factorial Complexity=2

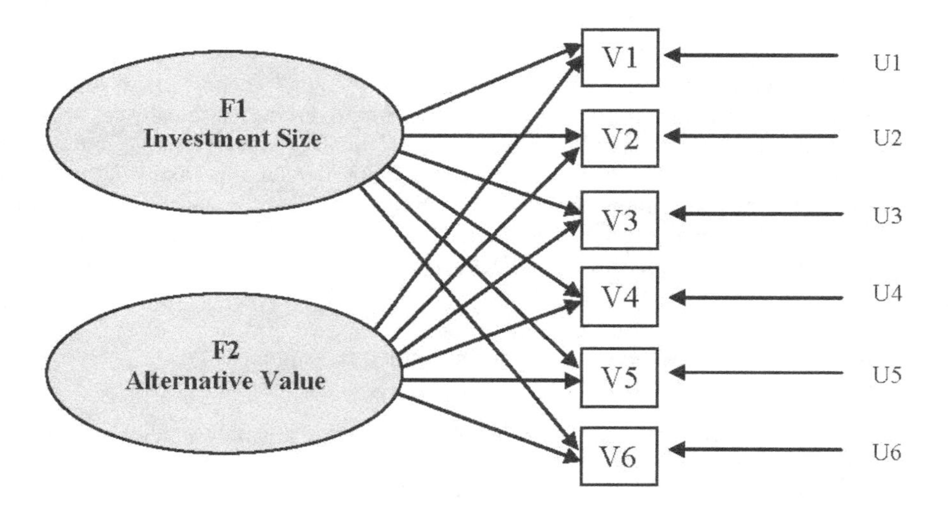

Observed Variables as Linear Combinations of Underlying Factors

It is possible to think of a given observed variable such as V1, as being a weighted sum of the underlying factors included in the factor model. For example, notice that in Figure 2.2, there are three factors that affect V1: Two common factors (F1 and F2), and one unique factor (U1). By multiplying these factors by the appropriate weights, it is possible to calculate any participant's score on V1. See the following equation:

$$V1 = b_{11}(F1) + b_{12}(F2) + d_1(U1)$$

In this equation, b_{11} is the regression weight for F1 (the amount of weight given to F1 in the prediction of V1), b_{12} is the regression weight for F2, and d1 is the regression weight for the unique factor associated with V1. You can see that a given person's score on V1 is determined by multiplying the underlying factors by the appropriate regression weights, and summing the resulting products. This is why, in factor analysis, the observed variables are viewed as linear combinations of underlying factors.

The preceding equation is therefore similar to the multiple regression equation as described in most statistics texts. In factor analysis, the observed variable (i.e., V1) is a counterpart to the criterion variable (Y) in multiple regression, and the latent factors (i.e., F1, F2 and U1) are counterparts to the predictor variables (i.e., the X variables) in multiple regression. We generally expect to obtain a different set of factor weights, and thus a different predictive equation, for each observed variable in a factor analysis.

Where does one find the regression weights for the common factors in factor analysis? These are found in the **factor pattern matrix**. An example of a pattern matrix is presented below:

Table 2.2

	Factor Pattern	
Variable	**Factor 1**	**Factor 2**
V1	.87	.26
V2	.80	.48
V3	.77	.34
V4	-.56	.49
V5	-.58	.52
V6	-.50	.59

You can see that the rows (running left to right) in the factor pattern represent the different observed variables such as V1 and V2. The columns in the factor pattern represent the different factors, such as F1 and F2. Where a row and column intersect, you will find a factor loading (or standardized regression coefficient). For example, when determining values of variable V1, F1 is given a weight of .87 and F2 is given a weight of .26; when determining values of V2, F1 is given a weight of .80 and F2 is given a weight of .48.

Communality versus the Unique Component

A **communality** is a characteristic of an observed variable. It refers to the variance in an observed variable that is accounted for common factors. If a variable exhibits a large communality, it means that this variable is strongly influenced by at least one common factor. The symbol for communality is h^2. The communality for a given variable is computed by squaring that variable's factor loadings for all retained common factors, and summing these squares. For example, using the factor loadings from the previous factor pattern, you may compute the communality for V1 in the following way:

$$h_1^2 = b_{11}^2 + b_{12}^2$$

$$= (.87)^2 + (.26)^2$$

$$= .755 + .068$$

$$= .82$$

So the communality for V1 is approximately 82. This means that 82% of the variance in V1 is accounted for by the two common factors. You can now compute the communality for each variable, and add these values to the table that contains the pattern matrix:

Table 2.3

	Factor Pattern		
Variable	**Factor 1**	**Factor 2**	**h^2**
V1	.87	.26	.82
V2	.80	.48	.87
V3	.77	.34	.71
V4	-.56	.49	.55
V5	-.58	.52	.61
V6	-.50	.59	.60

In contrast to the communality, the **unique component** refers to the proportion of variance in a given observed variable that is *not* accounted for by the common factors. Once communalities are computed, it is a simple matter to calculate the unique component: Simply subtract the communality from one. The unique component for V1 can be calculated in this fashion:

$$d_1^2 = 1 - h_1^2$$

$$= 1 - .82$$

$$= .18$$

And so, 18% of the variance in V1 is not accounted for by the common factors; alternatively, you could say that 18% of the variance in V1 is accounted for by the unique factor, U1.

If you then proceed to take the square root of the unique component, you can compute the coefficient "d." This should look familiar, because we earlier defined d as the weight given to a unique factor in determining values on the observed variable. For variable V1, the unique component was calculated as .18. The square root of .18 is approximately .42. Therefore, the unique factor U1 would be given a weight of .42 in determining values of V1 (i.e., $d_1 = .42$).

Exploratory Factor Analysis versus Principal Component Analysis

Some readers may be struck by the many similarities between exploratory factor analysis and principal component analysis. In fact, these similarities have even led some researchers to incorrectly report that they have conducted "factor analysis" when, in fact, they have conducted principal component analysis. Because of this common misunderstanding, this section will review some of the similarities and differences between the two procedures.

How Factor Analysis Differs from Principal Component Analysis

Purpose

Factor analysis is used to identify the factor or latent structure underlying a set of variables. In other words, if you wish to identify the number and nature of latent factors that are responsible for covariation in a dataset, then factor analysis, and not principal components analysis, should be used.

Principal Components versus Common Factors

A principal component is an artificial variable; it is a linear combination of (optimally weighted) observed variables. It is possible to calculate where a given participant stands on a principal component by simply summing that participant's (optimally weighted) scores on the observed variable being analyzed. For example, one could determine each participant's score on principal component 1 using the following formula:

$$C_1 = b_{11}(X_1) + b_{12}(X_2) + \ldots\ b_{1p}(X_p)$$

where

C_1 = the participant's score on principal component 1 (the first component extracted)

b_{1p} = the regression coefficient (or weight) for observed variable p, as used in creating principal component 1

X_p = the participant's score on observed variable p

In contrast, a common factor is a hypothetical latent variable that is assumed to be responsible for the covariation between two or more observed variables. Because factors are unmeasured latent variables, you may never know exactly where a given participant stands on an underlying factor (though it is possible to arrive at estimates, as you will see later).

In common factor analysis, the factors are not assumed to be linear combinations of the observed variables (as is the case with principal component analysis). Factor analysis assumes just the opposite: Observed variables are linear combinations of the underlying factors. This is illustrated in the following equation:

$$X_1 = b_1(F_1) + b_2(F_2) + \ldots\ b_q(F_q) + d_1(U_1)$$

where

X_1 = the participant's score on observed variable 1

b_q = the regression coefficient (or weight) for underlying common factor q, as used in determining the participant's score on X_1

F_q = the participant's score on underlying factor q

d_1 = the regression weight for the unique factor associated with X_1

U_1 = the unique factor associated with X_1

Because similar steps are followed in extracting principal components and common factors, it is easy to incorrectly assume that they are conceptually identical. Yet the preceding equations show that they differ in an important way. With principal components analysis, principal components are linear combinations of the observed variables; however, the factors of factor analysis are not viewed in this way. In factor analysis the observed variables are viewed as linear combinations of the underlying factors.

Some readers may be confused by this point because they know that it is possible to compute factor scores in exploratory factor analysis. Furthermore, they know that these factor scores are essentially linear composites of observed variables. In reality, however, these factor scores are merely *estimates* of where participants stand on the underlying factors. These so-called factor scores generally do not correlate perfectly with scores on the actual underlying factor. For this reason, they are referred to as **estimated factor scores**.

On the other hand, the principal component scores obtained in principal component analysis are not estimates; they are exact representations of the extracted components. Remember that a principal component is simply a mathematical transformation (a linear combination) of the observed variables. So a given participant's component score accurately represents where that participant stands on the principal component. It is therefore correct to discuss *actual* component scores rather than estimated component scores.

Variance Accounted For

Factor analysis and principal component analysis also differ with respect to the type of variance accounted for or explained. The factors of factor analysis account for common variance in a dataset, while the components of principal component analysis account for total variance in the dataset. This difference may be understood with reference to Figure 2.3.

Figure 2.3: Total Variation in Variable X_1 as Divided Into Common and Unique Components

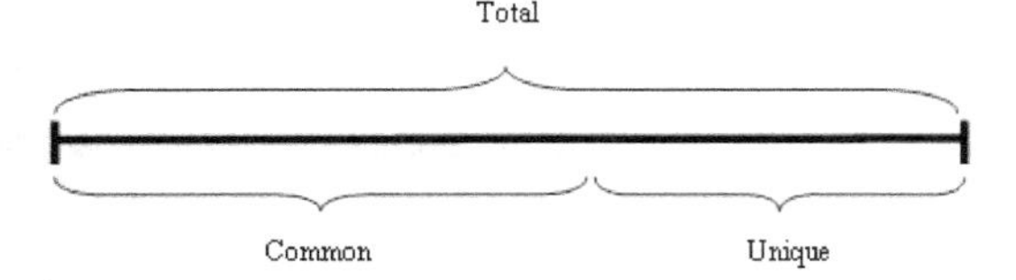

Assume that the length of the line in Figure 2.3 represents the total variance for observed variable X_1, and that variables X_1 through X_6 undergo factor analysis. The figure shows that the total variance in X_1 may be divided into two parts: Common variance and unique variance. **Common variance** corresponds to the communality of X_1: The proportion of total variance for the variable accounted for by the common factors. The remaining variance is the unique component: That variance (whether systematic or random) specific to variable X_1.

With factor analysis, factors are extracted to account only for the common variance; the remaining unique variance remains unanalyzed. This is accomplished by analyzing an **adjusted correlation matrix**: A correlation matrix with communality estimates on the diagonal. You cannot know a variable's actual communality prior to the factor analysis, and so it must be estimated using one of a number of alternative procedures. We recommend that squared multiple correlations be used as prior communality estimates. A variable's squared multiple correlation is obtained by using multiple regression to regress it on the remaining observed variables. (Later, you will find that these values can be obtained easily by using the PRIORS option with PROC FACTOR.) The adjusted correlation matrix that is analyzed in factor analysis has correlations between the observed variables off the diagonal and communality estimates on the diagonal.

With principal component analysis, however, components are extracted to account for **total variance** in the dataset, not just the common variance. This is accomplished by analyzing an **unadjusted correlation matrix**: A correlation matrix with ones (1.00) on the diagonal. Why ones? Since all variables are standardized in the analysis, each has a variance of one. Because the correlation matrix contains ones (rather than communalities) on the diagonal, 100% of each variable's variance will be accounted for by the combined components, not just the variance that the variable shares in common with other variables.

It is this difference that explains why only factor analysis—and not principal component analysis—can be used to identify the number and nature of the factors responsible for covariation in a dataset. Because principal component analysis makes no attempt to separate the common component from the unique component of each variable's variance, this procedure can provide a misleading picture of the factor structure underlying the data. Either procedure may be used to reduce a number of variables to a more manageable number; however, if one wishes to identify the factor structure of a dataset (such as that portrayed in Figure 2.1), only factor analysis is appropriate.

How Factor Analysis Is Similar to Principal Component Analysis

Purpose (in Some Cases)

Both factor analysis and principal component analysis may be used as **variable reduction procedures**; that is, both may be used to reduce a number of variables to a smaller, more manageable number. This is why both procedures are so widely used in analyzing data from multiple-item questionnaires the social science research; both procedures can be used to reduce a large number of survey questions into a smaller number of "scales."

Extraction Methods (in Some Cases)

This chapter shows how to use the principal axis method to extract factors. This is the same procedure used to extract principal components in the chapter on principal component analysis. (We will later show how to use the maximum likelihood method: An extraction method that is typically used only with factor analysis.)

Results (in Some Cases)

Principal component analysis and factor analysis often lead to similar conclusions regarding the appropriate number of factors (or components) to retain, as well as similar conclusions regarding how the factors (or components) should be interpreted. This is especially the case when the variable communalities are high (near 1.00). The reason for this should be obvious: When the principal axis extraction method is used, the only real difference between the two procedures involves the values that appear on the diagonal of the correlation matrix. If the communalities are very high (near 1.00), there is little difference between the matrix that is analyzed in principal component analysis and the matrix that is analyzed in factor analysis; hence the similar solutions.

Preparing and Administering the Investment Model Questionnaire

Assume that you are interested in measuring two constructs that constitute important components of Rusbult's investment model (1980). One construct is investment size: The amount of time or personal resources that the person has put into his or her relationship with a current partner; and the other construct is alternative value: The attractiveness of alternatives one's current romantic partner (Le and Agnew 2003).

Writing the Questionnaire Items

The questionnaire used discussed in the preceding chapter is again reproduced below. Note that items 1 to 3 were designed to assess investment size whereas items 4 to 6 were designed to assess alternative value.

Please rate each of the following items to indicate the extent to which you agree or disagree with each statement. Use a response scale in which 1 = "Strongly Disagree" and 7 = "Strongly Agree."

_____ 1. I have invested a lot of time and effort in developing my relationship with my current partner.
_____ 2. My current partner and I have developed interests in a lot of activities that I would lose if our relationship were to end.
_____ 3. My current partner and I have developed lot of mutual friendships that I would lose if our relationship were to end.
_____ 4. It would be more attractive for me to be involved in a relationship with someone else rather than continue a relationship with my current partner.
_____ 5. It would be more attractive for me to be by myself than to continue my relationship with my current partner.
_____ 6. In general, the alternatives to this relationship are quite attractive.

Number of Items per Factor

As mentioned in the previous chapter on principal component analysis, it is highly desirable to have at least three (and preferably more) variables loading on each factor when the analysis is complete. Because some items may be dropped during the course of the analysis, it is generally good practice to write at least five items for each construct that one wishes to measure; in this way, you increase the likelihood that at least three items per factor will survive the analysis. (You can see that preceding questionnaire violates this recommendation by including only three items for each factor at the outset.)

NOTE: Remember that the recommendation of three items per scale actually constitutes a *lower bound.* In practice, test and attitude scale developers normally desire that their scales contain many more than just three items to measure a given construct. It is not unusual to see individual scales that include 10, 20, or even more items to assess a single construct (e.g., O'Rourke and Cappeliez 2002). Other things being equal, the more items in a scale, the more reliable responses to that scale will be. The recommendation of three items per scale should therefore be viewed as a lower bound, appropriate only if practical concerns prevent you from including more items concerns (e.g., overall length of the questionnaire battery). For more information on scale construction, see Clark and Watson (1995), DeVellis (2012) and, Saris and Gallhofer (2007).

Minimal Sample Size Requirements

Exploratory factor analysis is a large-sample procedure, so it is important to use the following guidelines to determine the sample size which will be minimally adequate for an analysis as a general rule of thumb.

The minimal number of participants in the sample should be the larger of:

- 100 participants or
- 10 times the number of variables being analyzed (Floyd and Widaman 1995)

If questionnaire responses are being analyzed, then the number of variables is equal to the number of questionnaire items. To illustrate, assume that you wish to perform an exploratory factor analysis on responses to a 50-item questionnaire. Ten times the number of items on the questionnaire equals 500. Therefore, it would be best if your final sample provides usable (complete) data from at least 500 participants. It should be remembered, however, that any participant who fails to answer just one item will not provide usable data for the factor analysis, and will therefore be dropped from the final sample (unless you impute for missing responses; van Buuren, 2012). A certain number of participants can always be expected to leave at least one question blank; therefore, to insure that the final sample includes at least 500 usable responses, you would be wise to administer the questionnaire to perhaps 550 participants.

These rules regarding the number of participants per variable again constitute a lower bound, and some have argued that they should apply only under two optimal conditions for exploratory factor analysis: When many variables are expected to load on each factor; and when variable communalities are high. Under less optimal conditions, larger samples may be required. We again address the topic of sample size requirements in Chapter 4 "Path Analysis".

SAS Program and Exploratory Factor Analysis Results

This section provides instructions on writing the SAS program, along with an overview of the SAS output. A subsequent section will provide a more detailed treatment of the steps followed in the analysis, and the decisions to be made at each step.

Writing the SAS Program

The DATA Step

To perform an exploratory factor analysis, data may be input in the form of raw data, a correlation matrix, a covariance matrix, as well as other types of datasets (see Appendix A.2). In this example, raw data will be analyzed.

Assume that you administered your questionnaire to sample of 50 participants, and then entered their responses to each question. The SAS names given to these variables, and the format used in entering the data, are presented below:

Line	Column	Variable Name	Explanation
1	1–6	V1–V6	Participants' responses to survey questions 1 through 6. Responses were made using a 7-point scale, where higher scores indicate stronger agreement with the statement.
	8–9	COMMITMENT	Participants' scores on the commitment variable. Scores may range from 4 to 28, and higher scores indicate higher levels of commitment to maintain the relationship.

At this point, you are interested only in variables V1 to V6 (i.e., participant responses to the six questionnaire items). Scores on the commitment variable (COMMITMENT) are also included in the dataset because you will later compute correlations coefficients between estimated factor scores and COMMITMENT.

Below are the statements that will input these responses as raw data. The first three observations and the last three observations are reproduced here. For the entire (fictitious) dataset, see Appendix B, "Datasets."

```
data D1;
   input     #1     @1    (V1-V6)      (1.)
                    @8    (COMMITMENT)     (2.) ;

datalines;
776122 24
776111 28
111425  4
.
.
.
433344 15
557332 20
655222 13
;
run;
```

The dataset in Appendix B includes only 50 cases so that it will be relatively easy for interested readers to replicate these analyses. It should be restated, however, that 50 observations constitute an unacceptably small sample for an exploratory factor analysis (Floyd and Widaman 1995). Earlier it was said that a sample should provide usable data from the larger of either 100 cases or 10 times the number of observed variables. A small sample is being analyzed here for illustrative purposes only.

The PROC FACTOR Statement

The general form for the SAS program to perform an exploratory factor analysis with oblique rotation is presented below:

```
proc factor    data=dataset-name
               simple
               method=factor-extraction-method
               priors=prior-communality-estimates
               nfact=n
               plots=scree
               rotate=promax
               round
               flag=desired-size-of-"significant"-factor-loadings ;
   var variables-to-be-analyzed ;
run ;
```

Below is an actual program, including the DATA step that could be used to analyze some fictitious data from the investment model study.

```
data D1;
   input    #1    @1    (V1-V6)     (1.)
                  @8    (COMMITMENT)    (2.) ;
datalines;
776122 24
776111 28
111425  4
.
.
.
433344 15
557332 20
655222 13
;
run ;

proc factor    data=D1
               simple
               method=prin
               priors=smc
               nfact=2
               plots=scree
               rotate=promax
               round
               flag=.40    ;
   var V1 V2 V3 V4 V5 V6;
run;
```

Options Used with PROC FACTOR

The PROC FACTOR statement begins the factor procedure, and a number of options may be requested in this statement before it ends with a semicolon. Some options that are especially useful in social science research are presented below:

FLAG

causes the printer to flag (with an asterisk) factor loadings with absolute values greater than some specified size. For example, if you specify

```
flag=.35
```

an asterisk will appear next to any loading whose absolute value exceeds .35. This option can make it much easier to interpret a factor pattern. Negative values are not allowed in the flag option, and the flag option should be used in conjunction with the round option.

METHOD=factor-extraction-method
specifies the method to be used in extracting the factors. The current program specifies

```
method=prin
```

to request that the principal axis (principal factors) method be used for the initial extraction. Although the principal axis is a common extraction method, most researchers prefer the maximum likelihood method because it provides a significance test for solving the "number of factors" problem, and generally provides better parameter estimates. The maximum likelihood method may be requested with the option

```
method=ml
```

MINEIGEN=p
specifies the critical eigenvalue a factor must display if that factor is to be retained (here, p = the critical eigenvalue). Negative values are not allowed.

NFACT=n
allows you to specify the number of factors to be retained and rotated, where n = the number of factors.

OUT=name-of-new-dataset
creates a new dataset that includes all of the variables of the existing dataset, along with estimated factor scores for the retained factors. Factor 1 is given the variable name FACTOR1, factor 2 is given the name FACTOR2, and so forth. OUT= must be used in conjunction with the NFACT option, and the analysis must be based on raw data.

PRIORS=prior communality estimates
specifies prior communality estimates. The preceding specifies SMC to request that the squared multiple correlations between a given variable and the other observed variables be used as that variable's prior communality estimate.

ROTATE=rotation method
specifies the rotation method to be used. The preceding program requests a promax rotation that results in oblique (correlated) factors. This option is requested by specifying

```
rotate=promax
```

Orthogonal rotations may also be requested; Chapter 1 showed how to request an (orthogonal) rotation by specifying

```
rotate=varimax
```

ROUND
factor loadings and correlation coefficients in the matrices printed by PROC FACTOR are normally carried out to several decimal places. Requesting the ROUND option, however, causes all coefficients to be limited to two decimal places, rounded to the nearest integer, and multiplied by 100 (thus eliminating the decimal point). This generally makes it easier to read the coefficients.

PLOTS=SCREE
creates a plot that graphically displays the size of the eigenvalue associated with each factor. This can be used to perform a scree test to visually determine how many factors should be retained.

SIMPLE
requests simple descriptive statistics: The number of usable cases on which the analysis was performed and the means and standard deviations of the observed variables.

The VAR Statement

The variables to be analyzed are listed on the VAR statement, with each variable separated by at least one space. Remember that the VAR statement is a *separate* statement not an option within the factor statement, so do not forget to end the FACTOR statement with a semicolon before beginning the VAR statement.

Results from the Output

The preceding program would produce four pages of output. The following lists some of the information included in this output, and the page on which it appears:

- Page 1 presents simple statistics.
- Page 2 includes prior communality estimates, initial eigenvalues, scree plot of eigenvalues and cumulative variance, and final communality estimates.
- Page 3 includes the results of the orthogonal transformation matrix (varimax rotation), the rotated factor pattern matrix for the varimax solution, and final communality estimates.
- Page 4 includes results from the oblique rotation method (promax rotation) such as the inter-factor correlations, the rotated factor pattern matrix (standardized regression coefficients), the reference structure (semipartial correlations), the factor structure correlations and estimates of variance explained by each factor (ignoring other factors).

The following section reviews the steps by which exploratory factor analysis is conducted. Integrated into this discussion will be excerpts from the preceding output, along with guidelines for interpreting this output.

Steps in Conducting Exploratory Factor Analysis

Factor analysis is normally conducted in a sequence of steps, with somewhat subjective decisions being made at various steps. Because this is an introductory treatment of the topic, it will not provide a comprehensive discussion of all the options available to you at each step; instead, specific recommendations will be made, consistent with practices often followed in applied research. For a detailed discussion of exploratory factor analysis, see Kim and Mueller (1978a; 1978b), Loehlin (1987), and Rummel (1970).

Step 1: Initial Extraction of the Factors

The first step of the analysis involves the initial extraction of the factors. The preceding program specified the option

```
method=prin
```

which calls for the principal factors, or principal axis method. This is the same method used to extract the components of principal component analysis.

As with component analysis, the number of factors extracted will be equal to the number of variables being analyzed. Because six variables are being analyzed in the present study, six factors will be extracted. The first factor can be expected to account for a fairly large amount of the common variance. Each succeeding factor will account for progressively smaller amounts of variance. Although a large number of factors may be extracted in this way, only the first few factors will be sufficiently important to be retained for interpretation.

As with principal components, the extracted factors will have two important properties: (a) each factor will account for a maximum amount of the variance that has not already been accounted for by other previously extracted factors; and (b) each factor will be uncorrelated with all of the previously extracted factors. This second characteristic may come as a surprise, because earlier it was said that you were going to obtain an oblique solution (by specifying ROTATE=PROMAX) in which the factors would be correlated. In this analysis, however, the factors are in fact orthogonal (uncorrelated) at the time they are extracted. It is only later in the analysis that their orthogonality is relaxed, and they are allowed to become oblique. This will be discussed in more detail in a subsequent section on factor rotation.

These concepts will now be related to some of the results that appeared in the output created by the preceding program. Pages 1 and 2 of the output provided simple statistics, the eigenvalue table, and some additional information regarding the initial extraction of the factors. Those pages are reproduced here as Output 2.1.

Output 2.1: Simple Statistics, Prior Communalities, and Eigenvalue Table from Analysis of Investment Model Questionnaire (page 1)

The FACTOR Procedure

Input Data Type	Raw Data
Number of Records Read	50
Number of Records Used	50
N for Significance Tests	50

Means and Standard Deviations from 50 Observations		
Variable	Mean	Std Dev
V1	4.6200000	1.5371588
V2	4.3800000	1.5103723
V3	4.3600000	1.6383167
V4	2.7600000	1.2545428
V5	2.3600000	1.1021315
V6	2.5600000	1.3726185

Output 2.1 (page 2)

The FACTOR Procedure
Initial Factor Method: Principal Factors

Prior Communality Estimates: SMC					
V1	V2	V3	V4	V5	V6
0.78239483	0.81705605	0.67662145	0.47918877	0.52380277	0.49871459

Eigenvalues of the Reduced Correlation Matrix: Total = 3.77777847 Average = 0.62962975				
	Eigenvalue	Difference	Proportion	Cumulative
1	2.87532884	1.59874396	0.7611	0.7611
2	1.27658489	1.28903380	0.3379	1.0990
3	-.01244892	0.07484205	-0.0033	1.0957
4	-.08729097	0.03685491	-0.0231	1.0726
5	-.12414588	0.02610362	-0.0329	1.0398
6	-.15024950		-0.0398	1.0000

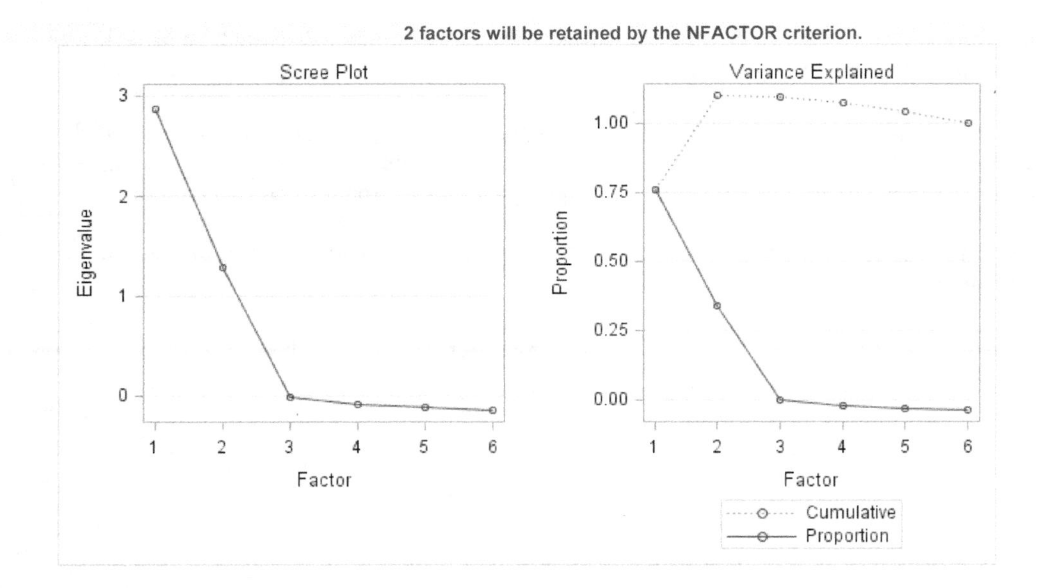

Factor Pattern				
	Factor1		Factor2	
V1	87	*	26	
V2	80	*	48	*
V3	77	*	34	
V4	-56	*	49	*
V5	-58	*	52	*
V6	-50	*	59	*

Printed values are multiplied by 100 and rounded to the nearest integer. Values greater than 0.4 are flagged by an '*'.

Variance Explained by Each Factor	
Factor1	Factor2
2.8753288	1.2765849

Final Communality Estimates: Total = 4.151914					
V1	V2	V3	V4	V5	V6
0.81677554	0.87417817	0.70443448	0.55882781	0.60705615	0.59064158

On page 1 of Output 2.1, the simple statistics section shows that the analysis was based on 50 observations. Means and standard deviations are also provided.

The first line of page 2 says "Initial Factor Method: Principal Factors." This indicates that the principal factors method was used for the initial extraction of the factors.

Next, the prior communality estimates are printed. Because the program included the PRIORS=SMC option, the prior communality estimates are squared multiple correlations.

Below that, the eigenvalue table is printed. An **eigenvalue** represents the amount of variance that is accounted for by a given factor. In the column labeled "Eigenvalue," the eigenvalue for each factor is presented. Each row in the matrix presents information about one of the six factors: The row labeled "1" provides information about the first factor extracted. The row labeled "2" provides information about the second factor extracted, and so forth.

Where the column headed "Eigenvalue" intersects with the rows labeled "1" and "2," you can see that the eigenvalue for factor 1 is approximately 2.88, while the eigenvalue for factor 2 is 1.28. This pattern is consistent with our earlier statement that the first factors extracted tend to account for relatively large amounts of variance, while the later factors account for relatively smaller amounts.

Step 2: Determining the Number of "Meaningful" Factors to Retain

As with principal component analysis, the number of factors extracted is equal to the number of variables analyzed, necessitating that you decide just how many of these factors are truly meaningful and worthy of being retained for rotation and interpretation. In general, we expect that only the first few factors will account for meaningful amounts of variance, and that the later factors will tend to account for relatively small amounts of variance (i.e., largely error variance). The next step of the analysis, therefore, is to determine how many meaningful factors should be retained for interpretation.

The preceding program specified NFACT=2 so that two factors would be retained; because this was the initial analysis, you had no empirical reason to expect two meaningful factors, and specified NFACT=2 on a hunch. If the empirical results suggest a different number of meaningful factors, the NFACT option may be changed for subsequent analyses.

The chapter on principal component analysis discussed four options that can be used to help make the "number of factors" decision; the first of these was the **eigenvalue-one** criterion or Kaiser-Guttman criterion (Kaiser 1960). When using this criterion, you retain any principal component with an eigenvalue greater than 1.00.

The eigenvalue-one criterion made sense in principal component analysis, because each variable contributed one unit of variance to the analysis. This criterion insured that you would not retain any component that accounted for less variance than had been contributed by one variable.

For the same reason, however, you can see that the eigenvalue-one criterion is less appropriate in common factor analysis. Remember that each variable does not contribute one unit of variance to this analysis but, instead, contributes its prior communality estimate. This estimate will be less than 1.00, and so it makes little sense to use the value of 1.00 as a cutting point for retaining factors. Without the eigenvalue-one criterion, you are left with the following three options.

The Scree Test

With the **scree test** (Cattell 1966), you plot the eigenvalues associated with each factor and look for a "break" between factors with relatively large eigenvalues and those with smaller eigenvalues. The factors that appear before the break are assumed to be meaningful and are retained for rotation; those appearing after the break are assumed to be unimportant and are not retained.

Specifying the PLOTS=SCREE option in the PROC FACTOR statement causes SAS to print an eigenvalue plot as part of the output. This scree plot is presented here as Output 2.2.

Output 2.2: Scree Plot of Eigenvalues from Analysis, and Proportion of Variance Explained, of Investment Model Questionnaire

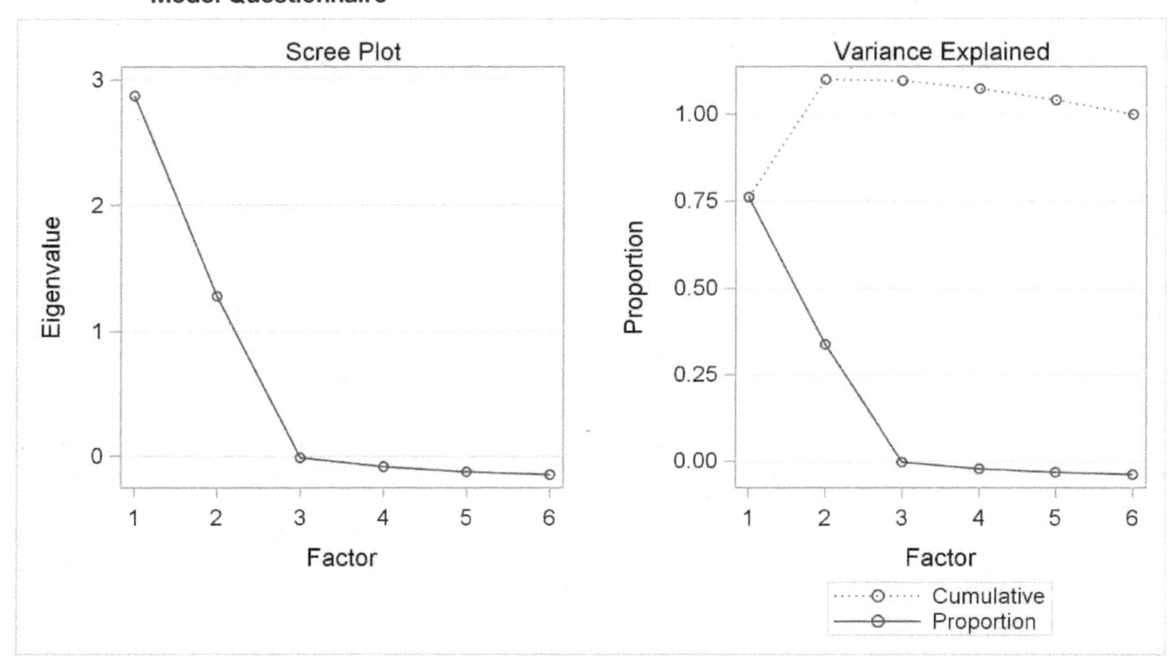

The scree plot appears on the left. You can see that the factor numbers are listed on the horizontal axis, while eigenvalues are listed on the vertical axis. With this plot, notice that there is a relatively large break between factors 1 and 2, another large break between factors 2 and 3, but that there is no break between factors 3 and 4, 4 and 5, or 5 and 6. Because factors 3 through 6 have relatively small eigenvalues, and the data points for factors 3 through 6 could almost be fitted with a straight line, they can be assumed to be relatively unimportant factors. Given this plot, a scree test would suggest that only factors 1 and 2 be retained because only these factors appear before the last big break. Factors 3 through 6 appear after the break, and thus will not be retained.

Proportion of Variance Accounted For

A second criterion in making the number of factors decision involves retaining a factor if it accounts for a certain **proportion (or percentage) the variance in the dataset**. For example, you may decide to retain any factor that accounts for at least 5% or 10% of the common variance. (See right-hand side graph, Output 2.2.) This proportion can be calculated with a simple formula:

```
                     Eigenvalue for the factor of interest
Proportion = ---------------------------------------------
                   Total eigenvalues of the correlation matrix
```

In principal component analysis, the "total eigenvalues of the correlation matrix" was equal to the total number of variables being analyzed (because each variable contributed one unit of variance to the dataset). In common factor analysis, however, the total eigenvalues will be equal to the sum of the communalities that appear on the main diagonal of the matrix being analyzed.

The proportion of common variance accounted for by each factor is printed in the eigenvalue table from output page 2 below the heading "Proportion." The eigenvalue table for the preceding analysis is presented again as Output 2.3.

Output 2.3: Eigenvalue Table from Analysis of Investment Model Questionnaire

Eigenvalues of the Reduced Correlation Matrix: Total = 3.77777847 Average = 0.62962975				
	Eigenvalue	Difference	Proportion	Cumulative
1	2.87532884	1.59874396	0.7611	0.7611
2	1.27658489	1.28903380	0.3379	1.0990
3	-.01244892	0.07484205	-0.0033	1.0957
4	-.08729097	0.03685491	-0.0231	1.0726
5	-.12414588	0.02610362	-0.0329	1.0398
6	-.15024950		-0.0398	1.0000

2 factors will be retained by the NFACTOR criterion.

From the "Proportion" column of the preceding eigenvalue table, you can see that the first factor alone accounts for 76% of the common variance, the second factor alone accounts for almost 34%, and the third factor accounts for less than 1%. (In fact, Factor 3 actually has a negative percentage; see the following box for an explanation.) If one were using, say, 10% as the criterion for deciding whether a factor should be retained, only Factors 1 and 2 would be retained in the present analysis. Despite the apparent ease of use of this criterion, however, remember that this approach has been criticized (Kim and Mueller 1978b).

How can you account for over 100% of the common variance? The final column of the eigenvalue table (labeled "Cumulative") provides the cumulative percent of common variance accounted for by the factors. Output 2.3 shows that factor 1 accounts for 76% of the common variance (the value in the table is 0.76), and factors 1 and 2 combined account for 110%. But how can two factors account for over 100% of the common variance?

In brief, this is because the prior communality estimates were not perfectly accurate. Consider this: If your prior communality estimates were perfectly accurate estimates of the variables' actual communalities, and if the common factor model was correctly estimated, then the factors that you retained would have to account for exactly 100% of the common variance, and the remaining factors would have to account for 0%. The fact that this did not happen in the present analysis is probably because your prior communality estimates (squared multiple correlations) were not perfectly accurate.

You may also be wondering why some of the factors seem to be accounting for a negative percent of the common variance (i.e., why they have negative eigenvalues). This is because the analysis is constrained so that the "Cumulative" proportion must equal 1.00 after the last factor is extracted. Since this cumulative value exceeds 1.00 at some points in the analysis, is was mathematically necessary that some factors have negative eigenvalues.

Interpretability Criterion

Perhaps the most important criterion to use when solving the "number of factors" problem is the **interpretability criteria**: Interpreting the substantive meaning of the retained factors and verifying that this interpretation "makes sense" in terms of what is known about the constructs under investigation. Below are four rules to follow when doing this. (A later section of this chapter will provide a step-by-step illustration of how to interpret a factor solution; the following rules will be more meaningful at that point.)

1. **Are there at least three variables (items) with significant loadings on each retained factor?** A solution is less satisfactory if a given factor is measured by less than three variables.
2. **Do the variables that load on a given factor share some conceptual meaning?** For example, if three questions on a survey all load on Factor 1, do all three of these questions seem to be measuring the same underlying construct?
3. **Do the variables that load on different factors seem to be measuring different constructs?** For example, if three questions load on Factor 1, and three other questions load on Factor 2, do the first three questions seem to be measuring a construct that is conceptually different from the construct measured by the last three questions?
4. **Does the rotated factor pattern demonstrate "simple structure?"** Simple structure means that the pattern possesses two characteristics: (a) most of the variables have relatively high factor loadings on only one factor, and near-zero loadings for the other factors; and (b) most factors have relatively high factor loadings for some variables, and near-zero loadings for the remaining variables. This concept of simple structure will be explained in more detail in a later section.

Recommendations

Given the preceding options, what procedure should you actually follow in solving the number of factors problem? This text recommends combining all three in a structured sequence. First, perform a scree test and look for obvious breaks in the plot. Because there will often be more than one break in the eigenvalue plot, it may be necessary to examine two or more possible solutions. Next, review the amount of common variance accounted for by each factor. We hesitate to recommend the rigid use of some specific but arbitrary cut off point, such as 5% or 10%. Still, if you are retaining factors that account for as little as 2% or 3% of the variance, it may be wise to take a second look at the solution and verify that these latter factors are of truly substantive importance. Finally, apply the interpretability criterion. If more than one solution can be justified on the basis of a scree test or the "variance accounted for" criteria, which of these solutions are the most interpretable? By seeking a solution that satisfies all three of these criteria, you maximize chances of correctly identifying the factor structure of the dataset.

Step 3: Rotation to a Final Solution

After extracting the initial factors, the computer will print an unrotated factor pattern matrix. The rows of this matrix represent the variables being analyzed, and the columns represent the retained factors. The entries in the matrix are factor loadings. In a factor *pattern* matrix, the observed variables are assumed to be linear combinations of the common factors, and the **factor loadings** are standardized regression coefficients for predicting the variables from the factors. (Later, you will see that the loadings have a different interpretation in a factor *structure* matrix.) With PROC FACTOR, the unrotated factor pattern is printed under the heading "Factor Pattern," and appears on output page 2. The factor pattern for the present analysis is presented as Output 2.4.

Output 2.4: Unrotated Factor Pattern from Analysis of Investment Model Questionnaire

Factor Pattern				
	Factor1		Factor2	
V1	87	*	26	
V2	80	*	48	*
V3	77	*	34	
V4	-56	*	49	*
V5	-58	*	52	*
V6	-50	*	59	*
Printed values are multiplied by 100 and rounded to the nearest integer. Values greater than 0.4 are flagged by an '*'.				

Variance Explained by Each Factor	
Factor1	Factor2
2.8753288	1.2765849

Final Communality Estimates: Total = 4.151914					
V1	V2	V3	V4	V5	V6
0.81677554	0.87417817	0.70443448	0.55882781	0.60705615	0.59064158

When more than one factor has been retained, an unrotated factor pattern is usually difficult to interpret. Factor patterns are easiest to interpret when some of the variables in the analysis have very high loadings on a given factor, and the remaining variables have near-zero loadings on that factor. Unrotated factor patterns often fail to display this type of pattern. For example, consider the loadings under the column heading "FACTOR1" in Output 2.4. Notice that variables V1, V2, and V3 do display fairly high loadings for this factor, which is good. Unfortunately, however, variables V4, V5, and V6 do not display near-zero loadings for this factor; the loadings for these three variables range from -.50 to -.58, which is to say that they are of moderate size. For reasons that will be made clear shortly, this state of affairs would make it difficult to interpret Factor 1.

To make interpretation easier, you will normally perform a linear transformation on the factor solution called a **rotation**. The previous chapter on principal component analysis demonstrated the use of an orthogonal rotation. It was explained that orthogonal rotations result in components (or factors) that are uncorrelated with one another.

In contrast, this chapter will illustrate the use of the promax rotation, which is a specific type of oblique rotation. Oblique rotations generally result in correlated factors (or components).

A promax rotation is actually conducted in two steps. The first step involves an orthogonal varimax prerotation. At this point in the analysis, the extracted factors are still uncorrelated. During the second step (the promax rotation), the orthogonality of the factors is relaxed, and they are allowed to correlate if warranted. Below, it will be seen that the interpretation of an oblique solution is more complicated than the interpretation of an orthogonal solution, though oblique rotations often provide better results (at least in those situations in which the actual, underlying factors truly are correlated).

Step 4: Interpreting the Rotated Solution

Orthogonal Solutions

During the prerotation step, SAS produces a rotated factor pattern similar to that which would be produced if you had specified ROTATE=VARIMAX. This matrix appears on output page 3 of the current output, and is presented as Output 2.5.

Output 2.5: Varimax (Orthogonal) Rotated Factor Pattern from Analysis of Investment Model Questionnaire

The FACTOR Procedure
Prerotation Method: Varimax

Orthogonal Transformation Matrix		
	1	2
1	0.82009	-0.57223
2	0.57223	0.82009

Rotated Factor Pattern				
	Factor1		Factor2	
V1	86	*	-28	
V2	93	*	-7	
V3	82	*	-16	
V4	-18		73	*
V5	-17		76	*
V6	-7		77	*

Printed values are multiplied by 100 and rounded to the nearest integer. Values greater than 0.4 are flagged by an '*'.

Variance Explained by Each Factor	
Factor1	Factor2
2.3518179	1.8000959

Final Communality Estimates: Total = 4.151914					
V1	V2	V3	V4	V5	V6
0.81677554	0.87417817	0.70443448	0.55882781	0.60705615	0.59064158

If you were interested in an orthogonal solution, it would be perfectly acceptable to interpret this rotated factor pattern in the manner described in the previous chapter on principal component analysis. Interested readers may turn to that chapter for a detailed discussion of how this is done. Because this chapter deals with oblique rotations, it will instead focus on how one interprets the results of the promax procedure.

Oblique Solutions

Before interpreting the meaning of the retained factors, you should first check the inter-factor correlations that appear on output toward the bottom of page 4. The results for the current analysis are presented here as Output 2.6.

Output 2.6: Inter-Factor Correlations from Analysis of Investment Model Questionnaire

Inter-Factor Correlations				
	Factor1		Factor2	
Factor1	100	*	-34	
Factor2	-34		100	*
Printed values are multiplied by 100 and rounded to the nearest integer. Values greater than 0.4 are flagged by an '*'.				

In Output 2.6, look in the section heading "Inter-Factor Correlations." Where the row heading "FACTOR1" intersects with the column heading "FACTOR2," you will find a correlation coefficient of -.34. This means that there is a correlation of -.34 between the two factors. At this point in the analysis, you do not know exactly what this correlations means, because you have not yet interpreted the meaning of the factors themselves. You will therefore return to this correlation after the interpretation of the factors has been completed.

In a sense, interpreting the nature of a given factor is relatively straightforward: You begin by looking for variables (survey items) that have high loadings on that factor. A high loading means that the variable is, in a sense, "measuring" that factor. You must review all of the variables with high loadings on that factor, and attempt to determine what the variables have in common. What underlying construct do all of the items seem to be measuring? In naming this construct, you name the factor.

As always, however, somewhat qualitative decisions must often be made. For example, how large must a factor loading be before you will conclude it is a "high" loading? As with the preceding chapter, we suggest that loadings equal to or greater than .40 be treated as meaningful loadings, and that loadings under .40 generally be ignored. As you gain expertise in performing factor analyses, you should explore the more sophisticated procedures for identifying "significant" loadings, such as those discussed by Stevens (2002).

With an orthogonal rotation, factor interpretation was fairly straightforward: You simply reviewed the factor pattern matrix to identify the variables with significant loadings on a given factor. With oblique rotations, however, the situation is somewhat more complex, because you must interpret two, and in some cases three different matrices, in order to fully understand the results. In all cases, the rotated factor pattern and factor structure matrices should be reviewed; in some cases, it may also be necessary to review the reference structure matrix.

First, you should review the **rotated factor pattern matrix**. This matrix appears on page 4 of the output for the current analysis. It is presented here as Output 2.7.

Output 2.7: Promax (Oblique) Rotated Factor Pattern from Analysis of Investment Model Questionnaire

Rotated Factor Pattern (Standardized Regression Coefficients)				
	Factor1		Factor2	
V1	85	*	-14	
V2	97	*	10	
V3	84	*	-1	
V4	-5		73	*
V5	-4		76	*
V6	7		79	*

Printed values are multiplied by 100 and rounded to the nearest integer. Values greater than 0.4 are flagged by an '*'.

Notice that "Standardized Regression Coefficients" appears in parentheses in the heading of this matrix. This should help remind you that the loadings appearing in this factor pattern are regression coefficients of the variables on the factors. In common factor analysis, the observed variables are viewed as linear combinations of the factors, and the elements of the factor pattern are regression weights associated with each factor in the prediction of these variables. The loadings in this matrix are also called **pattern loadings**, and may be said to represent the unique contribution that each factor makes to the variance of the observed variables (Rummel 1970).

You should rely most heavily on this rotated factor pattern matrix to interpret the meaning of each factor. The rotated factor pattern is more likely to display simple structure than the structure matrix (to be discussed below), and will be more useful in determining what names should be assigned to the factors.

The chapter on principal component analysis provided a structured procedure to follow in interpreting a rotated factor pattern. These guidelines are reproduced again below:

1. **Read across the row for the first variable**. All "meaningful loadings" (i.e., loadings greater than .40) have been flagged with an asterisk ("*"). This occurred because the FLAG=.40 option was specified in the preceding program. If a given variable has a meaningful loading on more than one factor (i.e., complex items), scratch that variable out and ignore it in your interpretation. In many situations, researchers wish to drop variables that load on more than one factor, because the variables are not "pure" measures of any one construct. In the present case, this means reviewing the row labeled V1, and reading to the right to see if it loads on more than one factor. In this case it does not, so you may retain this variable.
2. **Repeat this process for the remaining variables, scratching out any variable that loads on more than one factor**. In this analysis, none of the variables have high loadings for more than one factor, so none will have to be dropped.
3. **Review all of the surviving variables with high loadings on Factor 1 to determine the nature of this factor**. From the rotated factor pattern, you can see that only items 1, 2, and 3 load on Factor 1 (note the asterisks). It is now necessary to turn to the questionnaire itself and review the content of the questions in order to decide what a given factor should be named. What do questions 1, 2, and 3 have in common? What common construct do they seem to be measuring? For illustration, the questions being analyzed in the present case are again reproduced below. Remember that question 1 was represented as V1 in the SAS program, question 2 was V2, and so forth. To interpret Factor 1, you must read questions 1, 2, and 3 to see what they have in common.

Please rate each of the following statements to indicate the extent to which you agree or disagree with each using a response scale in which 1 = "Strongly Disagree" and 7 = "Strongly Agree."

_____ 1. I have invested a lot of time and effort in developing my relationship with my current partner.
_____ 2. My current partner and I have developed interests in a lot of fun activities that I would lose if our relationship were to end.
_____ 3. My current partner and I have developed lot of mutual friendships that I would lose if our relationship were to end.
_____ 4. It would be more attractive for me to be involved in a relationship with someone else rather than continue in a relationship with my current partner.
_____ 5. It would be more attractive for me to be by myself rather than to continue the relationship with my current partner.
_____ 6. In general, the alternatives to remaining in this relationship are quite attractive.

Questions 1, 2, and 3 all seem to be dealing with the size of the investment that the respondent has put into the relationship. It is therefore reasonable to label Factor 1 the "investment size" factor.

4. **Repeat this process to name the remaining retained factors**. In the present case, there is only one remaining factor to name: Factor 2. This factor has high loadings for questions 4, 5, and 6. In reviewing these items, it becomes clear that each seems to deal with the attractiveness of alternatives to one's current relationship. It is therefore reasonable to label this the "alternative value" factor.

5. **Determine whether this solution satisfies the "interpretability criteria."** An earlier section indicated that the overall results of a principal factor analysis are satisfactory only if they meet the following interpretability criteria:

   a. **Are there at least three variables (items) with significant loadings on each retained factor?** In the present example, three variables loaded on Factor 1, and three also loaded on factor 2, so this criterion was met.

   b. **Do the variables that load on a given factor share some conceptual meaning?** All three variables loading on Factor 1 are clearly measuring investment size, while all three loading on Factor 2 are clearly measuring alternative value. Therefore, this criterion is met.

   c. **Do the variables that load on different factors seem to be measuring different constructs?** Because the items loading on the "investment size" factor seem to be conceptually very different from the items loading on the "alternative value" factor, this criterion seems to be met as well.

   d. **Does the rotated factor pattern demonstrate "simple structure?"** Earlier, it was said that a rotated factor pattern demonstrates simple structure when it has two characteristics. First, most of the variables should have high loadings on one factor, and near-zero loadings on other factors. You can see that the pattern obtained here meets that requirement: Items 1 to 3 have high loadings on Factor 1, and near-zero loadings on Factor 2. Similarly, items 4 to 6 have high loadings on Factor 2, and near-zero loadings on Factor 1. The second characteristic of simple structure is that each factor should have high loadings for some variables, and near-zero loadings for the others. Again, the pattern obtained here also meets this requirement: Factor 1 has high loadings for items 1 to 3 and near-zero loadings for other items, while Factor 2 has high loadings for items 4 to 6, and near-zero loadings on the remaining items. In short, the rotated factor pattern obtained in this analysis does seem to demonstrate simple structure.

As stated earlier, the rotated factor pattern should be the first matrix reviewed in naming the factors. However, it does have one limitation: The pattern loadings of this matrix are not constrained to range between +1.00 and -1.00. In rare cases in which the factors are strongly correlated, some loadings may be as large as 10.00 or even larger. In such cases the interpretation of the pattern matrix may be difficult.

When faced with such a situation, it is generally easier to instead review the **reference structure matrix**. This appears under the heading "Reference Structure (Semipartial Correlations)" on output page 4. The reference structure for the current analysis of the investment model questionnaire is presented here as Output 2.8.

Output 2.8: Reference Structure (Semipartial Correlations) from Analysis of Investment Model Questionnaire

Reference Structure (Semipartial Correlations)				
	Factor1		Factor2	
V1	80	*	-13	
V2	91	*	10	
V3	78	*	-1	
V4	-5		68	*
V5	-4		72	*
V6	6		74	*
Printed values are multiplied by 100 and rounded to the nearest integer. Values greater than 0.4 are flagged by an '*'.				

The heading for the reference structure parenthetically includes the words "Semipartial Correlations." This is because the coefficients in this matrix represent the semipartial correlations between variables and common factors, removing from each common factor the effects of other common factors.

The steps followed in interpreting the reference structure are identical to those followed in reviewing the factor pattern. Notice that the size of the loadings in the above reference structure is very similar to those in the rotated factor pattern. It is clear that interpreting the reference structure in this study would have led to exactly the same interpretation of factors as was obtained using the rotated factor pattern.

In addition to interpreting the rotated factor pattern (and reference structure, if necessary), you should also review the **factor structure matrix**. The structure matrix for the present study also appears on page 5, and is presented here as Output 2.9.

Output 2.9: Factor Structure (Correlations) from Analysis of Investment Model Questionnaire

Factor Structure (Correlations)				
	Factor1		Factor2	
V1	89	*	-43	*
V2	93	*	-23	
V3	84	*	-30	
V4	-30		75	*
V5	-30		78	*
V6	-20		77	*
Printed values are multiplied by 100 and rounded to the nearest integer. Values greater than 0.4 are flagged by an '*'.				

The word "Correlations" appears in parentheses in the heading for this matrix, because the structure loadings that it contains represent the product-moment correlations between the variables and common factors. For example, where the row for V1 intersects with the column for FACTOR1, a structure loading of 89 appears. This indicates that the correlation between item 1 and Factor 1 is +.89.

The structure matrix is generally less useful for interpreting the meaning of the factors (compared to the rotated pattern matrix) because it often fails to demonstrate simple structure. For example, notice that the "low" loadings in this structure matrix are not really that low: The loading of V1 on Factor 2 is -.43; the corresponding loading from the rotated pattern matrix was considerably lower at -.14. Comparing the rotated pattern matrix to the structure matrix reveals the superiority of the former in achieving simple structure.

If this is the case, then why review the structure matrix at all? We do this because the pattern matrix and the structure matrix provide different information about the relationships between the observed variables and the underlying factors: The factor pattern reveals the unique contribution of each factor to the variance of the variable. The pattern loadings in this matrix are essentially standardized regression coefficients, comparable to those obtained in multiple regression.

The factor structure, on the other hand, reveals the correlation between a given factor and variable. It helps you understand the "big picture" of how the variables are really related to the factors. For example, consider the rotated factor pattern matrix which appeared on page 4 of the current output. It is presented again here as Output 2.10.

Output 2.10: Promax (Oblique) Rotated Factor Pattern from Analysis of Investment Model Questionnaire

Rotated Factor Pattern (Standardized Regression Coefficients)				
	Factor1		Factor2	
V1	85	*	-14	
V2	97	*	10	
V3	84	*	-1	
V4	-5		73	*
V5	-4		76	*
V6	7		79	*
Printed values are multiplied by 100 and rounded to the nearest integer. Values greater than 0.4 are flagged by an '*'.				

Notice that the pattern loading for V1 on Factor 2 is only -.14. Do not allow this very weak pattern loading to mislead you into believing that V1 and Factor 2 are completely unrelated. Because this is a *pattern* loading, its small value merely means that Factor 2 makes a very small *unique* contribution to the variance in V1.

For contrast, now consider the **structure loading** for V1 on Factor 2 (from Output 2.9). The structure loading reveals that V1 actually demonstrates a correlation with Factor 2 of -.43. Why would V1 be negatively correlated with Factor 2? Because V1 is directly related to Factor 1, and Factor 1, in turn, is negatively correlated with Factor 2. This negative correlation is illustrated graphically in Figure 2.4. Notice that there is a curved double-headed arrow that connects Factors 1 and 2. The arrow is identified with a negative sign. This curved arrow shows that these factors are negatively correlated with no assumed causation between them.

Figure 2.4: Path Model for a 6-Variable, 2-Factor Model, Oblique Factors, Factorial Complexity = 1

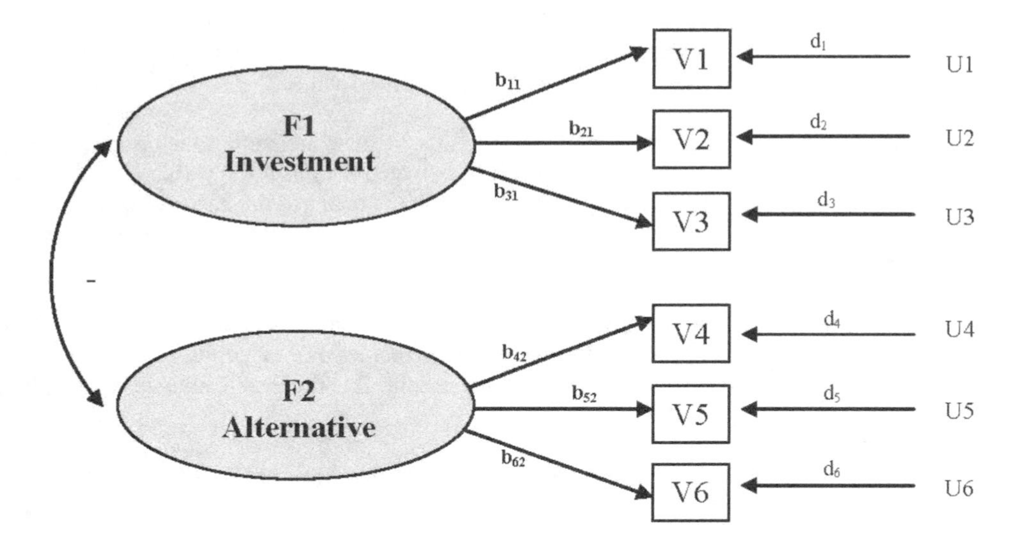

The model presented as Figure 2.4 is identical to Figure 2.1 with one exception. A curved double-headed arrow now connects Factors 1 and 2. This means that the factors are now oblique or correlated. This figure helps demonstrate how a variable could have a moderately large structure loading for a factor, but a small pattern loading. The structure loading for V1 and Factor 2 is -.43 because V1 is caused by Factor 1, and Factor 1 is negatively correlated with Factor 2. However, the pattern loading for V1 and Factor 2 is much smaller at -.14, because Factor 2 has essentially no direct effect on V1.

In summary, one should always review the pattern matrix to determine which groups of variables are measuring a given factor, for purposes of interpreting the *meaning* of that factor. One should then review the structure matrix to get the "big picture" concerning the simple bivariate relations between variables and factors.

If the structure matrix is so important, then why was it not discussed in the chapter on principal component analysis? This is because the pattern matrix and the structure matrix are one and the same in a principal component analysis with an orthogonal rotation. Technically, the loadings of the pattern matrix in principal component analysis can be viewed as regression coefficients, as in common factor analysis. Remember, however, that the principal components of this analysis are orthogonal, or uncorrelated. Because of this orthogonality, the regression coefficients for the components are equivalent to the correlation between the components and the variables. This is to say that the loadings of the pattern matrix can also be interpreted as correlations between the components and the variables. Hence, there is no difference between a factor pattern matrix and a factor structure matrix in principal component analysis with an orthogonal rotation. This is why only the pattern matrix is printed and interpreted.

Step 5: Creating Factor Scores or Factor-Based Scores

Once the analysis is complete, it is often desirable to assign scores to each participant to indicate where that participant stands on the retained factors. For example, the two factors retained in the present study were interpreted as an investment size factor and alternative value factor. You may wish to now assign one score to each participant to indicate where that participant stands on the investment size factor, and a different score to indicate where that participant stands on the alternative value factor. With this done, these factor scores could then be used either as predictor variables or criterion variables in subsequent analyses.

Before discussing the options for assigning these scores, it is necessary to first draw a distinction between factor scores versus estimated factor scores. A **factor score** represents a participant's actual standing on an underlying factor. An **estimated factor score**, on the other hand, is merely an estimate of a participant's standing on that underlying factor. In practice, researchers are never able to compute true "factor scores." This is because of a fundamental indeterminancy in factor analysis. In the end, factor scores are estimated by creating linear composites of the observed variables. That is, one computes factor scores by adding together optimally weighted scores on the observed variables. But remember that the common part of a given variable (that part influenced by the common factor) is inseparable from that variable's unique component. This means that there will always be some error associated with the computation of factor scores, and so it is therefore better to refer to them as *estimated* factor scores.

Estimated Factor Scores

Broadly speaking, two scoring approaches are available. The more sophisticated approach is to allow PROC FACTOR to compute estimated factor scores. An estimated factor score is a linear composite of the optimally weighted variables under analysis. For example, to calculate the participant's estimated factor score on factor 1, you would use the following equation:

$$F'_1 = b_{11}V_1 + b_{12}V_2 + b_{13}V_3 + \dots b_{1p}V_p$$

where

F'_1 = the estimated factor score for factor 1

b_{11} = the scoring coefficient for survey question 1 used in creating estimated factor score 1

V_1 = the participant's score on survey question 1

b_{12} = the scoring coefficient for survey question 2 used in creating estimated factor score 1

V_2 = the participant's score on survey question 2

b_{1p} = the scoring coefficient for survey question p (the last question), used in creating estimated factor score 1

V_p = the participant's score on survey question p

A different equation, with different scoring coefficients, would be used to calculate participants' scores on the remaining retained factors. In practice, you do not actually have to create equations such as those appearing here; instead, these estimated factor scores may be created automatically by requesting the creation of a new dataset within the SAS program. This is done by including the OUT and NFACT options in the FACTOR statement.

The general form for the NFACT option is

```
nfact=number-of-factors-to-be-retained
```

The general form for the OUT option is

```
out=name-of-new-SAS-dataset
```

The following SAS program incorporates these options:

```
proc factor   data=D1

                    simple
                    method=prin
                    priors=smc
                    nfact=2
                    rotate=promax
                    round
                    flag=.40
❶                   out=D2 ;
        var V1-V6 ;
    run;

❷   proc corr   data=D2;
❸    var COMMITMENT FACTOR1 FACTOR2;
❹   run;
```

Line ❶ of the preceding programs asks that an output dataset be created and given the name "D2." This name was arbitrary; any name consistent with SAS requirements would have been acceptable. The new dataset named D2 will contain (a) all of the variables contained in the previous dataset, as well as (b) new variables named FACTOR1 and FACTOR2. FACTOR1 will contain estimated factor scores for the first retained factor, and FACTOR2 will contain estimates for the second factor. The number of new "FACTOR" variables created will be equal to the number of factors retained by the NFACT statement.

The OUT option may be used only if the factor analysis has been performed on raw data (as opposed to a correlation or covariance matrix). The use of the NFACT statement is also required.

Having created the new estimated factor score variables named FACTOR1 and FACTOR2, you may be interested in seeing how they relate to some of the study's other variables (i.e., variables not analyzed in the factor analysis itself). You may therefore append a PROC CORR statement to your program following the last of the PROC FACTOR statements. In the preceding program, these statements appear on lines ❷ to ❹.

These PROC CORR statements request that COMMITMENT be correlated with FACTOR1 and FACTOR2. COMMITMENT represents participants' "commitment to the relationship." High scores on this variable indicate that participants intend to remain in the relationship with their current partners. (Assume that the variable COMMITMENT was also measured with the questionnaire, and that scores on COMMITMENT were entered as part of dataset D1.) These PROC CORR statements result in the SAS output that is presented here as Output 2.11.

Output 2.11: Correlations between COMMITMENT and Estimated Factor Score Variables FACTOR1 and FACTOR2

The CORR Procedure

3 Variables:	COMMITMENT Factor1 Factor2

Simple Statistics						
Variable	N	Mean	Std Dev	Sum	Minimum	Maximum
COMMITMENT	50	15.52000	6.67692	776.00000	4.00000	28.00000
Factor1	50	0	0.95720	0	-2.25877	1.68987
Factor2	50	0	0.88955	0	-1.29220	2.75565

Pearson Correlation Coefficients, N = 50 Prob > |r| under H0: Rho=0			
	COMMIT	Factor1	Factor2
COMMITMENT	1.00000	0.31881 0.0240	-0.29307 0.0389
Factor1	0.31881 0.0240	1.00000	-0.39458 0.0046
Factor2	-0.29307 0.0389	-0.39458 0.0046	1.00000

The correlations of interest appear in Output 2.11 below the heading "Pearson Correlations Coefficients, N = 50." Look at the first column of coefficients, under the heading "COMMITMENT." Where this column intersects with the row headed FACTOR1, you can see that FACTOR1 displays a correlation of approximately +.32 with commitment. This makes sense, because the first retained factor was interpreted as the "investment size" factor. It is logical that investment size would be positively correlated with commitment to maintain the relationship. The second estimated factor score variable, FACTOR2, has a correlation of -.29 with commitment; this too is logical. The second retained factor was interpreted as "alternative value." It makes sense that commitment would decrease as the attractiveness of one's alternatives increases. FACTOR1 and FACTOR2 may now be used as predictor or criterion variables in any other appropriate SAS procedure.

Factor-Based Scales

A second (and less sophisticated) approach to scoring involves the creation of factor-based scales. A **factor based scale** is a variable that estimates participant scores on the underlying factors, but does not use an optimally weighted formula to do this (as with the estimated factor scores created by PROC FACTOR).

Although a factor-based scale can be created in a number of ways, the following method has the advantage of being relatively straightforward:

- To calculate scores on factor-based scale 1, first determine which questionnaire items had high loadings on Factor 1.
- For a given participant, add together that participant's responses to these items. The result is that participant's score on the factor-based scale for Factor 1.
- Repeat these steps to calculate each participant's score for remaining retained factors.

Although this may sound like a cumbersome task, it is actually quite simple using SAS data manipulation statements. For example, assume that you have performed the factor analysis on your survey responses and have obtained the findings reported in this chapter. Specifically, it was found that survey items 1, 2, and 3 loaded on Factor 1 (the investment size factor), while items 4, 5, and 6 loaded on Factor 2 (the alternative value factor).

You would now like to create two new SAS variables. The first variable, called INVESTMENT, will include each participant's score on the factor-based scale for investment size. The second variable, called ALTERNATIVES, will include each participant's score on the factor-based scale for alternative value. Once these variables are created, you can use them as criterion variables or predictor variables in subsequent multiple regressions, ANOVAs, or other analyses. To keep things simple for the present example, assume that you are simply interested in determining whether there is a substantive correlation between COMMITMENT and INVESTMENT and between COMMITMENT and ALTERNATIVES.

At this point, it may be useful to review Chapter 4, "Working with Variables and Observations in SAS Datasets," from *A Step-By-Step Approach to Using SAS for Univariate and Multivariate Statistics* (O'Rourke,Hatcher, and Stepanski 2005), particularly the section on "Creating New Variables from Existing Variables." Such a review should make it easier to understand the data manipulation statements below.

Assume that earlier statements in the SAS program have already input participant responses to questionnaire items, including participants' scores on the variable COMMITMENT. These variables are included in a dataset called D2. Below are the subsequent lines that would create a new dataset called D3 which would include all of the variables in D2, as well as the newly created factor-based scales called INVESTMENT and ALTERNATIVES.

```
❶  data D3;
❷    set D2;

❸  INVESTMENT   = (V1 + V2 + V3);
❹  ALTERNATIVES = (V4 + V5 + V6);

❺    proc corr   data=D3;
❻   var COMMITMENT INVESTMENT  ALTERNATIVES;
❼     run;
```

Lines ❶ and ❷ request that a new dataset called D3 be created, and that this dataset be set up as a duplicate of existing dataset D2. On line ❸ the new variable called INVESTMENT is created. For each participant, his or her responses to items 1, 2, and 3 are added together. The result is the participant's score on the factor-based scale for the first factor. These scores are stored in a variable called INVESTMENT. The factor-based scale for the alternative value factor is created on line ❹, and these scores are stored in the variable called ALTERNATIVES. Lines ❺ to ❼ request correlation coefficients between COMMITMENT, INVESTMENT, and ALTERNATIVES be computed.

Fictitious results from the preceding program are presented as Output 2.12.

Output 2.12: Correlations between COMMITMENT and Factor-Based Scales INVESTMENT and ALTERNATIVES

The CORR Procedure

3 Variables:	COMMITMENT INVESTMENT ALTERNATIVES

Simple Statistics

Variable	N	Mean	Std Dev	Sum	Minimum	Maximum
COMMITMENT	50	15.52000	6.67692	776.00000	4.00000	28.00000
INVESTMENT	50	13.36000	4.36479	668.00000	3.00000	21.00000
ALTERNATIVES	50	7.68000	3.22895	384.00000	3.00000	18.00000

Pearson Correlation Coefficients, N = 50
Prob > |r| under H0: Rho=0

	COMMITMENT	INVESTMENT	ALTERNATIVES
COMMITMENT	1.00000	0.33798 0.0164	-0.26380 0.0642
INVESTMENT	0.33798 0.0164	1.00000	-0.30588 0.0308
ALTERNATIVES	-0.26380 0.0642	-0.30588 0.0308	1.00000

You can see that the correlations between COMMITMENT and the estimated factor scores (FACTOR1 and FACTOR2) discussed earlier are slightly different from the correlations between COMMITMENT and the factor-based scales (INVESTMENT and ALTERNATIVES) presented above. For example, the correlation between COMMITMENT and FACTOR1 (the estimated factor-score variable for investment size) was approximately .32, while the correlation coefficient between COMMITMENT and INVESTMENT (the factor-based scale for investment size) was slightly higher at approximately .34. These differences are to be expected as the estimated factor scores (FACTOR1 and FACTOR2) are *optimally weighted* linear composites, while the factor-based scales (INVESTMENT and ALTERNATIVES) are not optimally weighted. In fact, it would be instructive to create a single correlation matrix that includes both the estimated factor scores as well as the factor-based scales. This could be done with the following statements:

```
proc factor   data=D1
              simple
              method=prin
              priors=smc
              nfact=2
              rotate=promax
              round
              flag=.40
              out=D2 ;
   var V1-V6 ;
run;

data D3;
   set D2;

   INVESTMENT   = (V1 + V2 + V3);
   ALTERNATIVES = (V4 + V5 + V6);

proc corr   data=D3;
   var COMMITMENT FACTOR1 FACTOR2 INVESTMENT ALTERNATIVES;
run;
```

This program resulted in the correlation matrix presented here as Output 2.13.

Output 2.13: Correlations between COMMITMENT, Estimated Factor Score Variables FACTOR1 and FACTOR2, and Factor-Based Scales INVESTMENT and ALTERNATIVES

The CORR Procedure

5 Variables:	COMMITMENT Factor1 Factor2 INVESTMENT ALTERNATIVES

Simple Statistics						
Variable	N	Mean	Std Dev	Sum	Minimum	Maximum
COMMITMENT	50	15.52000	6.67692	776.00000	4.00000	28.00000
Factor1	50	0	0.95720	0	-2.25877	1.68987
Factor2	50	0	0.88955	0	-1.29220	2.75565
INVESTMENT	50	13.36000	4.36479	668.00000	3.00000	21.00000
ALTERNATIVES	50	7.68000	3.22895	384.00000	3.00000	18.00000

Pearson Correlation Coefficients, N = 50 Prob > |r| under H0: Rho=0					
	COMMITMENT	Factor1	Factor2	INVESTMENT	ALTERNATIVES
COMMITMENT	1.00000	0.31881 0.0240	-0.29307 0.0389	0.33798 0.0164	-0.26380 0.0642
Factor1	0.31881 0.0240	1.00000	-0.39458 0.0046	0.99431 <.0001	-0.32121 0.0229
Factor2	-0.29307 0.0389	-0.39458 0.0046	1.00000	-0.38401 0.0059	0.99043 <.0001
INVESTMENT	0.33798 0.0164	0.99431 <.0001	-0.38401 0.0059	1.00000	-0.30588 0.0308
ALTERNATIVES	-0.26380 0.0642	-0.32121 0.0229	0.99043 <.0001	-0.30588 0.0308	1.00000

The correlations of interested appear under the heading "Pearson Correlation Coefficients, N = 50." Remember that FACTOR1 contains the estimated factor scores for investment size, while INVESTMENT is the factor-based scale for investment size. Where the row for FACTOR1 intersects the row for INVESTMENT, you will find a correlation coefficient of .99, meaning that the estimated factor score variable and the factor-based scale for this construct are almost perfectly correlated. Similarly, the correlation of .99 between FACTOR2 and ALTERNATIVES shows that the estimated factor score variable and the factor-based scale for alternative value is also very strongly correlated.

Recoding Reversed Items Prior to Analysis

It is generally best to recode any reversed or negatively keyed items before conducting any of the analyses described here. In particular, it is essential that reversed items be recoded prior to the program statements that produce factor-based scales. The three questionnaire items designed to assess investment size are once again presented below:

Please rate each of the following statements to indicate the extent to which you agree or disagree with each using a response scale in which 1 = "Strongly Disagree" and 7 = "Strongly Agree."

_____ 1. I have invested a lot of time and effort in developing my relationship with my current partner.
_____ 2. My current partner and I have developed interests in a lot of fun activities that I would lose if our relationship were to end.
_____ 3. My current partner and I have developed a lot of mutual friendships that I would lose if our relationship were to end.

None of the above items are reversed; with each item, a response of "7" indicates a high level of investment. Below, however, item 1 is a reversed item. In contrast to the previous item, a response of "7" now indicates a *low* level of investment:

_____ 1. I have invested **very little** time and effort in developing my relationship with my current partner.
_____ 2. My current partner and I have developed interests in a lot of fun activities that I would lose if our relationship were to end.
_____ 3. My current partner and I have developed a lot of mutual friendships that I would lose if our relationship were to end.

If you were to perform a factor analysis on responses to these items, the factor loading for item 1 would have a sign that is the opposite of the sign of the loadings for items 2 and 3 (e.g., if items 2 and 3 had positive loadings, item 1 would have a negative loading). This would complicate the creation of a factor-based scale: With items 2 and 3, higher scores indicate greater investment whereas with item 1, lower scores indicate greater investment. Clearly, you would not wish to sum these three items together given the way they are presently coded. First, you will reverse item 1. Notice how this is done in the following program. (Assume that the data have already been input in a SAS dataset named D1.)

```
data D2 ;
   set D1 ;

❶   V1 = 8 - V1;

    INVESTMENT    = (V1 + V2 + V3) ;
    ALTERNATIVES = (V4 + V5 + V6) ;

proc corr    data=D2 ;
   var COMMITMENT INVESTMENT ALTERNATIVES ;
run ;
```

With line ❶, you are creating a new version of variable V1. Values on this new version of V1 will be equal to the quantity "8 minus the value of the old version of V1." Therefore, for participants whose score on the old version of V1 was 1, their value on the new version of V1 will be 7 (because 8 – 1 = 7). For participants whose score on the old version of V1 was 7, their value on the new version of V1 will be 1 (because 8 – 7 = 1), and so forth.

The general form of the formula used when recoding reversed items is:

```
Variable name = constant - variable name ;
```

In this formula, the "constant" is the following quantity:

(the number of points on the response scale used with the questionnaire item + 1)

Therefore, if you are using the 4-point response scale, the constant is 5; if you are using a 9-point scale, the constant is 10.

If you have prior knowledge about which items are going to appear as reversed items (with reversed factor loadings) in your results, it is best to place these recoding statements early in your SAS program, before the PROC FACTOR statements. This will make interpretation of the factors more straightforward, because it will eliminate significant loadings with opposite signs from appearing on the same factor. In any case, it is essential that the statements that recode reversed items appear before the statements that create any factor-based scales.

Step 6: Summarizing the Results in a Table

In some cases, you may wish to prepare a table presenting the rotated factor pattern and factor structure for the variables analyzed. One possible format is presented in Table 2.4, below.

Table 2.4: Questionnaire Items and Corresponding Factor Loadings from the Rotated Factor Pattern Matrix and Factor Structure Matrix, Decimals Omitted

Factor Pattern		Factor Structure		
1	**2**	**1**	**2**	**Questionnaire Item**
85	-14	89	-43	1. I have invested a lot of time and effort in developing my relationship with my current partner.
97	10	93	-23	2. My current partner and I have developed interest in a lot of fun activities that I would lose if our relationship were to end.
84	-1	84	-30	3. My current partner and I have developed lot of mutual friendships that I would lose if our relationship were to end.
-5	73	-30	75	4. It would be more attractive for me to be involved in a relationship with someone else rather than continue the relationship with my current partner.
-4	76	-30	78	5. It would be more attractive for me to be by myself than to continue my relationship with my current partner.
7	79	-20	77	6. In general, my alternatives to remaining in this relationship are quite attractive.

NOTE: N=50.

If feasible, it is ideal to include an additional column presenting the final communality estimates; the column heading would be "h^2," which is the symbol for communality. These final communality estimates appear in the output following the factor structure matrix. Table 1.2 from the previous chapter on principal component analysis shows how communalities may be presented in a table.

When many factors are retained or when the questionnaire items are long or numerous, it may not be possible to present the factor loadings, communalities, and questionnaire items all in a single table. In these instances, the loadings and communalities are presented in one table, and the items are presented in a second table (or within the text of the paper).

Step 7: Preparing a Formal Description of the Results for a Paper

The level of detail reported in research papers tends to be comparatively brief as factor analysis is often the first step in a series of analyses. The preceding analysis could be briefly summarized as follows:

> Responses to the 6-item questionnaire underwent exploratory factor analysis using squared multiple correlations as prior communality estimates. The principal factor method was used to extract factors, followed by a promax (oblique) rotation. A scree test suggested two meaningful factors so only these factors were retained for rotation.
>
> In interpreting the rotated factor pattern, an item was said to load on a given factor if the factor loading was .40 or greater for that factor, and was less than .40 for the other. Applying these criteria, three items were found to load on the first factor, which was subsequently labeled the investment size factor. Three items also loaded on the second factor, which was labeled the alternative value factor. Questionnaire items and corresponding factor loadings are presented in Table 2.4.

A More Complex Example: The Job Search Skills Questionnaire

The results presented in the preceding section were designed to be relatively simple to introduce the basic concepts of factor analysis. In conducting actual research, however, the results are seldom as clear cut. Very often you are forced to make somewhat qualitative decisions and are forced to choose between more than one interpretable solution. This section illustrates these problems by presenting a somewhat more complex analysis.

Assume that you are now conducting research in the area of college student career development. You have developed an instrument to assess student knowledge and ability in a wide variety of areas related to occupational choice and the job-search process. The instrument consists of 100 items, and the items are divided into 25 scales; each scale contains four items.

Below are the SAS variable names for each scale. Following the SAS variable name is the full name for the scale (in italics) and a sample item from the scale (in parentheses). Reviewing the scale names and sample items should make clear what type of knowledge or ability is assessed by each scale.

For example, the first scale is identified with the SAS variable name, "VALUES." The full name for this scale is "Clarifying values and interests," and the sample item is "My ability to describe just what are my work-related interests." Participants responded to each item on the questionnaire using a 7-point scale in which 1 = "Very Bad" and 7 = "Very Good."

1. VALUES: *Clarifying values and interests* (e.g., "My ability to describe just what are my work-related interests").
2. ABILITY: *Identifying work-related abilities and skills* (e.g., "My ability to describe just what are my strongest work-related skills and abilities").
3. ASSESS: *Using assessment instruments* (e.g., "My knowledge of what specific assessment instruments are available to help assess my interests").
4. STRATEGY: *Identifying effective job search strategies* (e.g., "My knowledge of effective job search strategies").
5. EXPERIENCE: *Getting job-related experience* (e.g., "My knowledge of how I could get relevant job experience in my field before I graduate").
6. ORGCHAR: *Identifying preferred organizational characteristics* (e.g., "My ability to clearly describe the exact characteristics an organization should have in order to satisfy my personal preferences").
7. RESOCCUP: *Researching potential occupations* (e.g., "My knowledge of what specific books, Internet sources, and other resources that provide useful information about specific occupations").
8. RESEMPLOY: *Researching specific employers* (e.g., "My ability to collect detailed information on a specific organization just before an employment interview").
9. GOALS: *Setting goals* (e.g., "My ability to clearly describe my career goals for the next five years").
10. BARRIER: *Dealing with occupational barriers* (e.g., "My knowledge of what types of occupational barriers that are likely to stand in my way of getting the job I really want").
11. MOTIVATED: *Staying motivated* (e.g., "My ability to maintain a high level of motivation throughout my job search").
12. RESUMES: *Using résumés* (e.g., "My ability to write a highly effective résumé").
13. RECOMMEND: *Using letters of recommendation* (e.g., "My knowledge of what I should do to insure that my referee writes a very effective letter of recommendation for me").
14. DIRECT: *Using the cover letter/direct mail approach* (e.g., "My knowledge of just what should be included in a cover letter used in the direct mail approach").
15. APPLICAT: *Completing application forms* (e.g., "My ability to complete an application form in such a way as to make the best possible impression on a prospective employer").
16. IDEMPLOY: *Identifying potential employers* (e.g., "My knowledge of exactly what books/references are available to help me identify organizations that might hire me").
17. CARDEVEL: *Using campus career development services* (e.g., "My ability to clearly describe exactly what services are offered by the career development office on this campus").
18. AGENCY: *Using employment agencies* (e.g., "My knowledge of how to make effective use of an employment agency").
19. FAIRS: *Using job fairs* (e.g., "My ability to make effective use of a job fair").
20. ADVERT: *Responding to advertised job openings* (e.g., "My knowledge of how to effectively respond to a job advertisement").
21. COUNSEL: *Using career counselors/consultants* (e.g., "My knowledge of how to use career counselors/consultants to make the most of the services they offer").
22. UNADVERT: *Applying directly for unadvertised jobs* (e.g., "My knowledge of the correct way to directly apply for an unadvertised position").

23. NETWORK: *Using the networking approach to job search* (e.g., "My knowledge of the most effective ways of producing job leads by asking for help from friends, relatives, past employers and other contacts).
24. INTERVIEW: *Managing the employment interview process* (e.g., "My knowledge of how to respond to tough interview questions").
25. SALARY: *Negotiating salary* (e.g., "My ability to successfully negotiate a fair and motivating salary").

Assume that you administered your scale to 258 college students and obtained usable responses from 220 of these. You determined each student's score on each of the 25 scales, meaning that there are 25 data points for each student. You now wish to perform an exploratory factor analysis to identify the latent structure underlying the data (from Ruddle, Thompson and Hatcher 1993).

Notice that the analysis will be performed on the 25 *scale scores*, not on the responses to each of the 100 individual questionnaire *items*. This approach is justifiable only if you have reason to believe that each of the 25 scales assesses just one construct. It would not be appropriate if, for example, items 1 and 2 within scale 1 assess one construct, and items 3 and 4 assess a different construct. In this latter case, it would be more appropriate to perform a factor analysis using all 100 of the individual items. (Of course, that analysis would require a large sample size in order to attain a good ratio of participants to variables.)

However, assume you have evidence that each scale does, in fact, assess just one construct. Assume that coefficient alpha exceeds .80 for each scale, and that the item-total correlations are quite high; these findings would suggest that the individual scales are unifactorial. Therefore, you will use scores on the 25 scales as observed variables in the factor analysis.

The SAS Program

The data analyzed here appear in Appendix B. Below is the SAS program (minus the DATA step) to perform an exploratory factor analysis on the data from your study.

```
     proc factor data=D1
                 simple
❶                method=ml
❷                priors=smc
❸                nfact=1
                  plots=scree
                  rotate=promax
                  round
❹                flag=.40 ;

       var V1-V25 ;
         /*  VALUES ABILITY ASSESS STRATEGY EXPERIENCE ORGCHAR
             RESOCCUP RESEMPLOY GOALS BARRIER MOTIVATED
             RESUMES RECOMMEND DIRECT APPLICAT IDEMPLOY
             CARDEVEL AGENCY FAIRS ADVERT COUNSEL UNADVERT
             NETWORK INTERVIEW SALARY ;   */
       run;
```

In most respects, the preceding program is similar to the other exploratory factor analysis presented previously in this chapter. The PRIORS option ❷ requests that squared multiple correlations again be used as prior communality estimates, and the FLAG option ❹ requests that factor loadings whose absolute values exceed .40 be flagged with asterisks. The NFACT option ❸ requests that one factor be retained. (Once again, you have no empirical evidence to expect any specific number of factors at this stage of the analysis; one factor was specified simply as a starting point.)

This program differs from the other analyses, however, in that the METHOD=ML option ❶ requests that the maximum likelihood method be used to extract factors. As previously indicated, most researchers prefer this method because it generally provides more accurate parameter estimates, and also provides a significance test to help solve the number of factors problem. Because of these advantages, the use of the maximum likelihood method will be described in this section.

The preceding program would produce three pages of output. Some of the information appearing on each page is summarized below:

- Page 1 presents the sample size and simple statistics.
- Page 2 includes the eigenvalue table, scree plot of eigenvalues, iteration history, the significance tests for the number of factors extracted, unrotated factor pattern matrix, and the final communality estimates.
- Page 3 simply includes a note reminding you that factor rotation is not viable with just one factor.

Portions of this output will be reproduced on the following pages as Output 2.14, Output 2.15, and Output 2.16.

Determining the Number of Factors to Retain

The Scree Plot

Because 25 scales were analyzed, you know that 25 factors will be extracted. The eigenvalue table for these factors, along with the scree plot, is reproduced here as Output 2.14.

Output 2.14: Preliminary Eigenvalues from Analysis of Job Search Skills Questionnaire

The FACTOR Procedure
Initial Factor Method: Maximum Likelihood

Preliminary Eigenvalues: Total = 40.1355388 Average = 1.60542155

	Eigenvalue	Difference	Proportion	Cumulative
1	33.4862182	30.2592664	0.8343	0.8343
2	3.2269518	1.6761612	0.0804	0.9147
3	1.5507906	0.1822790	0.0386	0.9534
4	1.3685116	0.3091964	0.0341	0.9875
5	1.0593152	0.2961206	0.0264	1.0139
6	0.7631945	0.1079114	0.0190	1.0329
7	0.6552831	0.2233701	0.0163	1.0492
8	0.4319130	0.1148151	0.0108	1.0600
9	0.3170979	0.0201290	0.0079	1.0679
10	0.2969689	0.1244940	0.0074	1.0753
11	0.1724749	0.0544826	0.0043	1.0796
12	0.1179923	0.0738060	0.0029	1.0825
13	0.0441863	0.0961917	0.0011	1.0836
14	-0.0520054	0.0594319	-0.0013	1.0823
15	-0.1114373	0.0174893	-0.0028	1.0795
16	-0.1289267	0.0361906	-0.0032	1.0763
17	-0.1651172	0.0548644	-0.0041	1.0722
18	-0.2199817	0.0027468	-0.0055	1.0667
19	-0.2227285	0.1178582	-0.0055	1.0612
20	-0.3405867	0.0203781	-0.0085	1.0527

Preliminary Eigenvalues: Total = 40.1355388 Average = 1.60542155				
	Eigenvalue	Difference	Proportion	Cumulative
21	-0.3609648	0.0438742	-0.0090	1.0437
22	-0.4048391	0.0013121	-0.0101	1.0336
23	-0.4061512	0.0431700	-0.0101	1.0235
24	-0.4493211	0.0439786	-0.0112	1.0123
25	-0.4932997		-0.0123	1.0000

1 factor will be retained by the NFACTOR criterion.

Output 2.15: Scree Plot from Analysis of Job Search Skills Questionnaire

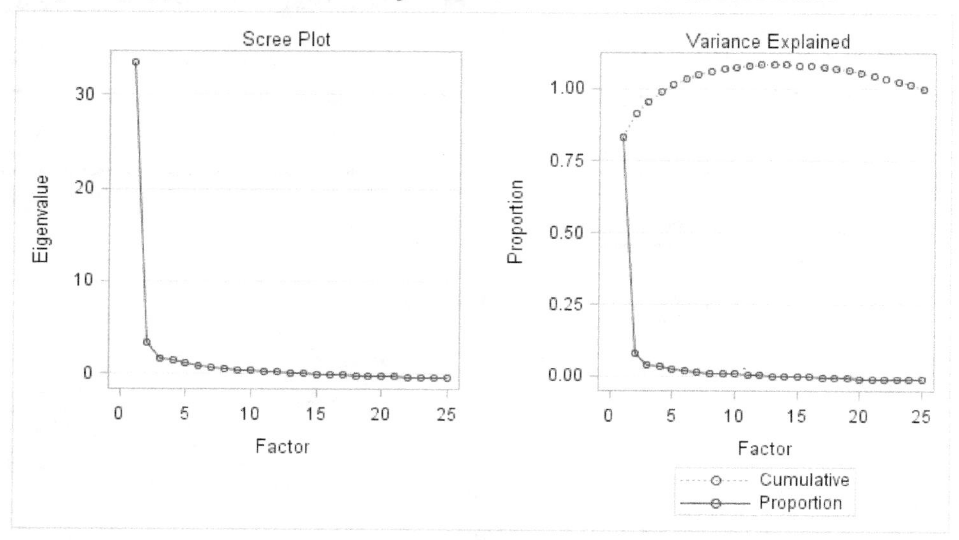

How many factors should you retain and rotate? Earlier, the scree test was used to help you make this decision. Remember that, with the scree test, you look for a major break in the eigenvalues. You hope that, following this break, the line will begin to "flatten out." Factors that appear before the break are retained whereas those appearing along the flat line after the break are assumed to account for only trivial variance and will not be retained.

In the scree plot of Output 2.15, there is clearly a major break following Factor 1. This *may* mean that this questionnaire is unifactorial (i.e., most of the scales may measure just one general "job search skills" factor). To assess the interpretability of this "one-factor" model, you will consult the factor pattern to determine which variables display the largest loadings for this factor. Identifying the variables with the highest loadings will help label the factor.

The interpretation of a one-factor solution is slightly different from the interpretation of multiple-factor models (as earlier presented). When only one factor is retained, rotation is not possible (whether orthogonal or oblique). This actually makes your task easier: When only one factor is retained, it is possible to review the unrotated factor pattern to interpret the factor. The unrotated factor pattern from your one-factor solution is presented here as Output 2.16.

Output 2.16: Factor Pattern from the One-Factor Solution, Analysis of Job Search Skills Questionnaire

Factor Pattern			
		Factor1	
V1	VALUES	48	*
V2	ABILITY	51	*
V3	ASSESS	63	*
V4	STRATEGY	73	*
V5	EXPERIENCE	69	*
V6	ORGCHAR	56	*
V7	RESOCCUP	76	*
V8	RESEMPLOY	81	*
V9	GOALS	53	*
V10	BARRIER	67	*
V11	MOTIVATED	57	*
V12	RESUMES	71	*
V13	RECOMMEND	76	*
V14	DIRECT	75	*
V15	APPLICAT	64	*
V16	IDEMPLOY	83	*
V17	CARDEVEL	69	*
V18	AGENCY	70	*
V19	FAIRS	68	*
V20	ADVERT	83	*
V21	COUNSEL	77	*
V22	UNADVERT	75	*
V23	NETWORK	73	*
V24	INTERVIEW	72	*
V25	SALARY	76	*
Printed values are multiplied by 100 and rounded to the nearest integer. Values greater than 0.4 are flagged by an '*'.			

Notice that every variable in Output 2.16 demonstrates a "meaningful" loading on Factor 1 (i.e., loading over .40). This is indicated by the fact that the loading for each variable is flagged with an asterisk. For example, the variable "VALUES" displays a loading of .48, the variable "ABILITY" displays a loading of .51, and so forth.

To interpret Factor 1 more effectively, it would be helpful to isolate those variables that demonstrate the largest loadings for it. Therefore, we will somewhat arbitrarily choose the value of .70 as a cut-off, and will construct a table that lists the scales that demonstrate a loading of .70 or greater for factor 1. These scales are listed in Table 2.5.

Table 2.5: Scales with Larger Factor Loadings from Output 2.16, Sorted by Size of Loadings

Factor Loading	Variable	Description
.83	IDEMPLOY	Identifying potential employers
.83	ADVERT	Responding to advertised job openings
.81	RESEMPLOY	Researching specific employers
.77	COUNSEL	Using career counselors/consultants
.76	RESOCCUP	Researching potential occupations
.76	RECOMMEND	Using letters of recommendation
.76	SALARY	Negotiating salary
.75	DIRECT	Using the cover letter/direct mail approach
.75	UNADVERT	Applying directly for unadvertised jobs
.73	STRATEGY	Identifying effective job search strategies
.73	NETWORK	Using the networking approach to job search
.72	INTERVIEW	Managing the employment interview process
.71	RESUMES	Using résumés
.70	AGENCY	Using employment agencies

Table 2.5 lists the scales that demonstrated a loading on Factor 1 of .70 or greater, and these scales are reordered according to the size of their loadings. Notice that most of the scales that loaded heavily on Factor 1 pertain to the "nuts-and-bolts" tasks associated with the job hunt itself (e.g., responding to job openings, learning about potential employers, negotiating salary). Because of this, if you ultimately decide that a one-factor solution is best, you will probably define this dimension as a *general job-search skills* factor.

Variance Accounted For

Before accepting the one-factor solution as your "final" solution, you will first review other criteria and consider some alternative solutions. The "Proportion" row of the eigenvalue table from Output 2.14 shows that the first factor accounts for approximately 83% of the common variance. Factor 2 accounts for an additional 8% of the variance: An amount that many researchers would consider meaningful. This information alone would probably warrant exploring a two-factor solution.

The Chi Square Test

As was mentioned earlier, one advantage of the maximum likelihood method of factor extraction is the fact that it provides a chi square test to help make the number of factors decision. The chi square test for the current analysis is presented as Output 2.17.

Output 2.17: Significance Tests For One-Factor Model, Job Search Skills Questionnaire

Significance Tests Based on 220 Observations			
Test	DF	Chi-Square	Pr > ChiSq
H0: No common factors	300	3737.4335	<.0001
HA: At least one common factor			
H0: 1 Factor is sufficient	275	775.0342	<.0001
HA: More factors are needed			

Chi-Square without Bartlett's Correction	811.47003
Akaike's Information Criterion	261.47003
Schwarz's Bayesian Criterion	-671.77754
Tucker and Lewis's Reliability Coefficient	0.84131

Squared Canonical Correlations
Factor1
0.96399635

The test you are most interested in appears to the right of the test "HO: 1 Factor is sufficient, HA: More factors are needed." This heading is self-explanatory; it tells you that the chi square statistic tests the null hypothesis that retaining one factor is sufficient. If you obtain a small p value for this test (i.e., $p < .05$), you are to reject this null hypothesis and consider the alternative hypothesis that more factors should be retained.

Output 2.17 shows that the obtained value of chi square for the test was large at approximately 775.03 (DF = 275). To the left of the chi square statistic and the degrees of freedom, the output provides the entry, "Pr > ChiSq < .0001." This is the p value for the obtained chi square statistic. Because this obtained p value is less than .05, you may reject the null hypothesis that one factor is adequate. This finding can be used as evidence that more factors should be retained. Under these circumstances, some researchers would sequentially add additional factors to the model until a nonsignificant chi square value is obtained.

However, we caution against the temptation to rely too heavily on the chi square test. Under circumstances that are often encountered in applied research, reliance on the chi-square test alone can lead you to retain too many factors. This is especially likely when the sample is large or there is even a minor misfit between the model and the data (Kim and Mueller 1978b). For this reason, use the chi square test as only one piece of information in making the number of factors decision; if the test suggests that additional factors are needed, consult other criteria before making a final decision (e.g., the scree test, proportion of variance accounted for, and interpretability criteria).

A Two-Factor Solution

So far you have obtained mixed support for a one-factor model. The scree test could be interpreted as supporting the retention of only one factor. One the other hand, the eigenvalue table showed that Factor 2 accounts for over 7% of the common variance, and the chi square test rejected the one-factor model. Combined, these findings justify exploring the possibility of a two-factor model.

The analysis was therefore repeated, this time specifying NFACT=2. This revised program again produced four pages of output, some of which is reproduced here as Output 2.18, Output 2.19, and Output 2.20. Some of the information appearing on these 4 pages of output is summarized below:

- Page 1 provides simple statistics.
- Page 2 includes the eigenvalue table, the factor pattern matrix, and significance tests for the number of factors extracted.
- Page 3 includes the orthogonal transformation matrix.
- Page 4 includes the rotated factor pattern matrix, variance explained by each factor (weighted and unweighted), the factor structure matrix, and final communality estimates.

The rotated factor pattern from the promax rotation is presented as Output 2.18.

Output 2.18: Rotated Factor Pattern from Promax Rotation, Two-Factor Solution, Job Search Skills Questionnaire

		Rotated Factor Pattern (Standardized Regression Coefficients)			
		Factor1		Factor2	
V1	VALUES	-9		78	*
V2	ABILITY	7		61	*
V3	ASSESS	51	*	17	
V4	STRATEGY	65	*	13	
V5	EXPERIENCE	46	*	32	
V6	ORGCHAR	0		77	*
V7	RESOCCUP	60	*	22	
V8	RESEMPLOY	69	*	18	
V9	GOALS	-1		74	*
V10	BARRIER	35		44	*
V11	MOTIVATED	8		67	*
V12	RESUMES	58	*	20	
V13	RECOMMEND	59	*	25	
V14	DIRECT	76	*	1	
V15	APPLICAT	47	*	23	
V16	IDEMPLOY	79	*	7	
V17	CARDEVEL	76	*	-7	
V18	AGENCY	76	*	-6	
V19	FAIRS	80	*	-14	
V20	ADVERT	78	*	9	
V21	COUNSEL	90	*	-14	
V22	UNADVERT	75	*	2	
V23	NETWORK	60	*	19	
V24	INTERVIEW	60	*	17	
V25	SALARY	74	*	4	

Printed values are multiplied by 100 and rounded to the nearest integer. Values greater than 0.4 are flagged by an '*'.

Remember that the pattern matrix reflects the unique contribution that each factor makes to the variance in a variable, so it is this matrix that you will first use to determine which variables load on which factor. For this analysis, you have flagged any loading over .40 with an asterisk, and will assume the flagged loadings are meaningful.

First, you should read across each row from left to right to see if any variable has a significant loading for more than one factor; these are known as *complex items*. These complex items need to be identified so that they will not be included in any factor-based scale that you will later create. It turns out that no variables load on both factors.

Next you should read down the first factor to see which variables demonstrated significant loadings for this factor. What do these variables have in common? What general construct do they all seem to be measuring? This process is then repeated in order to interpret Factor 2. While doing this, try to determine the way in which Factor 1 differs from Factor 2. In what way do the variables loading on Factor 1 (as a group) tend to differ from those loading on Factor 2?

To make this process easier, Table 2.6 sorts the scales according to the factors on which they load, and provides brief descriptions for the scales. (In this table, the variables have not been sorted according to the size of their loadings.)

Table 2.6: Variables Loading on Factors 1 and 2 According to Rotated Factor Pattern, Two-Factor Solution, Job Search Skills Questionnaire

Variables loading on Factor 1	
3. ASSESS	Using assessment instruments
4. STRATEGY	Identifying effective job search strategies
5. EXPERIENCE	Getting job-related experience
7. RESOCCUP	Researching potential occupations
8. RESEMPLOY	Researching specific employers
12. RESUMES	Using résumés
13. RECOMMEND	Using letters of recommendation
14. DIRECT	Using the cover letter/direct mail approach
15. APPLICAT	Completing application forms
16. IDEMPLOY	Identifying potential employers
17. CARDEVEL	Using campus career development services
18. AGENCY	Using employment agencies
19. FAIRS	Using job fairs
20. ADVERT	Responding to advertised job openings
21. COUNSEL	Using career counselors/consultants
22. UNADVERT	Applying directly for unadvertised jobs
23. NETWORK	Using the networking approach to job search
24. INTERVIEW	Managing the employment interview process
25. SALARY	Negotiating salary
Variables loading on Factor 2	
1. VALUES	Clarifying values and interests
2. ABILITY	Identifying work-related abilities and skills
6. ORGCHAR	Identifying preferred organizational characteristics
9. GOALS	Setting goals
10. BARRIER	Dealing with occupational barriers
11. MOTIVATED	Staying motivated

With Factor 1, you can see significant loadings for such variables as STRATEGY, RESEMPLOY, DIRECT, ADVERT, COUNSEL, UNADVERT, SALARY. In general, variables loading on Factor 1 seem to deal with the ability to perform tasks related to the job search process. People who score high on Factor 1 tend to be knowledgeable about which job search strategies are effective, how to conduct research on specific employers, how to make use of the services offered by career counselors, how to negotiate salary, and so forth. It therefore seems appropriate to label this the *job search skills* factor.

With Factor 2, on the other hand, you can see significant loadings for such scales as VALUES, ABILITY, ORGCHAR, GOALS, and MOTIVATED. People who score high on Factor 2 are able to clearly describe their work-related values and abilities. They know what their goals are, and feel that they will be able to stay motivated during their job search. Therefore, you might label Factor 2 the *goal clarity and motivation* factor. From the perspective of interpretability (at least) this two-factor solution appears to be acceptable.

Now that you have interpreted the meaning of the factors, it would be useful to know the nature of the relationship between Factor 1 and Factor 2. For this information, you may turn to the inter-factor correlations provided in the output. These are presented as Output 2.19.

Output 2.19: Inter-Factor Correlations from Two-Factor Solution, Job Search Skills Questionnaire

Inter-Factor Correlations				
	Factor1		Factor2	
Factor1	100	*	60	*
Factor2	60	*	100	*
Printed values are multiplied by 100 and rounded to the nearest integer. Values greater than 0.4 are flagged by an '*'.				

The inter-factor correlation of +.60 from Output 2.19 reveals a moderately strong positive correlation between the job search skills factor and the goal clarity and motivation factor; and this seems logical. It only makes sense that people who have a good deal of self-insight and motivation related to their careers would also have higher levels of the skills necessary to actually find a job. (In fact, this correlation is so high that some might wonder whether you are really justified in interpreting them as two separate factors.)

To understand the "big picture" concerning the relationship between factors, you will now review the factor structure matrix for your two-factor solution. This is presented as Output 2.20.

Output 2.20: Factor Structure from Promax Rotation, Job Search Skills Questionnaire

Factor Structure (Correlations)					
		Factor1		Factor2	
V1	VALUES	37		72	*
V2	ABILITY	43	*	65	*
V3	ASSESS	61	*	48	*
V4	STRATEGY	73	*	52	*
V5	EXPERIENCE	66	*	60	*
V6	ORGCHAR	46	*	77	*
V7	RESOCCUP	74	*	58	*
V8	RESEMPLOY	79	*	59	*
V9	GOALS	44	*	73	*
V10	BARRIER	61	*	65	*
V11	MOTIVATED	48	*	72	*
V12	RESUMES	70	*	55	*
V13	RECOMMEND	73	*	60	*
V14	DIRECT	77	*	47	*
V15	APPLICAT	61	*	52	*
V16	IDEMPLOY	84	*	55	*
V17	CARDEVEL	72	*	39	

Factor Structure (Correlations)					
		Factor1		Factor2	
V18	AGENCY	73	*	40	
V19	FAIRS	72	*	35	
V20	ADVERT	83	*	56	*
V21	COUNSEL	81	*	40	*
V22	UNADVERT	76	*	47	*
V23	NETWORK	71	*	55	*
V24	INTERVIEW	70	*	53	*
V25	SALARY	77	*	49	*
Printed values are multiplied by 100 and rounded to the nearest integer. Values greater than 0.4 are flagged by an '*'.					

Notice that almost all of the scales are flagged as having significant loadings for both factors. This finding only makes sense in light of the strong inter-factor correlation reported earlier. Remember that in a structure matrix, the loadings represent the correlation between a variable and a factor. Given the strong correlation between Factor 1 and Factor 2, it only makes sense that any variable that loads on Factor 1 will also be correlated with Factor 2, and that any variable that loads on Factor 2 will also be correlated with Factor 1. This is why it is necessary to review the structure matrix to fully understand an oblique solution: Reviewing only the pattern matrix would not reveal how strongly most variables are related to both factors.

A Four-Factor Solution

To illustrate that it is often possible to obtain more than one interpretable solution from a factor analysis, a four-factor solution will now be reviewed. Consider the eigenvalue table, reproduced once again here as Output 2.21 (originally shown in Output 2.14).

Output 2.21: Eigenvalue Table from Analysis of Job Search Skills Questionnaire

Preliminary Eigenvalues: Total = 40.1355388 Average = 1.60542155				
	Eigenvalue	Difference	Proportion	Cumulative
1	33.4862182	30.2592664	0.8343	0.8343
2	3.2269518	1.6761612	0.0804	0.9147
3	1.5507906	0.1822790	0.0386	0.9534
4	1.3685116	0.3091964	0.0341	0.9875
5	1.0593152	0.2961206	0.0264	1.0139
6	0.7631945	0.1079114	0.0190	1.0329
7	0.6552831	0.2233701	0.0163	1.0492
8	0.4319130	0.1148151	0.0108	1.0600
9	0.3170979	0.0201290	0.0079	1.0679
10	0.2969689	0.1244940	0.0074	1.0753
11	0.1724749	0.0544826	0.0043	1.0796
12	0.1179923	0.0738060	0.0029	1.0825
13	0.0441863	0.0961917	0.0011	1.0836
14	-0.0520054	0.0594319	-0.0013	1.0823
15	-0.1114373	0.0174893	-0.0028	1.0795

Preliminary Eigenvalues: Total = 40.1355388 Average = 1.60542155				
	Eigenvalue	Difference	Proportion	Cumulative
16	-0.1289267	0.0361906	-0.0032	1.0763
17	-0.1651172	0.0548644	-0.0041	1.0722
18	-0.2199817	0.0027468	-0.0055	1.0667
19	-0.2227285	0.1178582	-0.0055	1.0612
20	-0.3405867	0.0203781	-0.0085	1.0527
21	-0.3609648	0.0438742	-0.0090	1.0437
22	-0.4048391	0.0013121	-0.0101	1.0336
23	-0.4061512	0.0431700	-0.0101	1.0235
24	-0.4493211	0.0439786	-0.0112	1.0123
25	-0.4932997		-0.0123	1.0000

4 factors will be retained by the NFACTOR criterion.

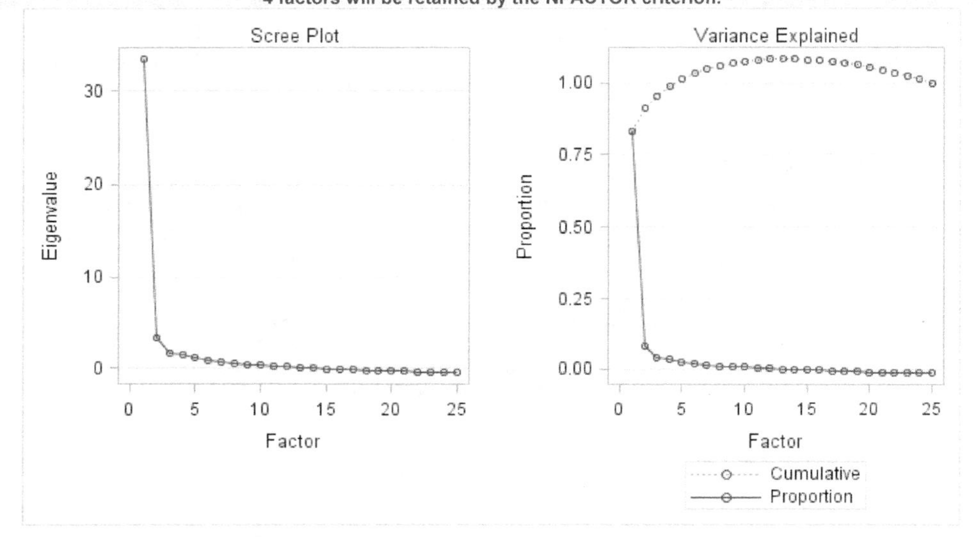

There is a clear break following Factor 1, another (much smaller) break following Factor 2, and from that point on the eigenvalues seem to "flatten out." On the basis of these "breaks" in the eigenvalues alone, it would be difficult to justify rotating four factors.

Still, some have argued that retaining and rotating too few factors has a more serious negative effect on the factor structure than rotating too many, and that it is probably best to err in the direction of over-factoring (Cattell 1952, 1958; Rummel 1970). In fact, one of Cattell's (1958) proposed solutions to the number of factors problem is to retain enough factors to account for 99% of the variance. With the preceding eigenvalue table, you can see that this would involve retaining the first four factors. This can be seen by reviewing the figures in the "Cumulative" column of Output 2.20. Notice that Factors 1 to 4 (combined) account for approximately 96% of the common variance in the dataset, while Factors 1 to 5 (combined) account for approximately 102% of the common variance. If you were to heed Cattell's recommendation, you would therefore retain and interpret Factors 1 to 4.

As an illustration, the results of this four-factor solution will be presented here. Output 2.22 provides the factor pattern matrix resulting from a promax rotation of four factors.

Output 2.22: Rotated Factor Pattern from Promax Rotation, Four-Factor Solution, Job Search Skills Questionnaire

Rotated Factor Pattern (Standardized Regression Coefficients)									
		Factor1		Factor2		Factor3		Factor4	
V1	VALUES	-6		76	*	1		-1	
V2	ABILITY	12		58	*	-6		3	
V3	ASSESS	9		18		33		20	
V4	STRATEGY	40		10		19		17	
V5	EXPERIENCE	-2		31		36		27	
V6	ORGCHAR	1		72	*	-3		8	
V7	RESOCCUP	4		6		5		83	*
V8	RESEMPLOY	52	*	7		-4		37	
V9	GOALS	3		72	*	2		-2	
V10	BARRIER	28		41	*	10		4	
V11	MOTIVATED	12		65	*	1		-1	
V12	RESUMES	51	*	21		15		-5	
V13	RECOMMEND	60	*	25		10		-8	
V14	DIRECT	64	*	3		22		-5	
V15	APPLICAT	55	*	22		2		-7	
V16	IDEMPLOY	41	*	-2		14		46	*
V17	CARDEVEL	7		2		86	*	-4	
V18	AGENCY	60	*	-5		25		-3	
V19	FAIRS	37		-10		48	*	5	
V20	ADVERT	76	*	4		5		5	
V21	COUNSEL	47	*	-11		48	*	4	
V22	UNADVERT	73	*	-2		6		3	
V23	NETWORK	45	*	17		16		7	
V24	INTERVIEW	81	*	10		-14		-1	
V25	SALARY	63	*	-7		-4		32	
Printed values are multiplied by 100 and rounded to the nearest integer. Values greater than 0.4 are flagged by an '*'.									

As before, you should begin by reviewing the rows of the factor pattern matrix to identify any variables with significant loadings for more than one factor. This process identifies that the variable IDEMPLOY loads on both Factors 1 and 4; also, the variable COUNSEL loads on Factors 1 and 3. These variables will therefore not be used in interpreting the factors.

To make it easier to interpret the meaning of these four factors, Table 2.7 groups together the scales according to the factors on which they load.

Table 2.7: Variables Loading on Factors 1, 2, 3, and 4 According to Rotated Factor Pattern, Four-Factor Solution, Job Search Skills Questionnaire

Variables loading on Factor 1	
4. STRATEGY	Identifying effective job search strategies
8. RESEMPLOY	Researching specific employers
12. RESUMES	Using résumés
13. RECOMMEND	Using letters of recommendation
14. DIRECT	Using the cover letter/direct mail approach
15. APPLICAT	Completing application forms
18. AGENCY	Using employment agencies
20. ADVERT	Responding to advertised job openings
22. UNADVERT	Applying directly for unadvertised jobs
23. NETWORK	Using the networking approach to job search
24. INTERVIEW	Managing the employment interview process
25. SALARY	Negotiating salary
Variables loading on Factor 2	
1. VALUES	Clarifying values and interests
2. ABILITY	Identifying work-related abilities and skills
6. ORGCHAR	Identifying preferred organizational characteristics
9. GOALS	Setting goals
10. BARRIER	Dealing with occupational barriers
11. MOTIVATED	Staying motivated
Variables loading on Factor 3	
17. CARDEVEL	Using campus career development services
19. FAIRS	Using job fairs
Variables loading on Factor 4	
7. RESOCCUP	Researching potential occupations

According to Table 2.7, the scales that loaded on Factor 1 all seem to deal with the finding and following-up on job leads. People who score high on this factor feel that they understand the best job-search strategies, are able to research employers and identify organizations that might hire them, are able to respond effectively to advertised and unadvertised job openings, are able to manage the interview process, and are able to successfully negotiate a good salary. This might therefore be labeled the *finding and pursuing job leads* factor.

Factor 2, on the other hand, should look familiar at this point: People who score high on this factor know what are their work-related values and abilities, and feel that they are able to set goals and stay motivated throughout their job search. This factor is similar to the *goal clarity and motivation* factor observed with the two-factor solution.

Only two variables loaded exclusively on Factor 3. Participants who scored high on this factor know how to use the campus career services office and also how to make effective use of job fairs (events that are typically coordinated by campus career development offices). This factor could be labeled the *using campus career services*.

Only 1 variable loads on Factor 4; a second variable was excluded as it loaded on both Factors 1 and 4 (i.e., complex item). The remaining variable pertains to researching potential occupations.

This factor solution proved to be fairly interpretable. For each factor, all variables that load on that factor seem to be measuring a similar underlying construct (i.e., all scales loading on Factor 1 seem to measure a "finding and pursuing job leads" construct). In addition, each factor seems to measure a conceptually different construct (i.e., the "finding and pursuing job leads" factor is conceptually different from the "goal clarity and motivation" factor).

Unfortunately, the solution is unsatisfactory because two of the factors are composed of less than three variables. Either new items need to be written focusing on the constructs that appear to be measured by Factors 3 and 4 (and another exploratory factor analysis later computed with a new dataset), a three-factor solution computed on the current data, or another scale selected.

Conclusion

Exploratory factor analysis is often an iterative process in which you begin with some *a priori* ideas regarding the nature of the factors to be investigated (hopefully based on theory and prior research), and then identify a number of variables that can be expected to be measure these factors. Performing an exploratory factor analysis on the obtained data will often teach you something that was not previously known: Perhaps a five-factor model emerges when a three-factor model was expected; or perhaps variables expected to load on Factor 1 instead load on Factor 4. These results should encourage you to return to the relevant literature, revise the initial model, and perhaps even find new ways of measuring your constructs of interest. A program of research that includes a number of exploratory factor analyses on different datasets, perhaps using improved measures at each step, stands the best chance of discovering the true nature of the factor structure which underlies your construct of interest.

Appendix: Assumptions Underlying Exploratory Factor Analysis

As with principal component analysis, a factor analysis is performed on a matrix of correlations, and this means that the data should satisfy the assumptions for the Pearson correlation coefficient. These assumptions are briefly reviewed below:

- **Interval-level measurement**. All analyzed variables should be assessed on an interval or ratio level of measurement.
- **Random sampling**. Each participant will contribute one score for each observed variable. These sets of scores should represent a random sample drawn from the population of interest.
- **Linearity**. The relationship between all observed variables should be linear.
- **Bivariate normal distribution**. Each pair of observed variables should display a bivariate normal distribution (e.g., they should form an elliptical scattergram when plotted). When the maximum likelihood method is used to extract factors, the output provides a significance test for the null hypothesis that the number of factors retained in the current analysis is sufficient to explain the observed correlations. The following assumption should be met for the probability value associated with this test to be valid:
- **Multivariate normality**. Responses obtained from participants should demonstrate an approximate multivariate normal distribution.

References

Cattell, R. B. (1952). *Factor analysis: An introduction and manual for the psychologist and social scientist.* New York: Harper and Row.

Cattell, R. B. (1958). Extracting the correct number of factors in factor analysis. *Educational and Psychological Measurement, 18,* 791–837.

Clark, L. A., and Watson, D. (1995). Constructing validity: Basic issues in objective scale development. *Psychological Assessment, 7,* 309–319.

DeVellis, R. F. (2012). *Scale development theory and applications* (3rd Ed.). Thousand Oaks, CA: Sage.

Floyd, F. J., and Widaman, K. F. (1995). Factor analysis in the development and refinement of clinical assessment instruments. *Psychological Assessment, 7,* 286–299.

Kim, J. O., and Mueller, C. W. (1978a). *Introduction to factor analysis: What it is and how to do it.* Beverly Hills, CA: Sage.

Kim, J. O., and Mueller, C. W. (1978b). *Factor analysis: Statistical methods and practical issues.* Beverly Hills, CA: Sage.

Le, B., and Agnew, C.R. (2003). Commitment and its theorized determinants: A meta-analysis of the investment model. *Personal Relationships, 10,* 37–57.Loehlin, J. C. (1987). *Latent variable models.* Hillsdale, NJ: Lawrence Erlbaum Associates.

O'Rourke, N., and Cappeliez, P. (2002). Development and validation of a couples measure of biased responding: The Marital Aggrandizement Scale. *Journal of Personality Assessment, 78,* 301–320.

O'Rourke, N., Hatcher, L., and Stepanski, E. J. (2005). *A Step-by-Step Approach to Using SAS for Univariate and Multivariate Statistics* (2nd Ed.). Cary, NC: SAS Institute Inc.

Ruddle, K., Thompson, J., and Hatcher, L. (1993). *Development of the Job Search Skills Inventory: A measure of job search abilities and knowledge.* Paper presented at the Carolinas Psychology Conference, North Carolina State University, Raleigh, NC.

Rummel, R. J. (1970). *Applied factor analysis.* Evanston, IL: Northwestern University Press.

Rusbult, C.E., (1980). Commitment and satisfaction in romantic associations: A test of the investment model. *Journal of Experimental Social Psychology, 16,* 172–186.

Saris, W. E., and Gallhofer, I. N., (2007). *Design, evaluation, and analysis of questionnaires for survey research.* Hoboken, NJ: Wiley InterScience.

Stevens, J. (2002). *Applied multivariate statistics for the social sciences* (4th Ed.). Mahwah, NJ: Lawrence Erlbaum.

Tabachnick, B. G., and Fidell, L. S. (2012). *Using Multivariate Statistics* (6th Ed.). Boston, MA: Allyn and Bacon.

van Buuren, S. (2012). *Flexible imputation of missing data.* Boca Raton, FL. Chapman and Hall.

Chapter 3: Assessing Scale Reliability with Coefficient Alpha

Introduction: The Basics of Response Reliability

You compute coefficient alpha when you have administered a multiple-item rating scale and want to determine the internal consistency of responses to the scale. Scale items may be scored dichotomously (scored as "right" or "wrong") or a multiple-point rating format (e.g., participants may respond to scale items using a 7-point scale).

This chapter shows how to use the CORR procedure to compute the coefficient alpha for the types of scales that are often used in social science research. However, this chapter will not show how to actually *develop* a multiple-item scale for use in research. To learn more about scale development, see DeVellis (2012), and Saris and Gallhofer (2007).

Example of a Summated Rating Scale

A **summated rating scale** usually consists of a short list of statements, questions, or other items to which participants respond. Very often, items that constitute the scale are statements, and participants indicate the extent to which they agree or disagree with each statement by selecting some response on a rating scale (e.g., a 7-point rating scale in which 1 = "Strongly Disagree" and 7 = "Strongly Agree").The scale is called a *summated* scale because the researcher typically sums responses to all selected responses to create an overall score on the scale. These scales are commonly referred to as **Likert-type** scales.

Imagine that you are interested in measuring job satisfaction in a sample of employees. To do this, you might develop a 10-item scale that includes items such as "in general, I am satisfied with my job." Employees respond to these items using a 7-point response format in which 1 = "Strongly Disagree," 4 = Neither Agree nor Disagree, and 7 = "Strongly Agree."

You administer this scale to 200 employees and compute a job satisfaction score for each by summing responses to the 10 items. Scores may range from a low of 10 (if the employee circled "Strongly Disagree" for each item) to a high of 70 (if the employee circled "Strongly Agree" for each item). Given the way these items were written, higher scores indicate higher levels of job satisfaction. With the job satisfaction scale now developed and administered to a sample, you hope to use it as a predictor or criterion variable in future research. However, the people who later read about your research are going to have questions about the psychometric properties of responses to your scale. At the very least, they will want to see empirical evidence that responses to the scale are reliable. This chapter discusses the meaning of scale reliability and shows how SAS can be used to obtain an index of internal consistency for summated rating scales.

True Scores and Measurement Error

Most observed variables measured in the social sciences (e.g., scores on your job satisfaction scale) actually consist of two components: a **true score** that indicates where the participant actually stands on the variable of interest, and a **measurement error**. Almost all observed variables in the social sciences contain at least some measurement error, even variables that seem to be objectively measured.

Imagine that you assess the observed variable "age" in a group of participants by asking them to indicate their age in years. To a large extent, this observed variable (what the participants wrote down) is influenced by the true score component. To a large extent, what they write will be influenced by how old they actually are. Unfortunately, however, this observed variable will also be influenced by measurement error. Some will write down the wrong age because they do not want to disclose how old they are, and other participants will write the wrong age because they did not understand the question. In short, it is likely that there will not be a perfect correlation between the observed variable (what the participants write down) and their true scores on the underlying construct (i.e., their actual age).

This can occur even though the "age" variable is relatively objective and straightforward. If a question such as this is going to be influenced by measurement error, imagine how much more error might result when more subjective constructs are measured such as items that constitute your job satisfaction scale.

Underlying Constructs versus Observed Variables

In applied research, it is useful to draw a distinction between underlying constructs versus observed variables. An **underlying construct** is the variable that you wish to measure. In the job satisfaction study, for example, you wanted to measure the underlying construct of job satisfaction within a group of employees. The **observed variable**, on the other hand, consists of the responses that you measured. In that example, the observed variable consisted of scores on the 10-item measure of job satisfaction. These scores may or may not be a good measure of the underlying construct.

Reliability Defined

With this understanding, it is now possible to provide some definitions. A **reliability coefficient** may be defined as the percent of variance in an observed variable that is accounted for by true scores on the underlying construct. For example, imagine that in the study just described, you were able to obtain two scores for the 200 employees in the sample: their observed scores on the job satisfaction questionnaire; and their true scores on the underlying construct of job satisfaction. Assume that you compute the correlation between these two variables. This correlation coefficient squared represents the reliability of responses to your job satisfaction scale; it is the percent of variance in observed job satisfaction scores that is accounted for by true scores on the underlying construct of job satisfaction (i.e., shared variance).

The preceding was a technical definition for reliability, but this definition is of little use in practice because it is generally not possible to obtain true scores for a variable. For this reason, reliability is estimated in terms of the **consistency** of scores that are obtained on the observed variable. Responses to an instrument are said to be reliable if consistent scores are obtained upon repeated administration, upon administration by alternate forms, and so forth. A variety of methods of estimating scale reliability are used in practice.

Test-Retest Reliability

Assume that you administer your measure of job satisfaction to a group of 200 employees at two points in time: once in January and again in March. If responses to the instrument are indeed reliable, you would expect that participants who provided high scores in January will tend to provide high scores again in March; conversely, those who provided low scores in January will likely provide low scores in March. These results would support the test-retest reliability of responses to the scale. Test-retest reliability is assessed by administering the same instrument to the same sample of participants at two points in time and then computing the correlation between sets of scores.

But what is an appropriate interval over which questionnaires should be administered? Unfortunately, there is no hard-and-fast rule of thumb here; this interval will depend on what is being measured. For enduring constructs such as personality variables, test-retest reliability has been assessed over several decades. For other constructs such as depressive symptomatology, the interval tends to be much shorter (e.g., weeks) due to the fluctuating course of depression and its symptoms. Generally speaking, the test-retest interval should not be too short so that respondents recall their responses to specific items (e.g., less than a week) but not as long as to measure natural variability in the construct (e.g., bona fide change in depressive symptoms). The former will lead to an overstatement of test-retest reliability, whereas the latter will lead to understatement of test-retest reliability.

Internal Consistency

A further problem with the test-retest reliability procedure is the time that it requires. What if you do not have time to administer the scale two times? In such situations, you are likely to turn to reliability indices that may be obtained with only one administration. In research that involves the use of questionnaire data, the most popular of these are the internal consistency indices of reliability. Briefly, **internal consistency** is the extent to which the individual items that constitute a test correlate with one another or with the test total. In the social sciences, the most widely used index of internal consistency is the coefficient alpha, also known as Cronbach's alpha, symbolized by the Greek letter α (Cronbach 1951).[1]

Reliability as a Property of *Responses* to Scales

You may have already noticed that in this chapter we describe reliability as a property of responses to scales, not of scales themselves. In other words, scales are not reliable, only responses to scales. This is because reliability estimates such as internal consistency often vary from population to population, and in some instances, even within subpopulations (Wilkinson and The American Psychological Association Task Force on Statistical Inference 1999). In other words, estimates such as internal consistency may be very different for one group (e.g., young adults) as compared to another (e.g., older adults) even for responses to the same scale. By keeping this difference in mind, you will avoid making this error and recognize that it is important to calculate reliability estimates for each study and not rely upon coefficients previously reported by other researchers.

Coefficient Alpha

Formula

Coefficient alpha is a general formula for scale reliability based on internal consistency. It provides the lowest estimate of reliability that can be expected for an instrument.

The formula for coefficient alpha is as follows:

$$r_{xx} = \left(\frac{N}{N-1}\right)\left(\frac{S^2 - \Sigma S_i^2}{S^2}\right)$$

where

r_{xx} = coefficient alpha

N = number of items constituting the instrument

S^2 = variance of summated scale scores (e.g., compute a total score for each participant by summing responses to the items that constitute the scale; the variance of this total score variable would be S^2)

$\sum S^2_i$ = the sum of the variances of the individual items that constitute this scale

When Will Coefficient Alpha Be High?

Other factors held constant, coefficient alpha will be high to the extent that many items are included in the scale, and the items that constitute the scale are highly correlated with one another.

To understand why a coefficient alpha is high when the items are highly correlated with one another, consider the second term in the preceding formula:

$$\left(\frac{S^2 - \Sigma S_i^2}{S^2}\right)$$

This term shows that the variance of the summated scales scores is (essentially) divided by itself to compute coefficient alpha. However, the combined variance of the individual items is first subtracted from this variance before the division is performed. This part of the equation shows that if combined variance of the individual items is a small value, then coefficient alpha will be a relatively larger value.

This is important because (with other factors held constant) the stronger the correlations between items, the smaller the $\sum S^2_i$ term. This is why coefficient alpha for responses to a given scale is likely to be large to the extent that the variables constituting that scale are strongly correlated.

Assessing Coefficient Alpha with PROC CORR

Imagine that you have conducted research in the area of prosocial behavior and have developed an instrument designed to measure two separate underlying constructs: helping others and financial giving. **Helping others** refers to prosocial activities performed to help coworkers, relatives, and friends; whereas **financial giving** refers to giving money to charities or the homeless. (See Chapter 1, “Principal Component Analysis” for a more detailed description of these constructs.) In the following questionnaire, items 1 to 3 were designed to assess helping others, and items 4 to 6 were designed to assess financial giving.

Instructions: Below are a number of activities in which people sometimes engage. For each item, please indicate how frequently you have engaged in this activity over the past six months. Provide your response by circling the appropriate number to the left of the item, and use the following response key:

7 = Very Frequently
6 = Frequently
5 = Somewhat Frequently
4 = Occasionally
3 = Seldom
2 = Almost Never
1 = Never

1 2 3 4 5 6 7 1. Went out of my way to do a favor for a coworker.
1 2 3 4 5 6 7 2. Went out of my way to do a favor for a relative.
1 2 3 4 5 6 7 3. Went out of my way to do a favor for a friend.
1 2 3 4 5 6 7 4. Gave money to a religious charity.
1 2 3 4 5 6 7 5. Gave money to a charity not associated with a religion.
1 2 3 4 5 6 7 6. Gave money to a panhandler.

Assume that you have administered this 6-item questionnaire to 50 participants. For the moment, we are concerned only with the reliability of responses to items 1 to 3 (the items that assess helping others).

Let us further assume that you have made a mistake in assessing the reliability of responses to this scale. Assume that you erroneously believed that the helping others construct was assessed by items 1 to 4 (whereas, in reality, the construct was assessed by items 1 to 3). It will be instructive to see what you learn when you mistakenly include item 4 in the analysis.

General Form

Here is the general form for the SAS statements to estimate this coefficient or Cronbach's alpha (internal consistency) for a summated rating scale:

```
proc corr    data=dataset-name    alpha    nomiss;
   var  list-of-variables;
run ;
```

In the preceding program, the ALPHA option requests that Cronbach's or coefficient alpha be computed for the variables included in the VAR statement; the NOMISS option is required. The VAR statement should list only those items that constitute the scale (or subscale) in question. You must perform a separate PROC CORR for each scale whose reliability you want to estimate.

A 4-Item Scale

Here is an actual program, including the DATA step, to analyze fictitious data from your study. Only a few sample lines of data appear here. The complete dataset appears in Appendix B. Ordinarily, one would not compute Cronbach's alpha in this case as internal consistency is often underestimated with so few items. The following examples are provided simply to illustrate the computation and meaning of Cronbach's alpha.

```
data D1;
   input    #1    @1    (V1-V6)    (1.)   ;

datalines ;
556754
567343
777222
.
.
.
767151
455323
455544
;
run ;

proc corr   data=D1   alpha   nomiss ;
   var V1 V2 V3 V4 ;
run ;
```

The results of this analysis appear as Output 3.1. These results provide the means, standard deviations, and other descriptive statistics that you should review to verify that the analysis proceeded as expected. The results below these descriptive statistics pertain to the reliability of scale responses.

Output 3.1: Simple Statistics and Coefficient Alpha Results for Analysis of Scale That Includes Items 1 to 4, Prosocial Behavior Study

The CORR Procedure

4 Variables:	V1 V2 V3 V4

Simple Statistics						
Variable	N	Mean	Std Dev	Sum	Minimum	Maximum
V1	50	5.18000	1.39518	259.00000	1.00000	7.00000
V2	50	5.40000	1.10657	270.00000	3.00000	7.00000
V3	50	5.52000	1.21622	276.00000	2.00000	7.00000
V4	50	3.64000	1.79296	182.00000	1.00000	7.00000

Cronbach Coefficient Alpha	
Variables	Alpha
Raw	0.490448
Standardized	0.575912

Cronbach Coefficient Alpha with Deleted Variable				
	Raw Variables		Standardized Variables	
Deleted Variable	Correlation with Total	Alpha	Correlation with Total	Alpha
V1	0.461961	0.243936	0.563691	0.326279
V2	0.433130	0.318862	0.458438	0.420678
V3	0.500697	0.240271	0.546203	0.342459
V4	-.037388	0.776635	-.030269	0.773264

Pearson Correlation Coefficients, N = 50 Prob > |r| under H0: Rho=0				
	V1	V2	V3	V4
V1	1.00000	0.49439 0.0003	0.71345 <.0001	-0.10410 0.4719
V2	0.49439 0.0003	1.00000	0.38820 0.0053	0.05349 0.7122
V3	0.71345 <.0001	0.38820 0.0053	1.00000	-0.02471 0.8648
V4	-0.10410 0.4719	0.05349 0.7122	-0.02471 0.8648	1.00000

To the right of the heading "Cronbach Coefficient Alpha (Raw)" you see that the reliability coefficient for responses to items 1 to 4 is only .49 (rounded to two decimal places). Reliability estimates for raw variables are normally reported in published reports as opposed to the standardized alphas.

How Large Must a Reliability Coefficient Be to Be Considered Acceptable?

As a rule of thumb, Nunnally (1978) suggested that coefficient alpha values .70+ are acceptable. You should remember, however, that this is only a rule of thumb. Some social scientists accept coefficient alphas under .70; in most social science disciplines, however, alpha values .80+ are seen as ideal. When alpha values are less than .70, less than half of all variance is shared among items (i.e., $70^2 < .50$).

Is a larger alpha coefficient always better than a smaller one? Not necessarily. An ideal estimate of internal consistency is between .80 and .90 (i.e., $.80 \leq \alpha \leq .90$; Clark and Watson 1995; DeVellis 2012). This is because coefficients in excess of .90 suggest item redundancy or excessive scale length. The number of items within scales is also a consideration when interpreting coefficient alpha values, however. This is because Cronbach's alpha underestimates internal consistency with fewer than eight items, and overestimates internal consistency with more than 30 items (Henson 2001).

Back to our example, the coefficient alpha of .49 reported in Output 3.1 is not acceptable; but how is it possible to significantly improve this coefficient?

In some situations, the reliability of responses to a multiple-item scale is improved by deleting those items with poor item-total correlations. An **item-total correlation** is the correlation between an individual item and the sum of the remaining items that constitute the scale. If an item-total correlation is small, this may be seen as evidence that the item is not measuring the same construct as other scale items. You may therefore choose to discard items with small item-total correlation values (assuming that data have been entered correctly).

Consider Output 3.1. Under the "Correlation with Total" (Raw Variables) heading, you can see that items 1 to 3 are each moderately correlated with the sum of the remaining items on the scale. Item V4, however, has an item-total correlation less than -.04. This suggests that item V4 is not measuring the same underlying construct as items V1 to V3.

In Output 3.1 under the "Alpha" heading, you find an estimate of what alpha would be if a given variable (item) were deleted from the scale. To the right of "V4," PROC CORR estimates that alpha would be approximately .78 if V4 were deleted. (This value appears where the row headed "V4" intersects with the column heading "Alpha" in the "Raw Variables" section.) This makes sense because variable V4 exhibits a correlation with the remaining scale items of only -.04. You could substantially improve internal consistency by removing the item that appears to be measuring a different construct than the other items.

A 3-Item Scale

Output 3.2 presents the results of PROC CORR when coefficient alpha is calculated for variables V1 to V3. This is done by specifying only V1 to V3 in the VAR statement.

Output 3.2: Simple Statistics and Coefficient Alpha Results for Analysis of Scale That Includes Items 1 to 3, Prosocial Behavior Study

The CORR Procedure

3 Variables:	V1 V2 V3

Simple Statistics						
Variable	N	Mean	Std Dev	Sum	Minimum	Maximum
V1	50	5.18000	1.39518	259.00000	1.00000	7.00000
V2	50	5.40000	1.10657	270.00000	3.00000	7.00000
V3	50	5.52000	1.21622	276.00000	2.00000	7.00000

Cronbach Coefficient Alpha	
Variables	Alpha
Raw	0.776635
Standardized	0.773264

Cronbach Coefficient Alpha with Deleted Variable				
	Raw Variables		Standardized Variables	
Deleted Variable	Correlation with Total	Alpha	Correlation with Total	Alpha
V1	0.730730	0.557491	0.724882	0.559285
V2	0.480510	0.828202	0.476768	0.832764
V3	0.657457	0.649926	0.637231	0.661659

Pearson Correlation Coefficients, N = 50 Prob > |r| under H0: Rho=0			
	V1	V2	V3
V1	1.00000	0.49439 0.0003	0.71345 <.0001
V2	0.49439 0.0003	1.00000	0.38820 0.0053
V3	0.71345 <.0001	0.38820 0.0053	1.00000

Output 3.2 provides a raw-variable coefficient alpha of .78 for the three items included in this analysis. This value is adequate only ($.70 \leq \alpha < .80$). This coefficient appears under the heading "Cronbach Coefficient Alpha" to the right of the heading "Raw." This coefficient exceeds the recommended minimum value of .70 (Nunnally 1978) and approaches the ideal range of $.80 \leq \alpha \leq .90$. Responses to the helping others subscale demonstrate a much higher level of reliability without item V4.

Summarizing the Results

Summarizing the Results in a Table

Researchers typically report the reliability of responses to a scale in a table with simple descriptive statistics for the study's variables such as means, standard deviations, and intercorrelations. In these tables, coefficient alpha estimates are often reported on the diagonal of the correlation matrix within parentheses. Such an approach appears in Table 3.1.

Table 3.1: Means, Standard Deviations, Intercorrelations, and Coefficient Alpha Estimates for the Study's Variables

Variable	M	SD	1	2	3
1. Authoritarianism	13.56	2.54	(.90)		
2. Helping others	15.60	3.22	.37	(.78)	
3. Financial giving	12.55	1.32	.25	.53	(.77)

NOTE: *N* = 200. Reliability estimates (Cronbach's alphas) appear on the diagonal.

In the preceding table, information for the authoritarianism variable is presented in both the row and the column heading "1." Where the row heading "1" intersects with the column heading "1," you will find Cronbach's alpha for responses to the authoritarianism scale; you can see that this coefficient is .90. In the same way, you can find coefficient alpha for acquaintance helping where row 2 intersects with column 2 ($\alpha = .78$), and you can find coefficient alpha for financial giving where row 3 intersects with column 3 ($\alpha = .77$).

Preparing a Formal Description of the Results for a Paper

When reliability estimates are computed for a relatively large number of scales, it is common to report them in a table (such as Table 3.1), and make only passing reference to them within the text of the paper when within acceptable parameters. For example, within the section on instrumentation, you might indicate:

> Estimates of internal consistency as measured by Cronbach's alpha are within acceptable limits for all study variables (i.e., $\alpha \geq .70$). These coefficients range from $.77 \leq \alpha \leq .90$ as reported along the diagonal of Table 3.1.

When reliability estimates are computed for only a few scales, it is possible to instead report these estimates within the body of the text itself. Here is an example of how this might be done:

> Internal consistency of scale responses was assessed by Cronbach's alpha. Reliability estimates were $\alpha = .90$, $\alpha = .78$, and $\alpha = .77$ for responses to the authoritarianism, helping others, and financial giving subscales, respectively.

Conclusion

Assessing scale reliability with Cronbach's or coefficient alpha (or some other reliability index) should be one of the first tasks you undertake when conducting questionnaire research. If responses to selected scales are not reliable, there is no point performing additional analyses. You can often improve suboptimal reliability estimates by deleting items with poor item-total correlations in keeping with the procedures described in this chapter. When several subscales on a questionnaire display poor reliability, it may be advisable to perform a principal component analysis or an exploratory factor analysis on responses to all questionnaire items to determine which tend to group together. If many items load on each retained factor and if the factor pattern obtained from such an analysis displays a simple structure, chances are good that responses will demonstrate adequate internal consistency.

Note

1. The Greek letter alpha (α) used to represent internal consistency (Cronbach's alpha) should not be confused with alpha (α) used to specify statistical significance threshold levels (e,g., $\alpha < .05$).

References

Clark, L. A., and Watson, D. (1995). Constructing validity: Basic issues in objective scale development. *Psychological Assessment, 7,* 309–319.

Cortina, J. M. (1993). What is coefficient alpha? An examination of theory and applications. *Journal of Applied Psychology, 78,* 98–104.

Cronbach, L. J. (1951). Coefficient alpha and the internal structure of tests. *Psychometrika, 16,* 297–334.

DeVellis, R. F. (2012). *Scale development theory and applications* (3rd Ed.).Thousand Oaks, CA: Sage.

Henson, R. K. (2001). Understanding internal consistency estimates: A conceptual primer on coefficient alpha. *Measurement and Evaluation in Counseling and Development, 34,* 177–189.

Nunnally, J. (1978). *Psychometric theory.* New York: McGraw-Hill.

Saris, W. E., and Gallhofer, I. N., (2007). *Design, evaluation, and analysis of questionnaires for survey research.* Hoboken, NJ: Wiley InterScience.

Wilkinson, L., and The American Psychological Association Task Force on Statistical Inference. (1999). Statistical methods in psychology journals: Guidelines and explanations. *American Psychologist, 54,* 594–604.

Chapter 4: Path Analysis

Introduction: The Basics of Path Analysis

Path analysis can be used to test theoretical models that specify directional relationships among a number of observed variables. Path analysis determines whether your model successfully accounts for the actual relationships observed in the sample data. The output of the CALIS procedure provides indices that specify whether the model, as a whole, fits the data as well as significance tests for specified directional paths. When a model provides a relatively poor fit to the data, additional results from CALIS can be used to modify the model and improve model fit.

This chapter deals only with models in which all variables are **manifest** (i.e., observed variables); it does not deal with models that specify directional relationships between **latent** or unobserved variables. For guidelines on how to test latent variable models, see Chapter 5, "Developing Measurement Models with Confirmatory Factor Analysis" and Chapter 6, "Structural Equation Modeling" of this text.

Some Simple Path Diagrams

When studying complex phenomena, a given outcome variable of interest may be influenced by a *variety* of other variables; in other words, few outcome variables are determined by just one variable. For example, imagine that you are an industrial/organizational psychologist who believes that employees' work performance (the outcome variable of interest) is influenced by the following four variables:

- employees' level of intelligence
- employees' level of work motivation
- workplace norms
- supervisory support

A very simple path diagram for this hypothetical directional model is presented in Figure 4.1.

Figure 4.1: Path Diagram: A Simple Model of the Determinants of Work Performance

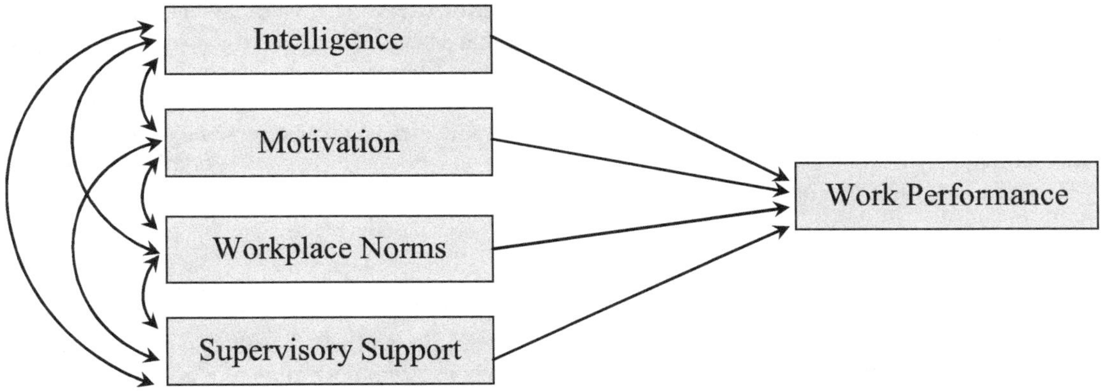

Intelligence, motivation, workplace norms, and supervisory support are antecedent or **independent variables** within this framework as each is assumed to predict to work performance. Similarly, work performance is the consequent or **dependent variable** in the model as it is predicted by independent variables. These terms are consistent with multiple regression; this is because path analysis is an extension or more complex form of multiple regression.

The boxes shown in Figure 4.1 are connected to one another by straight single-headed arrows and curved double-headed arrows. In path analysis and structural equation modeling, a **straight single-headed arrow** is used to represent a unidirectional path. The arrow originates at the variable exerting the influence (the independent variable), and points toward the variable being predicted (the dependent variable). For example, the straight single-headed arrow from intelligence to work performance represents the hypothesis that intelligence predicts work performance.

In contrast, a **curved double-headed arrow** connecting two variables represents covariance, or correlation, between variables. A curved arrow connecting two variables means that the two variables are expected to covary, but no hypothesis is made regarding any causal influence between them. For example, the two-headed curved arrow connecting intelligence to motivation means that no hypothesis is made as to which variable determines or predicts the other. Perhaps intelligence causes motivation, perhaps motivation causes intelligence, perhaps each has some bidirectional influence on the other, or perhaps their correlation is due to the influence of some shared but unmeasured variable.

The model presented in Figure 4.1 suggests that each of the four antecedent variables is expected to have a direct effect on work performance; notice that each arrow goes directly from the independent variable to work performance. This is the simplest type of path model. However, most models in social science research also predict that some variables have indirect effects on other variables. Figure 4.2 provides a model that includes indirect effects.

Figure 4.2: Path Diagram: A More Complex, Recursive Model of the Determinants of Work Performance

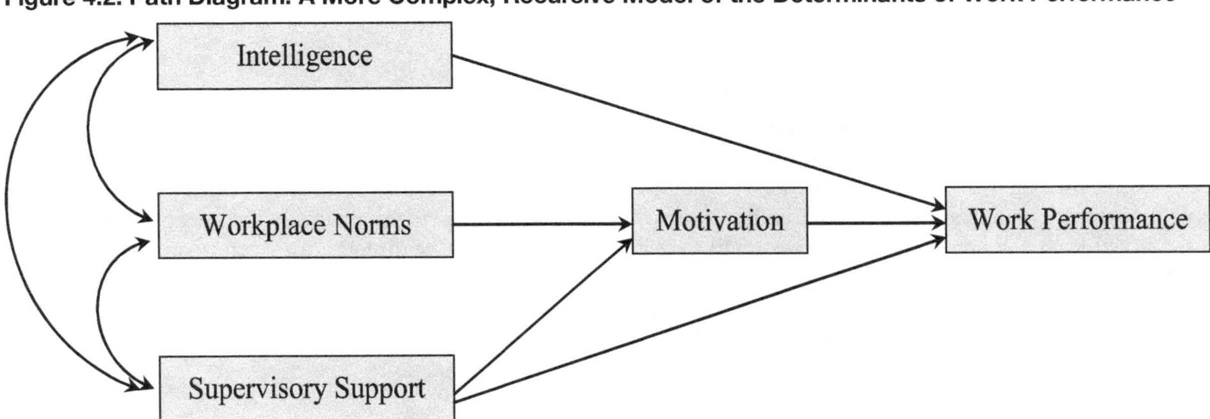

Figure 4.2 includes the same four variables previously discussed, but they are arranged in a somewhat different sequence. Most notably, motivation is now depicted as a **mediator variable**: A variable that mediates the effect of an independent variable onto a dependent variable. Notice that a single-headed arrow goes from workplace norms to motivation, and that a separate single-headed arrow goes from motivation to work performance. This indicates that workplace norms have only an indirect effect on work performance (i.e., workplace norms influence work performance by influencing motivation). This same idea can be expressed by saying that motivation completely mediates the effect of workplace norms on work performance. Workplace norms are not expected to have a direct effect on work performance as there is no direct path connecting workplace norms to work performance.

A variable may have both direct and indirect effects on a dependent variable. Figure 4.2 depicts such a relationship between supervisory support and work performance. A single-headed arrow goes directly from supervisory support to work performance, indicating the predicted direct effect. A path also goes from supervisory support to motivation, however, and a second path goes from motivation to work performance. This indicates that support is also assumed to indirectly impact work performance by first affecting motivation. (**Note:** These models are presented solely to describe the basic features of path analytic models and do not represent actual models of work performance.)

Important Terms Used in Path Analysis

Endogenous versus Exogenous Variables

In path analysis, a distinction is made between endogenous variables and exogenous variables. An **endogenous variable** is one whose variability is hypothesized to be determined by other variables in the model. Any variable that has a straight single-headed arrow pointing at it is an endogenous variable. In Figure 4.2, work performance is an endogenous variable as it is directly influenced by intelligence, motivation, and supervisory support. Motivation is also an endogenous variable; it is predicted by workplace norms and supervisory support.

Exogenous variables, on the other hand, are constructs that are influenced by variables outside of the directional model. Exogenous variables do not have any straight single-headed arrows pointing to them. In Figure 4.2, intelligence, norms, and support are all exogenous variables. These three variables are connected by curved arrows (indicating that they are expected to covary), but no single-headed arrows point toward them. This means that the researcher makes no predictions about what influences intelligence, norms, or supervisory support. In most models, exogenous variables will affect other variables; but, by definition, exogenous variables are never affected by other variables in the model.

Notice the straight arrow that runs from motivation to work performance. Does this mean that motivation is an exogenous variable? No, because there are also two arrows that point toward motivation. Any time a single-headed arrow points at a variable, that variable is an endogenous variable even if that variable is assumed to predict other constructs in that model.

Most of the figures in this chapter will follow the convention of having directional paths run from left to right. In general, exogenous variables will be presented toward the left side of the figure, and endogenous variables will be presented toward the middle and right.

Manifest versus Latent Variables

A **manifest variable** is one that is directly measured or observed, whereas a **latent variable** is a hypothetical construct that is not directly measured or observed. For example, a combined scale score on the "Weschler Adult Intelligence Scale (WAIS-IV)" is a manifest variable; it is possible to directly determine exactly where each participant stands on this variable. On the other hand, *intelligence* is generally thought of as a latent variable: It is a construct that is presumed to exist though it cannot be directly observed. You may administer an intelligence test to participants, but you know that the resulting IQ (Intelligence Quotient) scores are only estimates of the underlying construct of intelligence. You hope that the underlying latent variable of intelligence will influence participants' resulting WAIS-IV scores.

What do rectangles and ovals represent in diagrams? In path analysis, manifest variables are represented as squares or rectangles (see Figures 4.1 and 4.2). By definition, path analysis depicts assumed relationships among manifest variables only. In contrast, "confirmatory factor analysis" and "structural equation modeling" include both manifest and latent variables. In those models, latent variables are represented by circles or ovals. Chapters 5 and 6 will discuss models with latent or unobserved variables.

Recursive versus Nonrecursive Models

A **recursive path model** is one in which prediction is assumed to occur in one direction only. In a recursive model, a dependent variable never exerts influence (either directly or indirectly) on an independent variable that first exerts influence on it. In other words, recursive models are unidirectional. The model presented in Figure 4.2 is recursive; notice that prediction flows only in a left-to-right direction.

In contrast, in **nonrecursive path models** causation may flow in more than one direction, and a variable may have a direct or indirect effect on another variable that preceded it in the model. For example, a model may be nonrecursive because it predicts reciprocal causation between two variables (e.g., O'Rourke 2000). This is illustrated in Figure 4.3, which predicts reciprocal causation between the work performance and motivation variables. Here, motivation is assumed to predict work performance which, in turn, predicts motivation.

Figure 4.3: Path Diagram: Reciprocal Nonrecursive Model

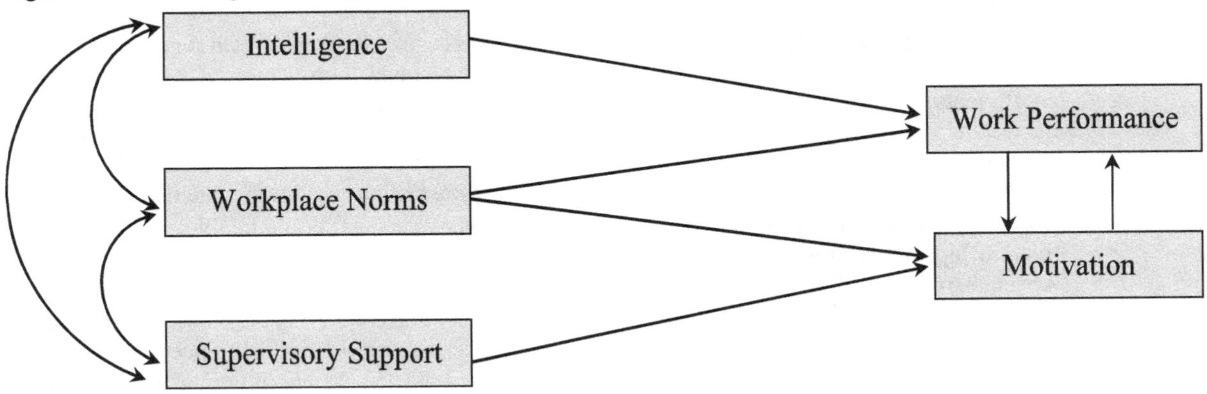

Models may also be nonrecursive because they contain feedback loop. For example, imagine a model in which variable A affects variable B, variable B affect variable C, and variable C, in turn, affects variable A. Variables A, B, and C can be said to constitute a feedback loop; the model that contains this loop is therefore nonrecursive.

Until recently, it was important to make a distinction between recursive and nonrecursive models because the other procedures for performing path analysis (multiple regression) did not easily lend themselves to the testing of nonrecursive models. Both types of model can be analyzed with PROC CALIS however.

NOTE: This chapter deals only with recursive models. This chapter will cover only the analysis of recursive (unidirectional) directional models. This will prepare you to deal with a wide variety of research problems that are commonly encountered in social science research. It will also serve as a good foundation for readers who want to move on to the somewhat more complex nonrecursive models. References for the analysis of nonrecursive models are provided later in the chapter.

Why Perform Path Analysis with PROC CALIS versus PROC REG?

As implied above, path analysis used to be performed with procedures such as PROC REG before the advent of PROC CALIS and other such programs. Each endogenous variable required the computation of a separate regression equation. With complex models, this required several regression equations be computed increasing the risk of *capitalization on chance* (i.e., incorrect rejection of the null hypothesis). In contrast, PROC CALIS requires only one model be computed no matter the number of endogenous variables within the model. Almost always it is preferable to compute a single multivariate equation versus multiple univariate equations.

As you will learn later in this chapter, PROC CALIS also enables you to estimate overall model goodness-of-fit indices not possible with PROC REG. Finally, PROC CALIS enables the researcher to compute modification indices. These will be discussed more fully later in this chapter; suffice to say at this point that PROC CALIS allows you to arrive at the best fitting model representing the relationships that exist among variables in a specific dataset. As a result of these features, we recommend that PROC CALIS be used instead of PROC REG when undertaking path analysis.

Necessary Conditions for Path Analysis

The use of path analysis assumes that a number of requirements have been met concerning the nature of the data as well as the theoretical model itself. Listed below are some important assumptions associated with the analysis of the simple recursive models to be covered in this chapter.

1. **Interval- or ratio-level measurement.** Endogenous variables should be assessed on an interval or ratio level of measurement. Alternative procedures for situations in which these assumptions are violated have been discussed elsewhere (e.g., Jöreskog and Sörbom 2001) but will not be covered in this chapter.
2. **Minimal number of values.** Endogenous variables should be continuous and should assume a minimum of five values (Byrne 1998).
3. **Normally distributed data.** Although parameter estimates may be correct with nonnormal data, the statistical tests used with PROC CALIS (e.g., chi-square test and significance tests for path coefficients) assume a multivariate normal distribution. It has been argued, however, that the maximum likelihood and generalized-least squares estimation procedures appear to be fairly robust against moderate violations of this assumption (Jöreskog and Sörbom 2001; Kline 2005). When data are markedly nonnormal, you should consider data transformations and/or the deletion of outliers.
4. **Linear and additive relationships.** Relationships among variables should be linear and additive (i.e., relationships between independent and dependent variables should not be curvilinear).
5. **Absence of multicollinearity.** Variables should be free of multicollinearity. Multicollinearity is a condition in which one or more variables are very strongly correlated (e.g., $r \geq .80$).Multicolinearity indicates redundancy of measurement (i.e., same construt measured more than once).
6. **Absence of measurement error.** All independent (antecedent) variables *should* be measured without error; this means that any analyzed independent manifest variable that is a perfectly reliable indicator of its underlying construct. Given the relatively high levels of measurement error associated with variables in the social science research, however, this is the most frequently violated assumption in path analysis. The following two chapters will show how this problem can be minimized through the analysis of structural equation models with multiple indicators and latent variables.
7. **Inclusion of all nontrivial causes.** All known nontrivial predictors of endogenous variables should be included in the model as independent variables. For example, if you perform a path analysis in which work performance is to be a dependent variable, the model should specify as independent variables all constructs that are known to have nontrivial effects on work performance (and, needless to say, these variables must be assessed in the study itself). If important antecedent variables are omitted, the path coefficients for the remaining antecedents are likely to be biased. If this requirement is met, the model is said to be **self-contained**, and all residual terms in the model should be uncorrelated. (The meaning of "residual terms" will be discussed below.)
8. **Overidentified model.** To be tested for goodness of fit, the model must be overidentified. The meaning of "just-identified" versus "overidentified" versus "underidentified" models will be discussed below.

Overview of the Analysis

Although it is relatively easy to perform path analysis using PROC CALIS, the process must be divided into a number of steps with important decisions to be made at each juncture. Here is an overview of the process.

Preparing the Program Figure

It is recommended that you always prepare a detailed figure that describes the predicted relationships among variables and identifies all parameters to be estimated. This program figure guides you when writing the SAS program. If carefully prepared, the program figure will make writing the program a much simpler task and reduce the possibility of specification errors.

Preparing the SAS Program

The analyses will be conducted using PROC CALIS. PROC CALIS will include statements to represent the model displayed in your program figure.

Interpreting the Results

The output of PROC CALIS provides significance tests for the null hypothesis that your theoretical model fits the data; these are known as "goodness-of-fit" statistics. It also provides estimates and significance tests for parameters such as path coefficients, variances, and covariances. If there is less than ideal fit between a model and data, modification indices may be used to determine how the model could be improved. Revised models can then be estimated to determine if they provide improved fit.

Sample Size Requirements for Path Analysis

There is no definitive consensus regarding sample size requirements for path analysis with PROC CALIS and other structural equation modeling (SEM) programs versus PROC REG. This is because analysis of covariance structures (e.g., SEM, confirmatory factor analysis) is based on large sample theory (Lehmann 1999). Although there are no latent variables in path analysis (by definition), one of the primary reasons to use path analyses versus SEM is the lower sample size requirement. We recommend that sample requirements be calculated similar to multiple regression and greater than 99; in other words, we recommend 100 as the minimal sample size required for path analyses (MacCallum 1986).

Irrespective of the number of dependent variables in one's path model, the number of independent variables is N–1 for the purposes of sample size estimation. Figure 4.3 is presented once again to help illustrate this point. Both work performance and motivation are predicted in this model (i.e., both are dependent variables), predicted by intelligence, workplace norms, supervisory support, and each other. As previously noted, this is a reciprocal, nonrecursive model because both dependent variables are hypothesized to be both dependent and independent variables. In this example, there are a total of five observed variables. Applying are N–1 rule, there are four independent variables for the purposes of estimating sample size requirements.

Figure 4.3: Path Diagram: Reciprocal Nonrecursive Model

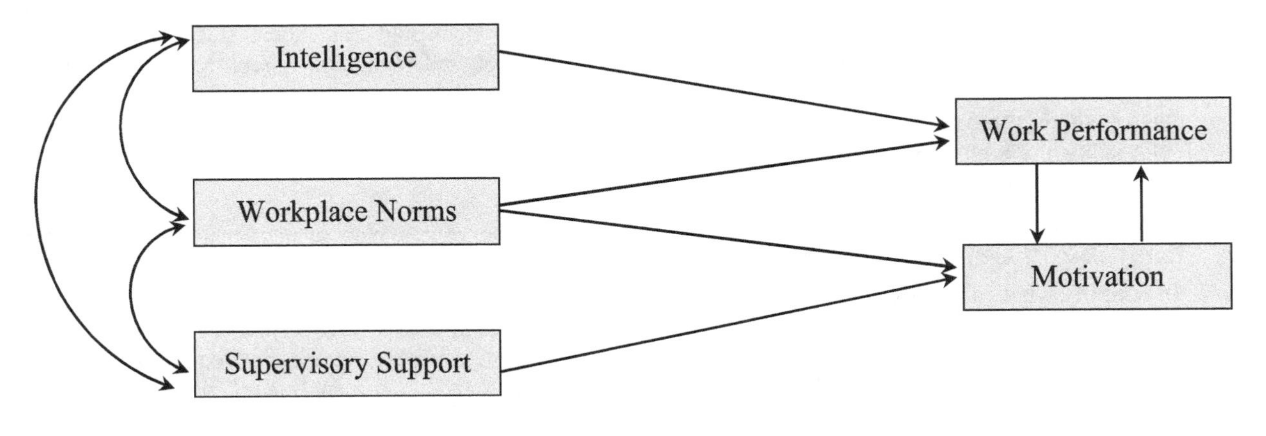

Statistical Power and Sample Size

The objective is to make sure that your sample is large enough to reveal what really exists in the population studied, all other things being equal. Stated another way, you need to have enough observations or data points to get an accurate snapshot of the phenomena examined. Such Type I or alpha (α) errors occur when the researcher fails to detect a phenomenon under study which, in fact, exists (i.e., would be detected with a larger sample size). In precise terms, this would be a false negative.

When working with small samples, you do not have sufficient *statistical power* (Cohen, 1998). In social science research, this usually means that you have too few participants to adequately test your research questions. Findings from such *underpowered* studies may apply only to that particular study; results may be misleading.

But is bigger always better? Not necessarily, at least when it comes to sample size. More precisely, excessively large samples can also lead to false findings. In such *overpowered* studies, very small between-group differences emerge as statistically significant (e.g., Giles, Ryan, and Anas, 2008). This risk is greater in research with population and other large data. Such overpowered analyses lead to Type II or beta (β) errors (i.e., statistically significant but meaningless findings). Type II error is a one reason for confusing findings commonly reported in the media that seemingly contradict previously research or common understanding.

Before beginning data collection, you should always identify what is an ideal, or at least adequate, sample size required for that particular study. An ideal sample size will include extra obsservations or participants to allow for skipped responses and unforeseen complications (e.g., 10% buffer). Your required sample size will differ depending on the research questions and selected statistical procedures. In other words, there is no one ideal sample size. You can have both too few and too many participants or data points. Bigger is not always better; sometimes bigger is worse.

Effect Sizes

It is also important to keep in mind that significant associations among phenomena range from strong to weak. Weak associations, for instance, are indirect and more distal (far apart). As an example, a strong association exists between having a parent with schizophrenia and the likelihood of developing schizophrenia oneself (Pulver, Liang, Brown, Wolyniec, McGrath, Adler et al. 1992). In contrast, a significant but small association exists between one's astrological sign and the likelihood of developing schizophrenia (i.e., greatest for those born under the signs of Pisces, Aries, and Taurus). More technically, a large effect size exists between genetics and schizophrenia; a smaller effect size exists between astrology and schizophrenia.[1]

A generally accepted rule of thumb is that the researcher should strive to identify medium to large effect sizes (e.g., $d > .14$ and $d > .34$, respectively; Cohen, 1992). This requires sufficiently large sample sizes in order to detect medium effect sizes. When performing procedures such as path analysis (and those described in Chapters 5 & 6). A sample of this size or larger is said to possess statistical power (i.e., sufficient to detect in this sample what exists in the population).

It is important to repeat that this does not necessarily mean finding statistically significant associations with a sample, but the associations that really exist in the population. Students and many social scientists commonly confuse the two. It is worth taking a moment to make sure that you understand the difference between the two and because these are not necessarily synonymous. Statistical significance pertains to associations within a dataset whereas statistical power pertains to the ability to identify what exists in the population sampled. The latter is the more important of the two though social scientists all too often focus only on the former (Cohen 1988).

Estimating Sample Size Requirements

To simplify what can easily become a complicated topic, Cohen (1992) provided a straightforward guide to assist social scientists when determining sample size requirements to test research hypotheses with a various statistical procedures. We have adapted this table to assist when determining sample size requirements for path analyses; these estimates are based on sample size requirements for regression analyses (Cohen 1992).

The left-hand column of Table 4.1 lists the number of independent variables that might be included in a path analysis (ranging from 2 to 15 independent variables). Remember, path analysis is an extension of multiple regression allowing 2+ dependent or predicted variables. This is performed as a single equation with PROC CALIS; whereas with PROC REG, separate calculations are computed for each dependent variable. A primary reason to perform path analyses with PROC CALIS is the reduced likelihood of detecting chance or random associations among variables. For this among other reasons, we recommend that path analysis be performed with PROC CALIS versus PROC REG in most instances.

Returning to our example, there are five variables in Figure 4.3; applying the N−1 guideline means that there are four independent variables for the purposes of sample size estimation (though really five IVs). Applying the recommendation that researchers aim to detect medium effect sizes, we focus on the middle column within the grouping shown in Table 4.1 (vs. small or large columns within each grouping). In keeping with the norm in social science research, we select the standard alpha (α) = .05 criterion. Where this column and row intersect (IV = 4), see in Table 4.1 that a sample of 100 is a recommended minimum. Analyses using smaller samples are at greater Type II error risk (i.e., fail to detect what exists in the population).

In order to further reduce the possibility of chance findings, the researcher may want to set a more stringent alpha level (i.e., to reduce of the likelihood of Type I errors). Where $\alpha \leq .01$, the researcher has greater confidence that findings are not random but genuinely reflect what exists in these data. Again, N = 5−1 so we now need 118 participants (vs. 100 participants where $\alpha \leq .05$).

There are comparatively few instances when the researcher would want to set a more liberal alpha level (e.g., $\alpha \leq .10$); this is not customary in social science research. However, assume the researcher wishes to try to replicate previous published findings, which suggests that large effect sizes exist between all variables; here, the researcher might argue that an *a priori* threshold value where $\alpha \leq .10$ is sufficient (and defensible). But note that this does not reduce your required sample size until you have 7+ independent variables as 100 is the minimum recommended sample size for path analysis.

Table 4.1: Sample Size Requirements for Path Analyses

	α = .01			α = .05			α = .10		
# Predictor Variables	Small	Medium	Large	Small	Medium	Large	Small	Medium	Large
2	698	100	100	481	100	100	217	100	100
3	780	108	"	547	100	"	233	"	"
4	841	118	"	599	100	"	242	"	"
5	901	126	"	645	100	"	256	"	"
6	953	134	"	686	100	"	267	"	"
7	998	141	"	726	102	"	272	"	"
8	1039	147	"	757	107	"	282	"	"
9	1076	152	"	779	112	"	297	"	"
10	1110	157	"	793	116	"	317	"	"
11	1142	161	"	803	119	"	339	"	"
12	1170	164	"	812	122	"	358	"	"
13	1195	165	"	820	124	"	375	"	"
14	1217	167	"	825	125	"	392	"	"
15	1237	168	"	830	125	"	407	"	"

Example 1: A Path-Analytic Investigation of the Investment Model

Rusbult's (1980) investment model will again be used for this example, this time to illustrate how path analysis is performed with PROC CALIS. Once again, the investment model identifies variables that are believed to affect satisfaction and commitment in romantic relationships (Le and Agnew 2003). One version of the model holds that **commitment** to a romantic relationship is determined by the following three constructs:

- **satisfaction** with the relationship

- one's **investments** in the relationship (e.g., the amount of personal time and resources put into the relationship)
- **alternative value**, or the attractiveness of alternatives to the relationship

Satisfaction, in turn, is said to be predicted by **rewards**, or the positive features that one associates with the relationship, and **costs**, or the negative features associated with the relationship.

It is possible to use some of the fundamental concepts of the investment model to develop the path model presented in Figure 4.4. The arrows presented in this figure represent the prediction that commitment is determined by satisfaction, investment size and alternative value, and that satisfaction is determined by rewards and costs.

Figure 4.4: The Basic Directional Model

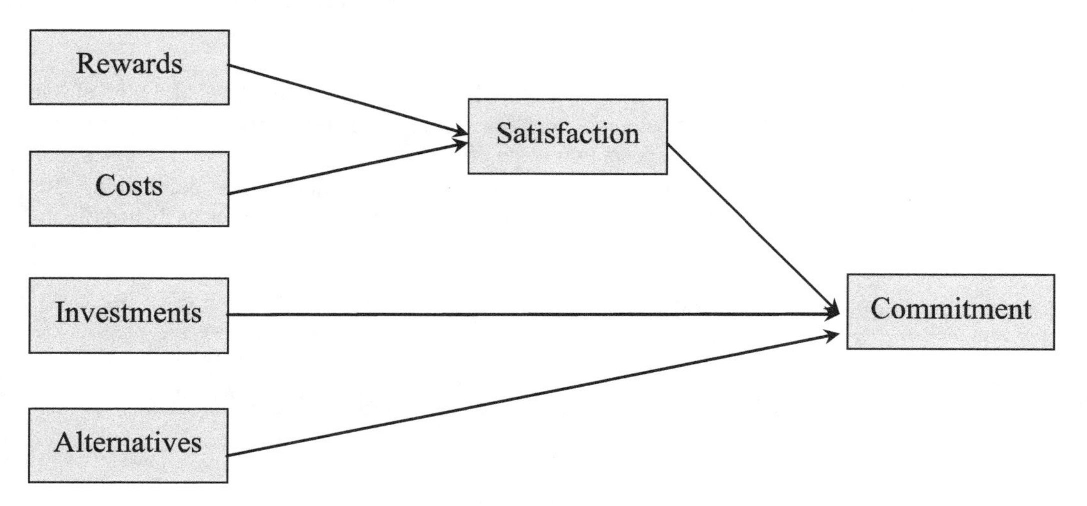

NOTE: The model presented in Figure 4.4 is based on Rusbult's investment model (1980) and should not necessarily be interpreted as an accurate representation of the theory. Note that the data analyses reported here are fictitious and are used only to illustrate statistical procedures. These results should not be viewed as valid tests of the investment model, or of any other theoretical framework. For those interested, see the review and meta-analysis by Le and Agnew (2003).

Overview of the Rules for Performing Path Analysis

The following sections will show you how to prepare a program figure and how to write the SAS program to test the path model presented in Figure 4.4. These sections will present 14 rules or guidelines to use when preparing figures and writing SAS programs to perform path analysis. The 14 rules are summarized below:

RULE 1: Only exogenous variables are supposed to have covariance parameters in the model.

RULE 2: A residual term is identified for each endogenous variable in the model.

RULE 3: Exogenous variables do not require residual terms.

RULE 4: Variance should be estimated for every exogenous variable in the model, including residual terms.

RULE 5: In most cases, covariance should be estimated for every possible pair of manifest exogenous variables; covariance is not estimated for endogenous variables.

RULE 6: For simple recursive models, covariance is generally not estimated for residual terms.

RULE 7: One equation should be created for each endogenous variable, with that variable's name to the left of the equals sign.

RULE 8:	Variables that have a direct effect on that endogenous variable are listed to the right of the equals sign.
RULE 9:	Exogenous variables, including residual terms, are never listed to the left of the equals sign.
RULE 10:	To estimate a path coefficient for a given independent variable, a unique path coefficient name should be created for the path coefficient associated with that independent variable.
RULE 11:	The last term in each equation should be the residual (disturbance) term for that endogenous variable; this term will have no name for its path coefficient.
RULE 12:	To *estimate* a path coefficient, create a name for that parameter.
RULE 13:	To *fix* a parameter at a given numerical value, insert that value in place of the parameter's name.
RULE 14:	To *constrain* two or more parameters to be equal, use the same name for those parameters.

The fact that these guidelines are assigned numbers (e.g., Rule 1, Rule 2) is not meant to indicate any hierarchy of importance. The numbers were merely assigned consecutively for ease of reference in this text.

It is not expected that the preceding rules will make sense to you at this point if you are just learning path analysis. As you read the following sections and gain experience conducting path analyses, however, these steps will become second nature. There are various ways or techniques to help learn path analysis and structural equation modeling; the rules or series of steps we describe are meant for the novice.

Once you become more familiar with PROC CALIS, you will discover in which instances you can skip over certain rules. See Appendix A.6 for an overview of how PROC CALIS handles these rules by default.

Preparing the Program Figure

Path analysis with PROC CALIS is somewhat more involved than path analysis with PROC REG. This is because the PROC CALIS program can include a relatively large number of statements that represent your directional model as a series of structural equations: An error in any statement, and your results may well be incorrect.

The likelihood of making mistakes is greatly reduced by first preparing a program figure. This **program figure** will display all of the important features of your model. Among other things, it will identify the endogenous and exogenous variables, identify the parameters to be estimated (e.g., path coefficients), and indicate which variables are free to covary. If you do a careful job preparing the program figure, writing the PROC CALIS program will be relatively easy.

This section discusses the steps to follow when preparing a program figure. A later section will show how the figure is then translated into a SAS program.

Step 1: Drawing the Basic Model

When preparing a program figure, try to adhere to the convention of placing the antecedent or independent variables toward the left side of the figure and the consequent of dependent variables to the right side of the figure. Draw straight single-headed arrows to indicate the predicted relationships. The basic model to be tested here was presented previously as Figure 4.4. Notice that the four exogenous variables appear on the left side of that figure; the reasons for this will later become clear.

Step 2: Assigning Short Variable Names to Manifest Variables

In Figure 4.5, short variable names have been assigned to the variables of the path diagram. With this system, manifest variables are represented by the letter "V" followed by a number. This text uses the convention of starting with the last dependent variable in the directional chain and naming it V1. We then trace backward (right to left), assigning the names V2, V3, and so forth, to variables that appear earlier in the chain. These short variable names are then written just above the variable's actual name in the path diagram. Following these conventions, Figure 4.5 shows that commitment has been named V1, satisfaction has been named V2, and so forth.

Figure 4.5: Assigning Short Variables Names to the Manifest Variables

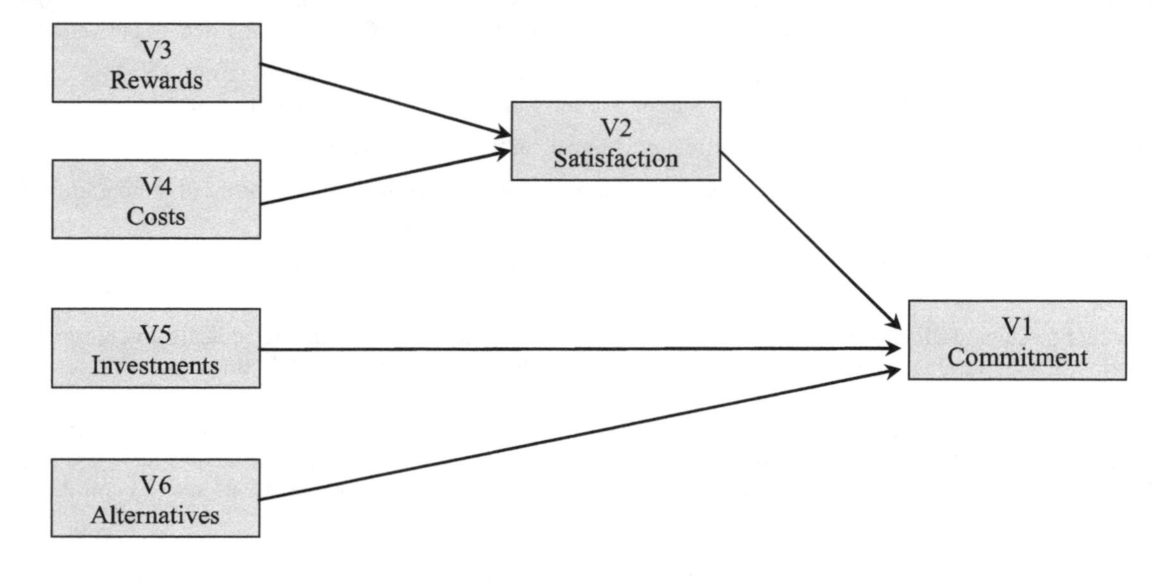

Step 3: Identifying Covariances Among Exogenous Variables

Remember that exogenous variables are those that do not have any single-headed arrows pointing at them. In the present model, the exogenous variables are rewards, costs, investments, and alternatives.

In Figure 4.6, curved double-headed arrows are used to connect all four of these exogenous variables to indicate that the four exogenous variables are expected to covary (i.e., correlated). This also means that you will ultimately prepare a program that calculates values for these covariance estimates.

Figure 4.6: Identifying Covariances Among Exogenous Variables and Residual Terms for Endogenous Variables

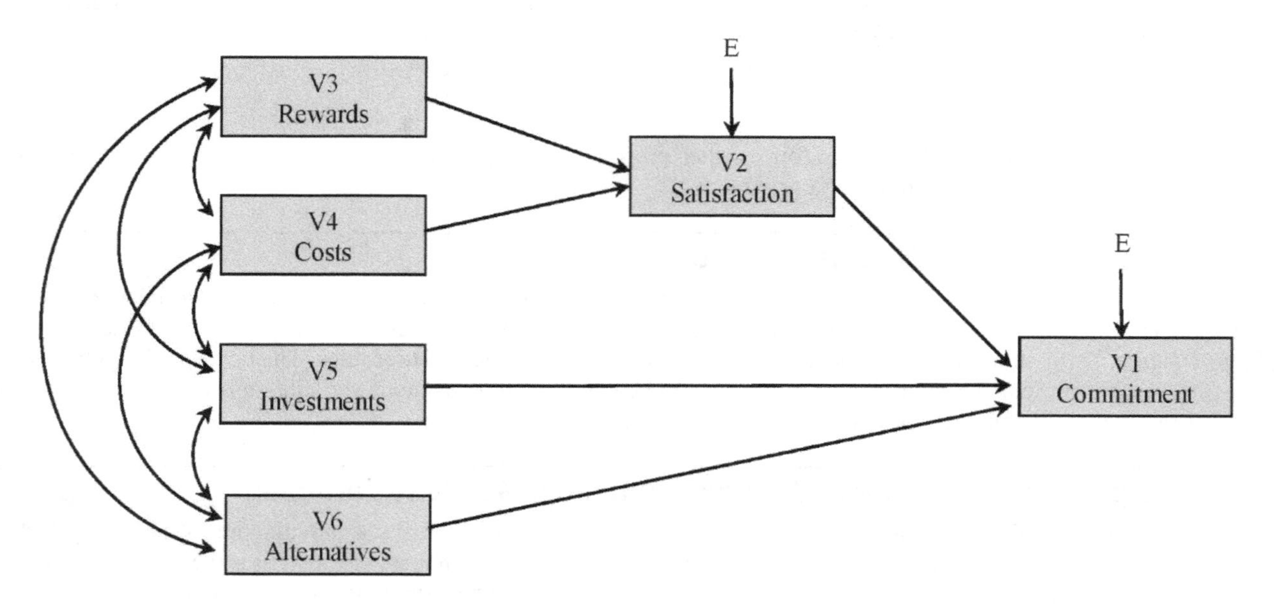

In preparing the program figure for your model, identify any variables that are expected to covary in the same manner by connecting them with curved double-headed arrows. You should heed the following rule, however:

RULE 1: Only exogenous variables are supposed to have covariance parameters in the model.

With the relatively simple models discussed in this chapter, endogenous variables do not covary with any other variable. This means that if a straight single-headed arrow points at a variable, there should not be a curved double-headed arrow pointing at that same variable.

With simple models, every exogenous variable is usually allowed to covary with every other exogenous variable. That is, you should estimate covariance for every possible pair of exogenous variables. This is not a strict rule, however. If you have good reason to believe that two exogenous variables are not correlated (e.g., contrary to theory), it is acceptable to constrain the covariance for this pair of constructs to zero. For example, you may have reason to believe that two exogenous variables are uncorrelated and you test a model that reflects this prediction. To do this in PROC CALIS, you explicitly specify that the covariance is 0. In most cases, however, it will be appropriate to estimate covariances for every pair of exogenous variables.

Step 4: Identifying Residual Terms for Endogenous Variables

After identifying the covariances for the exogenous variables, you should identify the residual terms for the endogenous variables in accord with Rule 2:

RULE 2: A residual term is identified for each endogenous variable in the model.

The residual term is sometimes referred to as either the **error term** or **disturbance term**. The **residual term** for a variable represents all the factors that influence variability in that variable but not included as antecedent variables in the model.

For example, the path diagram in Figure 4.6 hypothesizes that commitment is predicted by satisfaction, investments, and alternatives. It is highly unlikely, however, that these three antecedent variables will account for all variability in commitment. Therefore, commitment is expected to also be affected by a residual term. This residual term represents the effects on the dependent variable due to omitted independent variables, random shocks, measurement, or specification errors.

As per convention, residual terms are represented by the capital letter E (for **E**rror term). In Figure 4.6, the residual term for commitment is given the short name E1 because the corresponding name for the commitment variable is V1: The numerical suffix for the V and E terms ("1" in this case) should match. The figure shows that commitment (V1) is expected to be affected by the residual term, E1, while satisfaction (V2) is expected to be affected by its residual term, E2.

The remaining variables are exogenous variables, and therefore are not given residual terms, in accord with Rule 3:

RULE 3: Exogenous variables do not require residual terms.

Step 5: Identifying Variances to Be Estimated

Rule 4 indicates which variance terms should be estimated:

RULE 4: Variance should be estimated for every exogenous variable in the model, including residual terms.

We recommend that you identify every parameter to be estimated by placing a question mark (?) in the appropriate location in your program figure. More specifically, you can use the symbol "VAR?" to identify each variance to be estimated. Place this symbol directly under the name of that manifest variable or residual term. For example, Figure 4.7 shows that the VAR? symbol is placed directly under the word "Rewards" in the rectangle representing that variable. Similar symbols appear in the rectangles for the costs, investments, and alternatives variables. Because these are all exogenous variables, variance must be estimated.

Figure 4.7: Identifying the Variances and Covariances to Be Estimated

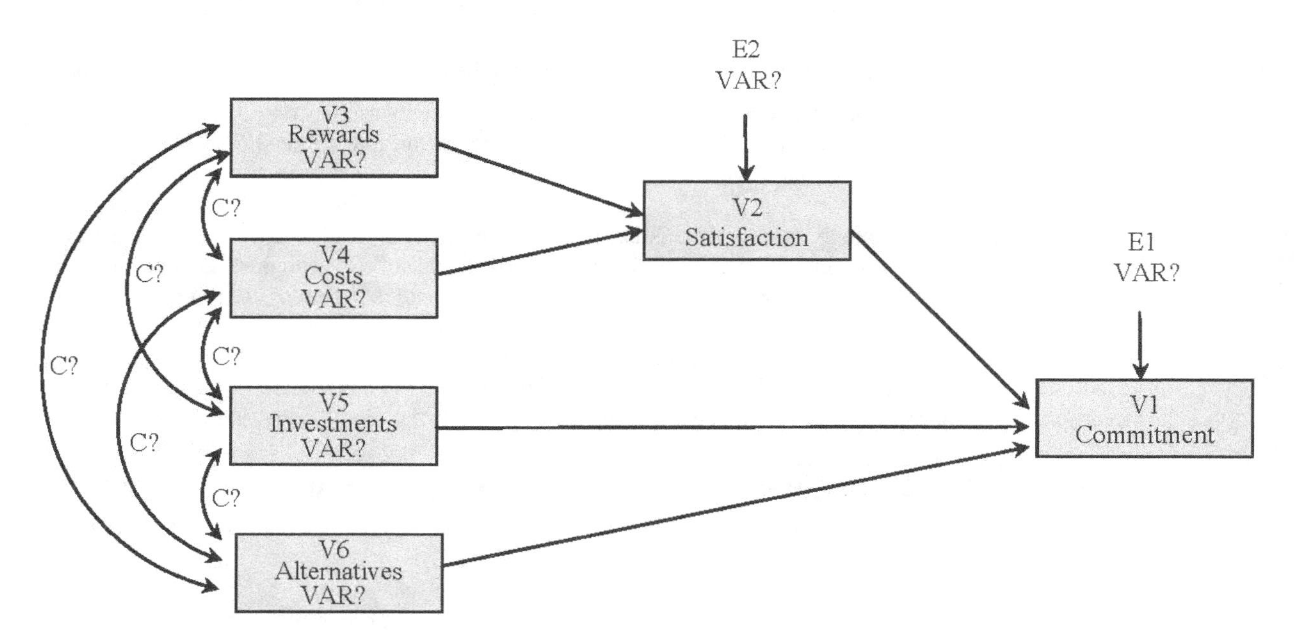

When identifying variances to be estimated, do not forget the residual terms! The residual terms of Figure 4.7 do not have single-headed arrows pointing toward them; this means that they are also exogenous variables. They therefore must have their variances estimated. Figure 4.7 shows that VAR? has been placed under the abridged names for residual terms E1 and E2.

Step 6: Identifying Covariances to Be Estimated

Figure 4.7 uses the symbol C? to identify the covariances to be estimated. In accord with Rule 5, these covariances will usually involve all manifest exogenous variables:

RULE 5: In most cases, covariance should be estimated for every possible pair of manifest exogenous variables; covariance is not estimated for endogenous variables.

Notice that a curved double-headed arrow connects the rewards variable (at the top of Figure 4.7) with the alternatives variable (at the bottom of the figure). This curved arrow is identified with the C? symbol indicating that the SAS program will estimate covariance between these variables. The remaining curved arrows connect the remaining pairs of manifest exogenous variables in this figure. Each is identified with the C? symbol; a total of six covariances are to be estimated for this model.

Notice, however, that no curved arrow connects either E1 or E2 with any other variable. This is in accord with Rule 6.

RULE 6: For simple recursive models, covariance is generally not estimated for residual terms.

Rule 6 does not necessarily apply to other types of models. For example, residual terms are allowed to covary in a time-series design in which the same variable is measured at more than one point in time. Also, in models with reciprocal causation, the residual terms for the two variables involved in the reciprocal relationship are often allowed to covary. But for the relatively simple models to be discussed in this chapter, Rule 6 will hold.

Step 7: Identifying the Path Coefficients to Be Estimated

Each straight, single-headed arrow in your figure represents a directional path. For each of these, the SAS program will estimate a **path coefficient**: A number that represents the amount of change in a dependent variable associated with a one-unit change in a given independent variable while holding constant the other independent variables. Path coefficients represent the size of the effect that a given independent variable has on a dependent variable.

This text uses the symbol P? to identify path coefficients to be estimated. In Figure 4.8, each path has been identified with this symbol. After the standardized path coefficients have been estimated by the SAS program, you will review their relative size to determine which independent variables had stronger effects on the dependent variables. The program results will also provide statistics testing the null hypothesis that a given path coefficient is zero in the population.

Figure 4.8: Identifying the Path Coefficients to Be Estimated

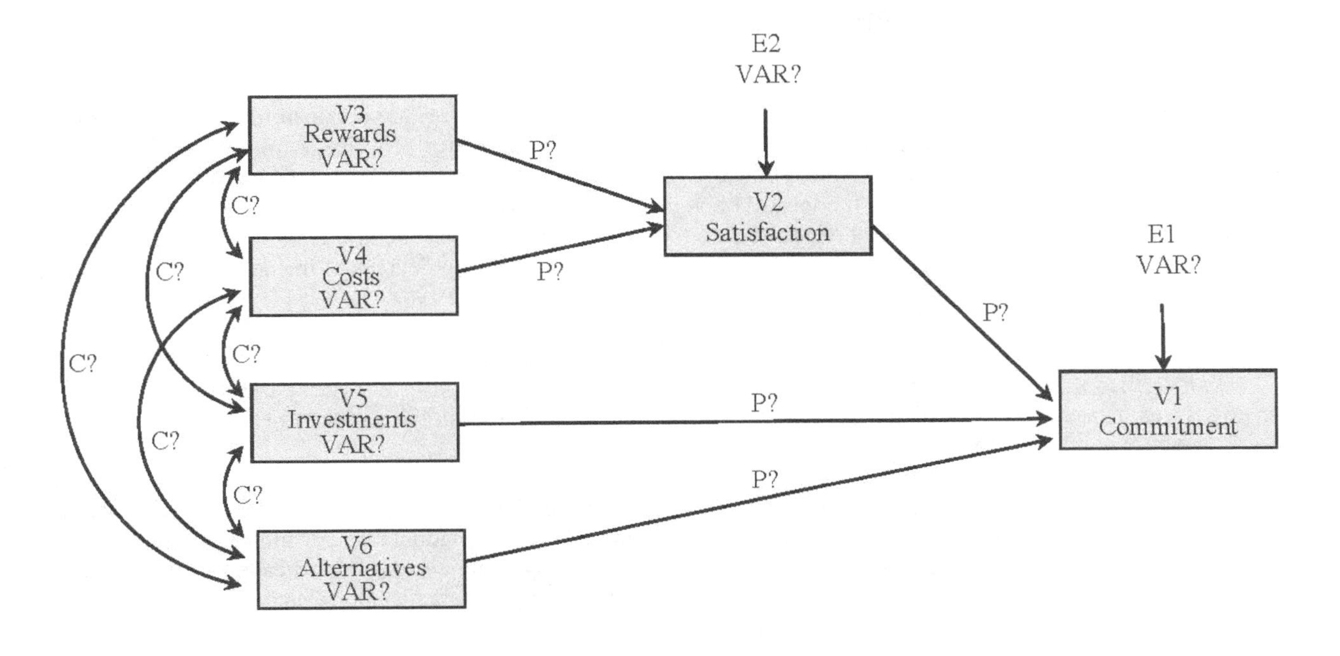

Step 8: Verifying that the Model Is Overidentified

The identification problem is one of the most important concepts to understand in path analysis and structural equation modeling. Because identification is less of a problem when testing simple recursive models such as those presented in this chapter, this section will merely introduce some basic concepts. Identification becomes a more serious problem when estimating nonrecursive models; for users interested in testing these more complicated models, this section also lists references that provide a more in-depth treatment of the topic.

To understand identification, it is necessary to first understand that path models may be represented as a system of functional equations. For example, the path diagram presented in Figure 4.8 predicts that the endogenous variable, satisfaction (V2), will be determined by rewards (V3), costs (V4), and its residual term (E2). A functional equation representing this part of the model could take the following form:

```
V2 = P23 V3 + P24 V4 + E2
```

Where P23 represents the path coefficient for the effect on V2 of V3, and P24 represents the effect on V2 of V4. This equation includes two unknowns: The two path coefficients. In performing the analysis, PROC CALIS will estimate values for these two path coefficients. A typical path model (such as that represented in Figure 4.8) will be represented by a number of equations and will usually include a variety of unknown parameters to be estimated, including variances, covariances and path coefficients.

It is highly desirable that a model not be **underidentified** prior to estimation. A model is said to be underidentified when it includes fewer linearly independent equations than unknown elements (Kline 2005). These unknowns are possible associations within path models where no hypotheses are specified. In Figure 4.8, for example, there are no arrows connecting investments (V5) or alternatives (V6) to satisfaction (V2); nor do direct paths appear from rewards (V3) or costs (V4) and commitment (V1). These unknowns (or unspecified associations) provide models with "degrees of freedom," which allow model fit statistics to be calculated.

When a model is underidentified, an infinite number of solutions can be generated for its parameters. For example, if PROC CALIS is used to estimate an underidentified model, performing the analysis with one set of starting values might generate one set of parameter estimates, while running the analysis a second time with a different set of starting values might generate a completely different set of parameter estimates (e.g., different values for the same path coefficients). Obviously, results obtained from the analysis of an underidentified model are completely unreliable.

Parameter estimates such as path coefficients are meaningful only if they are obtained from an **identified** model. A model may be identified either by being just-identified or overidentified. A **just-identified** model is one in which there are exactly as many linearly independent equations as unknowns (some texts refer to just-identified models as **saturated** models). Although a just-identified model has the advantage of allowing estimation of just one unique set of parameters for a given sample, it has the disadvantage of not allowing any tests for goodness of fit. This is because just-identified models have no degrees of freedom. It is for this reason that researchers typically ensure that models are overidentified (Byrne 1998).

A model is said to be **overidentified** when it includes more equations than unknown elements. As with a just-identified model, estimation of an overidentified model will result in only one set of parameter estimates from a given dataset. Overidentified models, however, have an additional desirable property: They can be tested for overall goodness of fit.

Fortunately, simple recursive models such as those covered in this chapter are just-identified or overidentified. When dealing with recursive models with uncorrelated residuals, the model may be said to be just identified if every variable in the model is related to every other variable by either a one- or two-headed arrow. Figure 4.9 presents an example of a just-identified model. Notice that every variable is interrelated with every other variable, either through a path or a covariance.

Figure 4.9: A Just-Identified, Fully Recursive Model

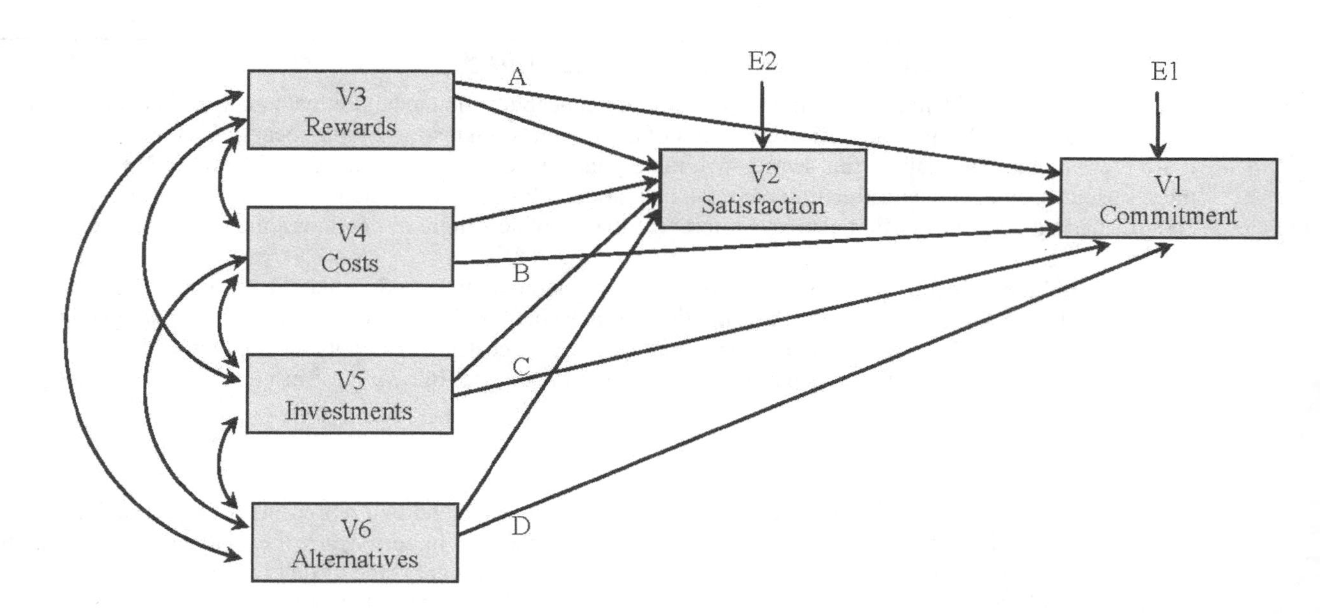

To create this just-identified model, four new paths were added to the investment model presented earlier. Paths A and B predict that rewards and costs will directly affect commitment, and paths C and D predict that investments and alternatives will affect satisfaction.

In performing path analysis, it is customary to test the "fit" between the model and data; the better the fit, the stronger the support for the model as hypothesized. It is important to remember, however, that a just identified model cannot be tested for goodness of fit. This is because a just-identified model just fits the data (i.e., is saturated). The reason for this should become clear when one considers that the data analyzed consist simply of a correlation (or variance-covariance) matrix; this matrix provides correlations between every pair of manifest variables. Now consider the just-identified model presented in Figure 4.9. Note that every variable is assumed to be related to every other variable. It comes as no surprise that such a saturated model is able to account for the covariation in the original matrix. A simpler or more parsimonious model has fewer interconnections; that model would be overidentified (Kline 2005).

A just-identified model becomes overidentified when you place restrictions on model parameters. Various restrictions may be imposed; for example, it is possible to constrain two path coefficients to the same value. But by far, the most common approach is to fix certain path coefficients to take on a value of zero. Fixing a path to zero has the same effect as eliminating that path from the model.

Figure 4.10 shows the model that resulted from fixing paths A, B, C, and D (from Figure 4.9) at zero. These paths have been eliminated, with the result that the path model in Figure 4.10 is now overidentified. This model can be tested for goodness of fit.

In summary, remember that a recursive model may only be tested for goodness of fit only when certain restrictions are placed on the just-identified model. This is usually achieved by eliminating possible paths (i.e., providing degrees of freedom).

Figure 4.10: An Overidentified Model

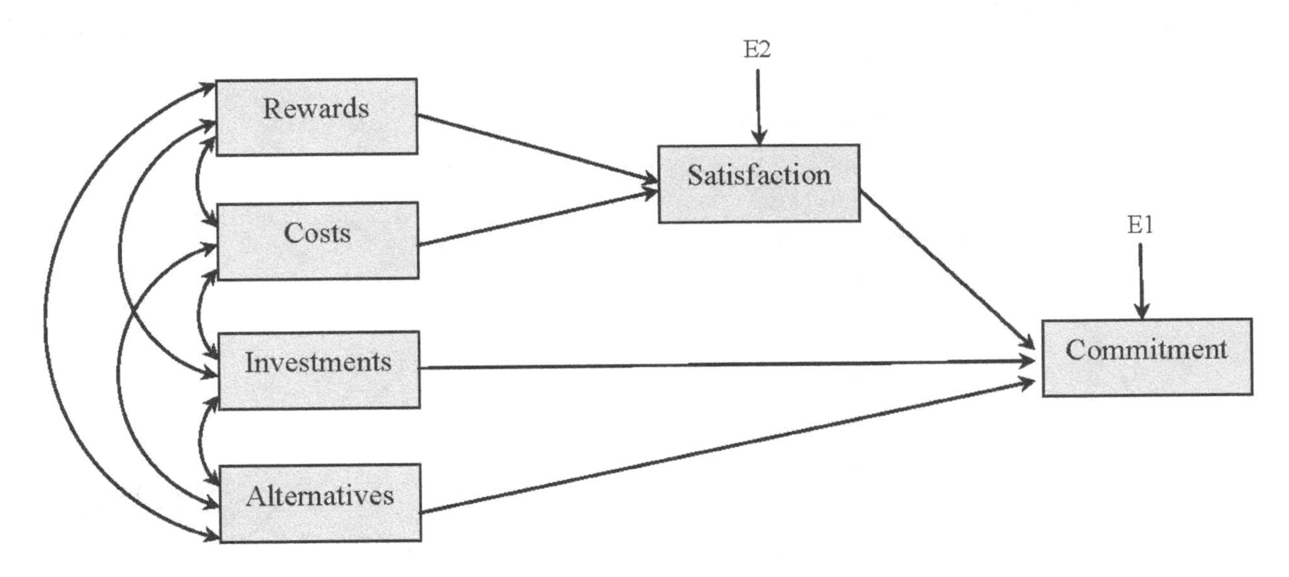

For more complex models, such as nonrecursive path models or structural equation models, the identification problem can be more troublesome; these models *can* be underidentified. The researcher who estimates an underidentified model may unknowingly obtain meaningless and misleading parameter estimates.

How can you be sure that these more complex path models are overidentified? When the variables are standardized, solving the equations for parameters is equivalent to solving the path equations. Kline (2005) describes techniques for ensuring a model meets these conditions.

In addition to checking log and output files, you should also routinely verify that the number of data points in the analysis is larger than the number of parameters to be estimated; when this is not the case, the model is not identified. The number of data points may be calculated with the following equation:

Number of data points = (p [p + 1]) / 2

where p = the number of manifest variables being analyzed. For example, in the current path analytic investment model, six variables are analyzed. Inserting this in the formula results in the following:

Number of data points = (6 [6 + 1]) / 2

= (6 [7]) / 2

= (42) / 2

= 21

Once it has been established that the current analysis involves 21 data points, it is necessary to determine the number of parameters to be estimated. This will be equal to the sum of the number of

- path coefficients
- variances
- covariances to be estimated

This will be easy to determine, as you have already identified these parameters in your program figure. This program figure is again reproduced as Figure 4.11. Remember that the symbol VAR? was used to represent each variance to be estimated, C? was used to represent each covariance, and P? was used to represent each path coefficient. By referring to Figure 4.11, you can see that the analysis will estimate six variances (for E1, E2, V3, V4, V5, and V6), six covariances (between variables V3, V4, V5, and V6), and five paths. The total number of parameters to be estimated is therefore equal to 17.

Figure 4.11: The Completed Program Figure

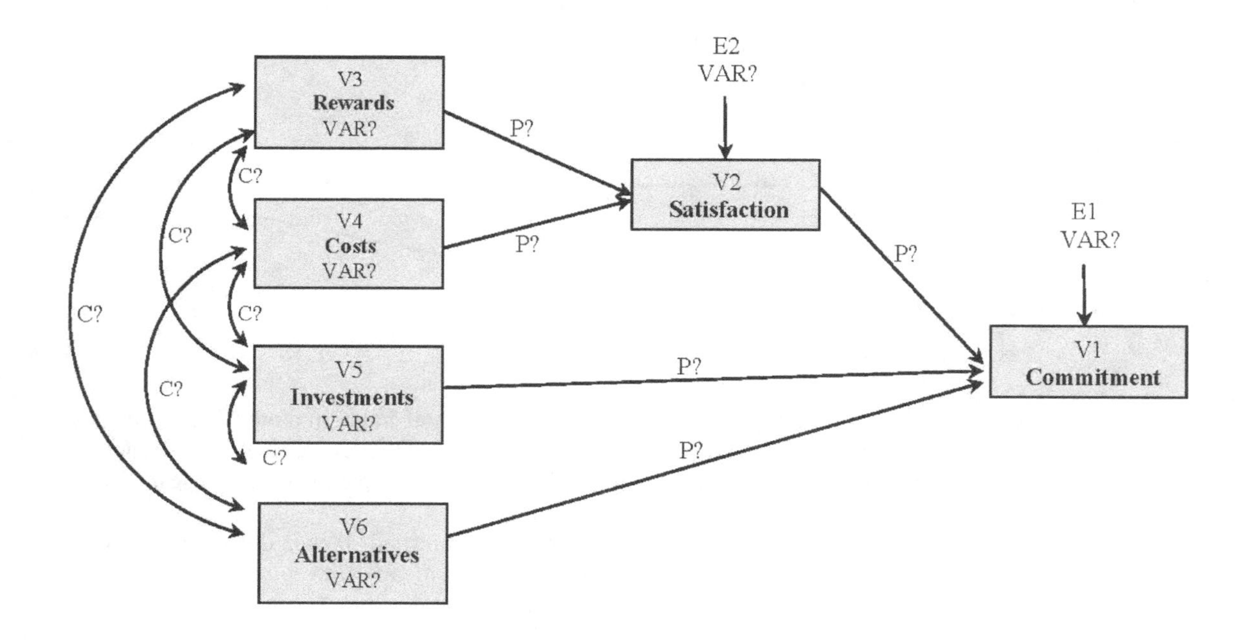

It has already been noted that one requirement for overidentification is that the number of data points should exceed the number of parameters to be estimated. Because the present analysis involves 21 data points but 17 parameters, you may conclude that the model presented in Figure 4.11 meets this criterion.

Is it necessary to ascertain the number of data points and parameter estimates in this way prior to each analysis? Technically, the answer is no; this is because this information is reported in the PROC CALIS output. On page 4 of the output (presented below), PROC CALIS indicates the number of data points to the left of the words "Functions (Observations)" and the number of parameter estimates to the right of the word "Parameter Estimates." Despite this, it is good practice to determine these values prior to computing the model to verify the model is, in fact, testable.

As a final test for identification, you may wish to repeat your analyses several times, each time using very different starting values for parameter estimates. If PROC CALIS arrives at the same final parameter estimates each time, it is likely (but not certain) that your model is identified.

IMPORTANT: Remember that the last two procedures—counting parameters and functions (observations) and conducting analyses with differing starting values—are necessary but not sufficient conditions for identification. In other words, if your model fails to pass these tests, they are clearly underidentified; but if they do pass these tests, it does not definitively prove that they are identified. The only way to be sure is to use one of the more time-consuming approaches, such as those described by Kline (2005). Remember that these more complicated approaches are necessary only with more complex models such as nonrecursive models, models with correlated residuals, and structural equation models. Simple recursive models without correlated residuals will always be overidentified so long as some of the paths are fixed to zero.

Preparing the SAS Program

Overview

The PROC CALIS program used to analyze a simple recursive path model (such as the one described here) is longer than most SAS programs, but is not especially complicated or difficult to understand. Preparing the program figure in advance will greatly facilitate the writing of this program; the instructions provided here should be helpful in leading you through the various PROC CALIS statements. Below is the entire PROC CALIS program, including the DATA step used to analyze the model presented in Figure 4.11:

```
      data D1(type=corr) ;
         input _type_ $ _name_ $ V1-V6 ;
            label
               V1 = COMMITMENT
               V2 = SATISFACTION
               V3 = REWARDS
               V4 = COSTS
               V5 = INVESTMENTS
               V6 = ALTERNATIVES   ;
      datalines;
      n      .    240     240     240     240     240     240
      std    .  2.3192  1.7744  1.2525  1.4086  1.5575  1.8701
      corr  V1  1.0000    .       .       .       .       .
      corr  V2   .6742  1.0000    .       .       .       .
      corr  V3   .5501   .6721  1.0000    .       .       .
      corr  V4  -.3499  -.5717  -.4405  1.0000    .       .
      corr  V5   .6444   .5234   .5346  -.1854  1.0000    .
      corr  V6  -.6929  -.4952  -.4061   .3525  -.3934  1.0000
      ;
      run;
❶     proc calis  modification ;
❷        lineqs
            V1 = PV1V2 V2 + PV1V5 V5 + PV1V6 V6 + E1,
            V2 = PV2V3 V3 + PV2V4 V4            + E2;
```

```
❸        variance
           E1 = VARE1,
           E2 = VARE2,
           V3 = VARV3,
           V4 = VARV4,
           V5 = VARV5,
           V6 = VARV6;
❹        cov
           V3 V4 = CV3V4,
           V3 V5 = CV3V5,
           V3 V6 = CV3V6,
           V4 V5 = CV4V5,
           V4 V6 = CV4V6,
           V5 V6 = CV5V6;
❺        var V1 V2 V3 V4 V5 V6 ;
    run;
```

All lines appearing before ❶ of the preceding program constitute the DATA step, here entered as a correlation matrix. Line ❶ includes the PROC CALIS statement that initiates this procedure will be performed. Equations included in the LINEQS statement (beginning line ❷) identify the predicted relationships between the model's variables (i.e., which variables are assumed to predict the model's endogenous variables). The VARIANCE statement, beginning at line ❸, identifies the variances to be estimated, while the subsequent COV statement (beginning line ❹) indicates which covariances are to be estimated. Finally, the VAR statement on line ❺ indicates which variables are to be analyzed. Each of these sections is discussed in detail below.

This section of the chapter is divided into five subsections. First, the options for data input are described. Next, the conventions and options used with the PROC CALIS, LINEQS, VARIANCE, and COV statements are reviewed. Once these statements are completed, your path model will have been converted into a series of equations and will be ready for analysis.

The DATA Input Step

With PROC CALIS, data may be input as raw data, or as either correlation or covariance matrices. Appendix A2, "Data Input," provides instruction on how to input these different types of data.

Inputting Raw Data

The advantage of inputting raw data is that this format allows use of the KURTOSIS option with PROC CALIS to compute a number of univariate and multivariate measures of kurtosis, and enables you to identify the observations that contribute most to kurtosis. This can be used to identify multivariate outliers to remove from the dataset.

Inputting a Correlation or Covariance Matrix

Inputting a correlation or covariance matrix has the advantage of usually requiring less computer time, but the disadvantage of not allowing for the computation of measures of kurtosis or the identification of outliers. In many cases, a good compromise is to use both methods: The first runs should involve the analysis of raw data (if available) to assess kurtosis and to identify outliers. Once outliers have been eliminated, the correlation or covariance matrix based on the resulting sample can be analyzed in subsequent runs. PROC CORR can be used to create the correlation matrix, if this is desired. It is usually desirable to specify the NOMISS option when computing these correlations, as this assures that every correlation will be based on the same set of observations. For example, the correlation matrix needed for the current model could have been produced with the following statements (the NOPROB option suppresses printing of p values for the correlations):

```
proc corr   data=D1 nomiss noprob ;
   var V1 V2 V3 V4 V5 V6;
run;
```

Problems with Large Differences in Standard Deviations

With PROC CALIS, problems can result if the standard deviations for some of the variables are considerably larger than the standard deviations for other variables. For example, if the largest standard deviation in an analysis is 150 for variable V1, and the smallest is 1.3 for variable V2, the CALIS procedure may encounter difficulty when estimating the model. This may be noted on the SAS output itself, with a message stating that not all parameters were identified. Near-zero standard errors for parameter estimate t tests are another warning sign.

When inputting raw data, these problems can be avoided by rescaling variables so that they are all on approximately the same scale. For example, if all scores on V1 were divided by 100, the standard deviation for that variable would decrease from 150 to 1.5, which is comparable to the standard deviation of 1.3 for V2. Despite the new standard deviation, all correlations between the variables remain unchanged.

When data are input in the form of a correlation matrix with standard deviations, it is not necessary to actually divide all scores by some constant. When you encounter a problem with standard deviations of this nature, simply move the decimal point for the troublesome standard deviation to achieve a standard deviation of the desired magnitude.

The PROC CALIS Statement

The CALIS procedure can perform a variety of analyses, including confirmatory factor analysis and structural equation modeling. This chapter, however, focuses only on how the CALIS procedure can be used to perform path analysis. Confirmatory factor analysis and structural equation modeling are covered in Chapters 5 and 6 of this text, respectively.

The general form for the PROC CALIS statement is as follows:

```
proc calis   options ;
```

The words PROC CALIS invoke the procedure, and these should be followed by at least one blank space and a list of options (if desired) with the name of each option separated by at least one blank space. The statement ends with a semicolon.

The preceding program uses the following PROC CALIS statement. This statement requests the modification option.

```
proc calis  modification ;
```

The modification option requests a number of modification indices that can be useful in identifying changes that would improve model fit. Often, your original model will not provide a satisfactory fit to the data, and it will be necessary to change it to better reflect the actual relationships among model variables. The modification option prints the **Lagrange Multiplier or LM test**, which identifies paths or covariances that might be *added* to the model. It also prints results of the **Wald test**, which identifies paths and covariances that may be *deleted* from the model (i.e., not statistically significant). Although these indices can be quite valuable in developing an improved model, it is emphasized that changes made to the model should be driven by your understanding of theory and existing research on the topic, not simply by the results of these statistical tests (Byrne 1998). This point will be further discussed in the "Modifying the Model" section of this chapter.

The PROC CALIS statement can have additional options that may be used with the procedure. A few that may be particularly useful in social science research are presented below:

ALL
: prints all optional output.

DATA= dataset-name
: specifies the input dataset to be analyzed. If the DATA=option is omitted, the most recently created SAS dataset will be analyzed.

GCONV= p
: specifies the absolute gradient convergence criterion. Smaller values may be specified to obtain more precise parameter estimates, but this will significantly increase the time required for computation. If not specified, the default value is 1 E–8.

KURTOSIS or KU
: prints coefficients of univariate kurtosis and skewness along with various coefficients of multivariate kurtosis. This option also prints the numbers of the observations that make the greatest contribution to normalized multivariate kurtosis. The KURTOSIS option can help identify outliers and should be requested during the first run. The dataset must be a raw dataset, however; it cannot be requested if the input dataset is a correlation or covariance matrix.

METHOD=name
: specifies the method of parameter estimation. Maximum likelihood estimation is the default. Below are the most common optimization techniques that may be requested:

GLS
: requests generalized least-squares parameter estimation and requires a nonsingular correlation matrix. This method performs a statistical test of the goodness of fit of the model to the data, but assumes multivariate normality of all variables and independence of observations.

LSGLS
: requests unweighted least-squares estimation followed by generalized least-squares estimation.

LSML
: requests unweighted least-squares parameter estimation followed by normal-theory maximum-likelihood estimation.

ML
: requests normal-theory maximum-likelihood parameter estimation. Requires a nonsingular correlation matrix. This method performs a statistical test of the goodness of fit of the model to the data but assumes multivariate normality of all variables and independence of observations. This is the default method.

NONE
: request that no estimation method be used.

ULS
: requests unweighted least-squares parameter estimation.

MODIFICATION or MOD
: requests Lagrange Multiplier (LM indices) along with univariate and multivariate Wald test indices.

RESIDUAL or RES
: requests that the absolute and normalized correlation (or covariance) matrix be printed, along with the rank order of the largest residuals and a bar chart of the residuals.

SIMPLE or S
: requests means, standard deviations, skewness, and univariate kurtosis of manifest variables.

SUMMARY or PSUM or PSUMMARY
: requests that only the fit assessment table be printed.

TOTEFF or TE
: requests that total effects, indirect effects, and latent variable regression score coefficients be printed.

The LINEQS Statement

PROC CALIS provides a number of ways for describing the model to be analyzed. This chapter describes the LINEQS approach to model specification. The LINEQS statement is used to identify the variables that have direct effects on the endogenous variables in the path model. This is done with a series of equations with a separate equation for each endogenous variable. This chapter will present a system of notation to use with the LINEQS statement. Although it might seem most logical to use path-style input for path analyses, our intent is to provide a foundation or building blocks for the subsequent forms of analysis presented in this text such a confirmatory factor analysis and structural equation modeling, which also use LINEQS-style input.

Below is the general form for the LINEQS statement:

```
lineqs
   v = p v + p v + p v ..... + e,
   v = p v + p v + p v ..... + e,
   v = p v + p v + p v ..... + e ;
```

where

```
v = manifest variables
p = path coefficients
e = residual term for corresponding endogenous variables
```

An asterisk, *, can also be specified in place of a path coefficient name. This enables PROC CALIS to name the parameter automatically.

Although the preceding displays the general form for three equations, any number of equations is actually possible. Also, any number of independent variables (to the right of the equals sign) is possible.

To make this a bit more concrete, below is the LINEQS statement for the present path model:

```
lineqs
   V1 = PV1V2 V2 + PV1V5 V5 + PV1V6 V6 + E1,
   V2 = PV2V3 V3 + PV2V4 V4            + E2;
```

The LINEQS statement begins with the word LINEQS and ends with a semicolon. Each equation in the statement must be separated by a comma. The preceding statement includes two equations. The first equation identifies the variables that predict endogenous variable V1. This is evident as “V1” appears to the left of the equals sign in the first equation. The second equation identifies the variables that predict V2.

The names of the independent variables that predict an endogenous variable appear to the right of the equals sign in the equation for that variable. With the above equations, you can see that V1 is predicted by V2, V5, V6 with a residual term, E1. At the same time, V2 is predicted by V3 and V4, along with its residual term, E2.

To the immediate left of each independent variable is the name assigned to that variable’s path coefficient or an asterisk, *. In the present case, the path coefficient for V2 is given the name “PV1V2,” the coefficient for V5 is given the name “PV1V5,” and the coefficient for V6 is given the name “PV1V6.” The conventions used in naming these coefficients are discussed below.

SAS requires certain conventions when creating names for variables included in the LINEQS statement. For example, it requires that names for the residual terms of manifest variables begin with the letter E (for **E**rror term), names for latent factors begin with the letter F, and names for the disturbance terms of these factors begin with the letter D. (Latent factors and their disturbance terms are not relevant to path analysis; F and D terms will not be discussed in this chapter.) We recommend some additional conventions to simplify your task.

Naming Manifest Variables

In the previous section on preparing the program figure, we advised that the manifest variables be named using “string variables.” Each variable name consists of the prefix V and a numerical suffix. Thus you have variable names such as V1 and V2. Technically, manifest variables can actually be given any name so long as their names adhere to the usual SAS conventions (e.g., begins with a letter). The conventions we recommend will help create more meaningful names for path coefficients.

Naming Residual Terms

Similarly, the earlier section advised that residual terms be given names such as E1 and E2. The numerical suffix should always match the suffix of the corresponding manifest variable so that E1 is the residual term for V1, E2 is the residual name for V2, and so forth.

Naming Path Coefficients

Technically, path coefficients can be given any name that meets the usual SAS conventions. We, however, recommend a system with path coefficient names such as PV1V2, PV1V5, and PV2V4. With this convention, the name for a path coefficient:

- begins with the prefix P (to identify this as the name of a **P**ath coefficient)
- continues with the name of the dependent variable being predicted (e.g., V1 or V2)
- concludes with the name of the independent variable where the path originates (e.g., V4, V5, or V6)

Thus, the name PV1V5 tells you that this is the name of the path coefficient for the path to V1 from V5. The name PV2V4 indicates that this is the name of the coefficient for the path to V2 from V4. This approach has the advantage of making each coefficient's name both unique and meaningful. For example, path coefficients sometimes appear in sections of output such as Wald test results (to be described later) that provide no independent information to indicate which variables are associated with that coefficient. By using this system, you can identify the relevant variables simply by looking at the path coefficient's name.

Identifying the Variables to Be Included in Each Equation

It is now possible to show how the program figure (previously presented) can be used to construct the equations to be included in the LINEQS statement. That program figure for the investment model is presented here as Figure 4.12.

Figure 4.12: The Completed Program Figure

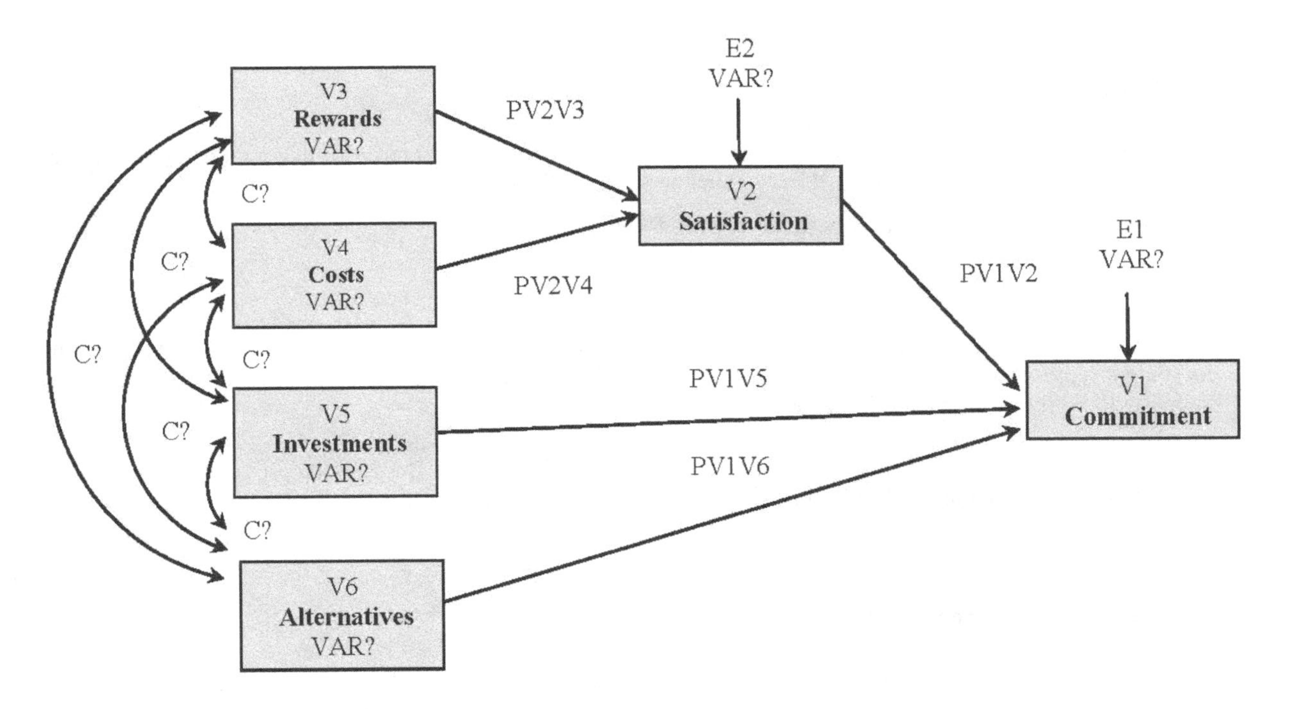

The process begins by determining how many equations should be included in the LINEQS statement. (There is one equation for each endogenous variable.) Remember that a variable is an endogenous variable if any straight single-headed arrow points at it. Figure 4.12 shows two endogenous variables: V1 and V2. One equation will be created for both variables.

Next, review the figure to determine which variables should be listed as independent variables for a given endogenous variable. In this context, **independent variables** are those variables that have arrows pointing *directly* at the endogenous variable of interest. For V1, the independent variables are V2, V5, and V6. Once these independent variables have been identified, you are now ready to write the equation for V1, while adhering to Rules 7, 8, and 9. (These rules pick up where Rule 6 left off in the previous section on "Preparing the Program Figure.")

RULE 7: One equation should be created for each endogenous variable with that variable's name to the left of the equals sign.

RULE 8: Variables that have a direct effect on that endogenous variable are listed to the right of the equals sign.

RULE 9: Exogenous variables, including residual terms, are never listed to the left of the equals sign.

In compliance with Rule 7, your equation for V1 would therefore begin with:

```
V1 =
```

In accordance with Rule 8, the following independent variable names would then be added:

```
V1 =        V2  +      V5  +        V6
```

Next, Rule 10 makes provision for path coefficient names:

RULE 10: To estimate a path coefficient for a given independent variable, a unique path coefficient name should be created for the path coefficient associated with that independent variable.

And so the following path coefficient names would be added to the equation:

```
V1 = PV1V2 V2 + PV1V5 V5 + PV1V6 V6
```

Finally, Rule 11 indicates how each equation should end:

RULE 11: The last term in each equation should be the residual term for that endogenous variable; this E term will have no name for its path coefficient.

The completed equation for V1 takes the following form:

```
V1 = PV1V2 V2 + PV1V5 V5 + PV1V6 V6 + E1,
```

Notice that, given the conventions for creating path coefficient names, just looking at PV1V2 tells you that this coefficient represents the strength of the effect of V2 on V1. Notice also that the equation ends with a comma, since it will be followed by another equation.

Following the same rules, it is now possible to create the equation for V2. Because the program figure shows that it is affected by V3 and V4, the equation will take on the following form:

```
V2 = PV2V3 V3 + PV2V4 V4 + E2;
```

The preceding equation ends with a semicolon as it is the last equation.

The program figure tells you which variables should be included in each LINEQS equation. You need only find the endogenous variable of interest and determine which variables have direct effects on it as indicated by straight, single-headed arrow. Figure 4.12 shows that V1 (commitment) is predicted by V2, V5, and V6 along with its residual term, E1. Similarly, you can see that V2 (satisfaction) is predicted by V3 and V4 along with its residual term, E2. It is in this way that the figure tells you which terms should be included in the equations for V1 and V2.

Once again, the full LINEQS statement appears as follows:

```
lineqs
   V1 = PV1V2 V2 + PV1V5 V5 + PV1V6 V6 + E1,
   V2 = PV2V3 V3 + PV2V4 V4            + E2;
```

Is it really necessary to line up the residual terms? The above statement was prepared so that the residual variables (E1 and E2) are vertically aligned. This is not required by the CALIS procedure but is recommended to reduce the chance of errors. Experience has taught us that it is easy to forget a residual term for a given equation if they are not lined up in this manner!

Estimating, Fixing, and Constraining Paths

The CALIS procedure estimates three different types of parameters:

- path coefficients, which represent the amount of change in a dependent variable associated with a one-unit change in the relevant independent variable while holding constant the effects of the remaining independent variables
- variances, which represent the variability in exogenous variables
- covariances, which represent the covariation between pairs of exogenous variables

All three types of parameters can be either estimated, fixed, or constrained. The following rules explain how this is done in the LINEQS statement; this section provides specific examples of how path coefficients may be estimated, fixed, or constrained.

RULE 12: To *estimate* a parameter, create a name for that parameter (or let PROC CALIS create the name for you).

When PROC CALIS estimates a path coefficient, it simply determines the "optimal" value for that coefficient in much the same way that PROC REG determines the "optimal" value for regression coefficients in a multiple regression equation. Whenever your LINEQS statement includes a name for a coefficient, the procedure's output will provide a numerical estimate for that coefficient. For example, consider the following equation:

```
V2 = PV2V4 V4 + E2;
```

The preceding equation asks PROC CALIS to estimate just one coefficient, the one representing the effect of V4 on V2. The output of this fictitious program may indicate that this path coefficient is equal to 0.23.

RULE 13: To fix a parameter at a given numerical value, insert that value in the place of the parameter's name.

For example, the following equation fixes the path for the effect of V4 on V2 at .50 while estimating the coefficient for the effect of V5 on V2:

```
V2 = .50 V4 + PV2V5 V5 + E2;
```

While it is unlikely that researchers new to path analysis will want to fix path coefficients at specific numbers, it is nonetheless easy to do as the preceding rule suggests.

It is sometimes desirable to fix the value of a path coefficient at 1.00; this is easy to do with PROC CALIS. You may have noticed that there are no path coefficient names to represent the effects of the E terms on the endogenous variables. This is because these coefficients are usually "fixed" to be equal to 1.00. With PROC CALIS, leaving off the name of a path coefficient just before the independent variable's name has the effect of fixing that coefficient at 1.00. In the following equation, both the path for the E term as well as the path for the V4 term are fixed at 1.00. The path coefficient for V5 is again estimated:

```
V2 = V4 + PV2V5 V5 + E2;
```

Be warned that this convention applies only to fixing *path coefficients* at 1.00. It will not apply to variances or covariances as described below.

When you fix a path coefficient at zero, this has the effect of eliminating the corresponding path from the model. In practice, it is not necessary to include a zero (0) in the equation. To fix that path at zero, merely omit the name of the relevant independent variable from the equation. For example, consider the following equation:

```
V2 = PV2V4 V4 + E2;
```

This equation indicates that V2 is affected only by V4; the paths from all other variables to V2 are automatically fixed at zero.

Finally, Rule 14 indicates how two or more paths may be constrained to be equal to each other:

RULE 14: To *constrain* two or more parameters to be equal, use the same name for those parameters.

Again, it is unlikely that learning path analysis at an introductory level will require this, but it is a useful to know for future reference. The two (or more) path coefficients may be given any name that complies with the usual SAS conventions. However, we recommend that coefficients be given a "PEQ" prefix (for "**P**aths constrained to be **Eq**ual") followed by a numerical suffix such as PEQ1 or PEQ2. For example, assume that you want the path from V4 to V2 be constrained to be equal to the path from V5 to V1. This means that, in the printed output, these two paths will take on exactly the same value (whatever that will be). Therefore, both paths will be given exactly the same name: PEQ1. The following LINEQS statements show how this constraint may be requested:

```
lineqs
   V1 = PV1V2 V2 + PEQ1 V5 + E1,
   V2 = PV2V3 V3 + PEQ1 V4 + E2;
```

Starting Values for Path Coefficients

Usually it is unnecessary to provide PROC CALIS with starting values for path coefficients or any other parameters to be estimated. However, if you provide starting figures that are close to the parameter estimates at which the procedure will ultimately arrive, this may allow the program to converge more quickly (i.e., with fewer iterations). Also, as previously mentioned, one test to determine whether a model is identified involves running the program several different times with different starting values.

When used, the starting value for a given path coefficient should appear in parentheses immediately to the right of that coefficient. Below, for example, .55 is used as the starting value for the coefficient named PV1V2, and .3 is used as the starting value for the coefficient named PV2V3:

```
lineqs
   V1 = PV1V2 (.55) V2 + PV1V5 V5 + PV1V6 V6 + E1,
   V2 = PV2V3 (.3)  V3 + PV2V4 V4            + E2;
```

The VARIANCE Statement

The primary purpose of the VARIANCE statement is to specify constraints on variances in an analysis.. Below you will see that the statement can also be used to name estimated variances, though this is not necessary. (It is important not to confuse the VARIANCE statement with the VAR statement; these are distinct SAS syntax. In other words, VAR is not an abridged version of VARIANCE!)

The VARIANCE statement follows a format very similar to the LINEQS statement. It begins with the letters VARIANCE, and ends with a semicolon. Each equation in the statement is separated by a comma. Below is the VARIANCE statement with explicitly named variance parameters. (Naming the variance parameters is not required.)

```
variance
   E1 = VARE1,
   E2 = VARE2,
   V3 = VARV3,
   V4 = VARV4,
   V5 = VARV5,
   V6 = VARV6;
```

The above lines request that variance be estimated for the residual terms E1 and E2, as well as the exogenous manifest variables V3, V4, V5, and V6. Recall that Rule 4 (described earlier) states that variance should be estimated for every exogenous variable in the model including both exogenous manifest variables and residual terms.

Fortunately, by completing the program figure in advance, the variances to be estimated have already been identified. This was done in Figure 4.12 by inserting the symbol VAR? below the name of every exogenous variable (again, including the residual variables). The program figure shows that variances should be estimated for V3, V4, V5, V6, E1, and E2.

Naming Variance Estimates

A variance is a parameter just like a path coefficient. Like path coefficients, actual numerical estimates for these variances will be calculated by PROC CALIS. Technically, any name that conforms to SAS conventions for variable names can be used. We, however, recommend an approach that provides meaningful names to facilitate understanding of the output. With this system, you will begin the name with a VAR prefix (to identify this as the name of a **VAR**iance estimate), and complete the name with the name of the variable whose variance is being estimated such as V3 or E2.

This results in variance estimate names such as VARV3 and VARE2. With this system, you need only observe the name VARV3 to realize that this is the variance estimate for V3, and VARE2 is the variance estimate for E2.

Estimating, Fixing, and Constraining Variances

An earlier in this chapter, we provided rules for estimating, fixing and constraining parameters. This section will demonstrate how these rules apply to variance estimates.

To estimate variance estimate for an exogenous variable, you need only create an equation within the VARIANCE statement that contains the variable's name to the left of the equals sign, and the variance estimate's name to the right of the equals sign. The following statement requests variance estimates for exogenous variables V3 and E2:

```
variance
   V3 = VARV3,
   E2 = VARE2;
```

To fix a variance estimate at a specific numerical value, provide that numerical value in place of the variance estimate name. Here, the variance of V3 is fixed at .93, while the variance of E2 is estimated:

```
variance
   V3 = .93,
   E2 = VARE2;
```

To constrain two or more variance estimates to be equal, use the same variance estimate name for those variables. We recommend that the name for this estimate begin with the VAR prefix (for "**Var**iances constrained to be **Eq**ual"), and conclude with a numerical suffix so that the resulting name takes on the form of VAREQ1, VAREQ2, etc. Below, the variances for V3 and V4 are constrained to be equal; the name for the resulting variance estimate is VAREQ1:

```
variance
   V3 = VAREQ1,
   V4 = VAREQ1,
   V5 = VARV5,
   E2 = VARE2;
```

Starting Values for Variance Estimates

Starting values are usually not required. If used, however, starting values should appear in parentheses to the immediate right of the variance estimate name but before the comma or semicolon. Below, VARV3 is given a starting value of 1.3, and VARE2 is given a starting value of .65:

```
variance
   V3 = VARV3 (1.3),
   E2 = VARE2 (.65);
```

The COV Statement

The COV statement can be used to constrain pairs of variables that are expected to covary (i.e., correlated) or to specify correlated errors. The statement begins with the letters COV, and ends with a semicolon; commas are used to separate the equations included in the statement. Within each equation, the pairs of covarying variables are presented to the left of the equals sign, and the name for the corresponding covariance estimate appears to the right of the equal sign. The COV statement used in the present program is presented below:

```
cov
        V3 V4 = CV3V4,
        V3 V5 = CV3V5,
        V3 V6 = CV3V6,
        V4 V5 = CV4V5,
        V4 V6 = CV4V6,
        V5 V6 = CV5V6;
```

The preceding statement requests that covariances be estimated for all possible combinations for the four manifest exogenous variables displayed in Figure 4.12. Note that no endogenous variables appear in these statements consistent with Rule 5 (presented earlier), which states that covariances are normally estimated for pairs of exogenous variables but not for endogenous variables.

Naming Covariance Estimates

To create meaningful covariance names, we recommend that the name begin with the C prefix (to identify this as the name of a **C**ovariance estimate), and conclude with the names of the two variables that are covarying. For example, the name CV3V4 is assigned to the covariance estimate between V3 and V4. This approach creates meaningful names. You need only see the name CV5V6 to know that it represents the covariance between V5 and V6.

NOTE: Be consistent when referring to these covariance estimates. The name given to a covariance estimate must always be typed the same way each time it appears in the program. For example, if it appears once as CV3V4 and later as CV4V3, errors will result. To avoid confusion, we recommend that variable name with the lower numerical value always appear first followed by the higher value. That is, the covariance name should be created as CV3V4 and not as CV4V3.

Estimating, Fixing, and Constraining Covariances

The general rules previously specified for estimating, fixing, and constraining parameters apply to covariances as well. To specify the name of a covariance estimate, the two variables expected to covary are listed on the left side of the equals sign (separated by at least one blank space), and the name for that covariance is listed to the right of the equals sign. To illustrate, the following COV statement requests that covariance between V4 and V5 be estimated:

```
cov
   V4 V5 = CV4V5;
```

Which covariances should be estimated? You have already identified these by preparing your program figure in advance. In Figure 4.12, manifest variables V3, V4, V5, and V6 have all been interconnected by curved double-headed arrows indicating that they are all expected to covary. Each of these arrows has been labeled with the C? symbol. There will be one equation in the COV statement for every C? symbol in the figure.

All manifest exogenous variables are usually allowed to covary in a path analysis. Although they are manifest variables, the residual terms (E variables) are usually not allowed to covary. This can be seen in the COV statement used for this program:

```
cov
   V3 V4 = CV3V4,
   V3 V5 = CV3V5,
   V3 V6 = CV3V6,
   V4 V5 = CV4V5,
   V4 V6 = CV4V6,
   V5 V6 = CV5V6;
```

To fix a covariance at a specific numerical value, provide that numerical value after an equals sign in the COV statement. Below, the covariance between V4 and V5 is fixed at .45, and the variance between V4 and V6 and between V5 and V6 is estimated:

```
cov
   V4 V5 = .45,
   V4 V6 = CV4V6
   V5 V6 = CV5V6;
```

Fixing a covariance estimate at zero has the effect of eliminating any curved double-headed arrow between the relevant variables. To constrain two or more covariance estimates to be equal, use the same covariance estimate name for those pairs of variables. This approach involves beginning the common name with the CEQ prefix (for "**C**ovariances constrained to be **Eq**ual") and concluding with a numerical suffix so that the resulting name takes the form of CEQ1, CEQ2, and so forth. Below, the covariance between for V4 and V5 is constrained to be equal to the covariance between V4 and V6:

```
cov
   V4 V5 = CEQ1,
   V4 V6 = CEQ1,
   V5 V6 = CV5V6;
```

Starting Values for Covariance Estimates

Starting values for covariance estimates are optional. If used, starting values should appear in parentheses immediately to the right of the equals sign (or, if you name them, the name of the covariance estimate), before the comma or semicolon. Starting values are provided below for all covariance estimates in the current example:

```
cov
   V3 V4 = CV3V4 (.50),
   V3 V5 = CV3V5 (.40),
   V3 V6 = CV3V6 (.40),
   V4 V5 = CV4V5 (.50),
   V4 V6 = CV4V6 (.75),
   V5 V6 = CV5V6 (.20);
```

The VAR Statement

The VAR statement identifies the manifest variables to be analyzed in the path analysis. Technically, the VAR statement was not necessary in the present program because all variables in the dataset were analyzed. It was included as a matter of good practice, however.

The general form for the VAR statement is

```
var  list-of-manifest-variables-to-be-analyzed ;
```

For the current program, the following VAR statement was used:

```
var  V1 V2 V3 V4 V5 V6 ;
```

Interpreting the Results of the Analysis

When the analysis has been completed, SAS creates two new files. The log file contains lines from the original SAS program along with notes and error messages, and the output file contains the actual results of the analysis. As always, the log file should be reviewed first to check for any errors or warnings. Next, the output file should be reviewed for signs of other problems; the following section describes how to do this.

Making Sure That the SAS Output File "Looks Right"

The SAS output file contains the results of the analyses. Below is a brief description of the information contained on each page. (In the interest of brevity, some of the information appearing in the output file has not been listed here.)

- a list of the endogenous variables and exogenous variables specified in the LINEQS statement.
- simple statistics.
- initial estimation methods.
- the number of parameters estimated and the model's iteration history.
- a variety of goodness-of-fit indices (to be discussed below).
- unstandardized equation parameter estimates and their standard errors corresponding to those specified in the LINEQS statement. This page also reports unstandardized covariance estimates, squared multiple correlation values and R^2 values for exogenous variables.
- standardized equation parameter estimates and their associated t values (significance estimates). This page also reports standardized covariance estimates and their associated t values.
- modification indices: Lagrange Multiplier (LM) and Wald test results (to be discussed).

The output itself is reproduced here as Output 4.1 through 4.8. Several tables of the output must be reviewed to verify that the program ran as expected. First, the structural equations should be reviewed to verify that the LINEQS statements were written correctly. The information indicates that your program has specified two endogenous variables (V1 and V2) and six exogenous variables (V3, V4, V5, V6, E1, and E2). This is consistent with your model.

Output 4.1: CALIS Output (Pages 1 and 2), Analysis of Initial Model for Investment Model Study

The CALIS Procedure
Covariance Structure Analysis: Model and Initial Values

Modeling Information	
Maximum Likelihood Estimation	
Data Set	WORK.D1
N Obs	240
Model Type	LINEQS
Analysis	Covariances

Variables in the Model		
Endogenous	Manifest	V1 V2
	Latent	
Exogenous	Manifest	V3 V4 V5 V6
	Latent	
	Error	E1 E2
Number of Endogenous Variables = 2 Number of Exogenous Variables = 6		

Initial Estimates for Linear Equations															
V1	=		.	*	V2	+	.	*	V5	+		.	*	V6	+ 1.0000 E1
					PV1V2				PV1V5					PV1V6	
V2	=		.	*	V3	+	.	*	V4	+	1.0000			E2	
					PV2V3				PV2V4						

Initial Estimates for Variances of Exogenous Variables			
Variable Type	Variable	Parameter	Estimate
Error	E1	VARE1	.
	E2	VARE2	.
Observed	V3	VARV3	.
	V4	VARV4	.
	V5	VARV5	.
	V6	VARV6	.

Initial Estimates for Covariances Among Exogenous Variables			
Var1	Var2	Parameter	Estimate
V3	V4	CV3V4	.
V3	V5	CV3V5	.
V3	V6	CV3V6	.
V4	V5	CV4V5	.
V4	V6	CV4V6	.
V5	V6	CV5V6	.

Simple Statistics			
Variable		Mean	Std Dev
V1	COMMITMENT	0	2.31920
V2	SATISFACTION	0	1.77440
V3	REWARDS	0	1.25250
V4	COSTS	0	1.40860
V5	INVESTMENTS	0	1.55750
V6	ALTERNATIVES	0	1.87010

The equations in the middle of output page 1 (under the heading "Initial Estimates for Linear Equations") correspond to the equations in the LINEQS statements. Each parameter to be estimated is identified by an asterisk ("*"); the information on this page shows that five path coefficients are to be estimated (e.g., PV1V2, PV1V5). The number 1.00 appears before E1 and E2 indicating that these paths are fixed at 1. Finally, the tables on the lower half of the page can be reviewed to verify that you have correctly specified which variances and covariances are to be estimated. Page 2 reports mean and standard deviation values (Std Dev) for all observed variables. These should be checked to verify that no errors have been made when inputting the data.

Page 3 reports initial estimates for all parameters, variances, and covariances. Page 4 of Output 4.2 provides the iteration history; this page should be reviewed to verify that the program converged. In this case, you can see that convergence was achieved after eight iterations. If the model is overidentified and the program has been written correctly but still has not converged within 50 iterations, it may be necessary to provide starting values and/or increase the allowed number of iterations with the MAXITER option in the PROC CALIS statement (or question the viability of one's model).

The final note on page 4 tells you that the convergence criterion has been satisfied. For most models, the optimization technique used by CALIS requires the repeated computation of two values: The function value (optimization criterion), and the gradient element (first-order partial derivatives).

Also appearing at the top of page 5 are the number of data points associated with the analysis or the amount of independent information in the data matrix. This appears in Output 4.2 to the right of the Model Info (Observations) heading. Previously, it was shown that this value may be calculated with the following equation:

$$\text{Number of data points} = (p\,[\,p + 1\,])\,/\,2$$

where p = number of manifest variables being analyzed. Because p = 6 in this analysis (i.e., the number of observed variables in the path model), the resulting number of data points is 21. Earlier, it was noted that a necessary (but not sufficient) condition for model identification is that the number of functions (observations) must exceed the number of parameters to be estimated. This page reports that 17 parameters are to be estimated so this condition has been met.

Output 4.2: CALIS Output (Pages 3 and 4), Analysis of Initial Model for Investment Model Study

Covariance Structure Analysis: Optimization

Initial Estimation Methods	
1	Observed Moments of Variables
2	McDonald Method
3	Two-Stage Least Squares

Optimization Start Parameter Estimates			
N	Parameter	Estimate	Gradient
1	PV1V2	0.35345	-0.08618
2	PV1V5	0.49848	-7.893E-16
3	PV1V6	-0.52990	6.193E-16
4	PV2V3	0.73873	-3.886E-16
5	PV2V4	-0.43082	1.3878E-16
6	VARE1	1.62060	-2.419E-16
7	VARE2	1.42946	4.4811E-18
8	VARV3	1.56876	2.5939E-18
9	VARV4	1.98415	8.9752E-19
10	VARV5	2.42581	4.2251E-17
11	VARV6	3.49727	9.4999E-18
12	CV3V4	-0.77716	-2.776E-17
13	CV3V5	1.04288	-1.11E-16
14	CV3V6	-0.95121	6.9389E-18
15	CV4V5	-0.40675	4.8572E-17
16	CV4V6	0.92856	-6.939E-17
17	CV5V6	-1.14585	-3.643E-17
Value of Objective Function = 0.1579645379			

Covariance Structure Analysis: Optimization

Levenberg-Marquardt Optimization

Scaling Update of More (1978)

Parameter Estimates	17
Functions (Observations)	21

Optimization Start			
Active Constraints	0	Objective Function	0.1579645379
Max Abs Gradient Element	0.0861824737	Radius	1

Iteration	Restarts	Function Calls	Active Constraints	Objective Function	Objective Function Change	Max Abs Gradient Element	Lambda	Ratio Between Actual and Predicted Change
1	0	4	0	0.15659	0.00137	0.0177	0	1.199
2	0	6	0	0.15646	0.000137	0.00613	0	1.311
3	0	8	0	0.15644	0.000016	0.00166	0	1.340
4	0	10	0	0.15644	1.885E-6	0.000645	0	1.346
5	0	12	0	0.15644	2.265E-7	0.000214	0	1.347
6	0	14	0	0.15644	2.723E-8	0.000076	0	1.347
7	0	16	0	0.15644	3.276E-9	0.000026	0	1.347
8	0	18	0	0.15644	3.94E-10	9.045E-6	0	1.347

Optimization Results			
Iterations	8	Function Calls	21
Jacobian Calls	10	Active Constraints	0
Objective Function	0.156439117	Max Abs Gradient Element	9.0449671E-6
Lambda	0	Actual Over Pred Change	1.3468218043
Radius	0.0000581107		

Convergence criterion (GCONV=1E-8) satisfied.

Assessing the Fit between Model and Data

When conducting path analysis, you will usually begin with some theoretical model that hypothesizes a set of associations among observed variables. You then obtain raw data from a sample, which may be recomputed as a covariance matrix or, as in this case, a correlation matrix. A theoretical model provides a good fit to data when it successfully accounts for observed covariances in this matrix.

A number of procedures and statistics have been developed to assess the extent to which a model fits data, and you will usually refer to several of these in the course of the analysis. This is because there is no single index of goodness of fit that is universally accepted (Byrne 1998); each index provides somewhat different information. This section describes a few indices that should be particularly useful to you if you are learning about path analysis for the first time. Consistent with the format used throughout this text, these indices are introduced by leading you through a number of structured steps.

Step 1: Reviewing the Chi-Square Test

The original goodness-of-fit index used in path analysis was the chi-square (χ^2) statistic or likelihood ratio. This statistic tests for significance between the actual covariance matrix among variables, and the estimated covariance matrix based upon the model. The chi-square statistic tests the hypothesis that a specified model holds exactly in the population from which data are derived (MacCallum, Browne, and Sugarawa 1996). If the null hypothesis is correct, then the obtained chi-square value should be small and the associated p value should be relatively large. The p value (or probability value) associated with the test indicates the likelihood of obtaining a chi-square value this large or larger if the null hypothesis were true (i.e., if the model fits the population perfectly).

Remember that the null hypothesis in path analysis is a hypothesis of good fit; think about what this means in practical terms. For instance, if you have developed a theoretical model and hope to obtain support for it (as most often will be the case), then you hope *not* to reject this null hypothesis. In other words, you hope that the chi-square value will be small (near zero) and, as a result, the associated p (or probability) value will be large (e.g., $p > .05$).

The chi-square test for the present analysis is presented here as Output 4.3. This section of the CALIS output provides a large number of goodness-of-fit indices, some of which will be discussed below.

Output 4.3: CALIS Output (Page 5), Analysis of Initial Model for Investment Model Study

Covariance Structure Analysis: Maximum Likelihood Estimation

Fit Summary		
Modeling Info	Number of Observations	240
	Number of Variables	6
	Number of Moments	21
	Number of Parameters	17
	Number of Active Constraints	0
	Baseline Model Function Value	2.9937
	Baseline Model Chi-Square	715.4838
	Baseline Model Chi-Square DF	15
	Pr > Baseline Model Chi-Square	<.0001
Absolute Index	Fit Function	0.1564
	Chi-Square	37.3889
	Chi-Square DF	4
	Pr > Chi-Square	<.0001
	Z-Test of Wilson & Hilferty	4.9302
	Hoelter Critical N	61
	Root Mean Square Residual (RMR)	0.2258
	Standardized RMR (SRMR)	0.0645
	Goodness of Fit Index (GFI)	0.9538

Parsimony Index	Adjusted GFI (AGFI)	0.7574
	Parsimonious GFI	0.2543
	RMSEA Estimate	0.1869
	RMSEA Lower 90% Confidence Limit	0.1350
	RMSEA Upper 90% Confidence Limit	0.2438
	Probability of Close Fit	<.0001
	ECVI Estimate	0.3030
	ECVI Lower 90% Confidence Limit	0.2354
	ECVI Upper 90% Confidence Limit	0.4027
	Akaike Information Criterion	71.3889
	Bozdogan CAIC	147.5598
	Schwarz Bayesian Criterion	130.5598
	McDonald Centrality	0.9328
Incremental Index	Bentler Comparative Fit Index	0.9523
	Bentler-Bonett NFI	0.9477
	Bentler-Bonett Non-normed Index	0.8213
	Bollen Normed Index Rho1	0.8040
	Bollen Non-normed Index Delta2	0.9531
	James et al. Parsimonious NFI	0.2527

The chi-square test appears in the middle of the page and is repeated again below:

	Chi-Square	37.3889
	Chi-Square DF	4
	Pr > Chi-Square	<.0001

Remember that, if the model provides a good fit, you expect to see a small value of chi-square and a large p value. In the present analysis, chi-square value is 37.39 with 4 degrees of freedom, which was highly significant (i.e., $p < .01$). Because it was significant, *technically* you reject your null hypothesis of good model fit. In other words, this test did not support your model.

How are the degrees of freedom determined? For a simple recursive path analytic model without correlated residuals (such as the one considered here), the degrees of freedom are equal to the number of data points used in the analysis, minus the number of parameters to be estimated. Page 4 of Output 4.2 indicates that the current analysis has 21 functions (or observations) and 17 parameters; hence 21 minus 17 equals the 4 degrees of freedom associated with this test.

It is also useful to think of the degrees of freedom as being equal to the number of restrictions that are placed on the data. Remember that, if your model were a fully recursive (or saturated) model, then no restrictions would have been placed on any of the relationships; every variable would be connected to every other variable by either a directional path or a double-headed arrow. You did, however, impose some restrictions in order to create the model tested here. Specifically, you fixed the following four paths to be zero: the path from V3 to V1; the path from V4 to V1; the path from V5 to V2 and; the path from V6 to V2. Fixing these paths to be equal to zero had the effect as eliminating them from your path model. Imposing these four restrictions resulted in four degrees of freedom and made it possible to assess your model with the chi-square test.

As previously noted, chi-square values should be nonsignificant suggesting that derived data do not differ from the population from which they were drawn. This criterion, however, is rarely met. In part, this is due to sample size sensitivity (i.e., small differences emerge as significant with large samples). Today, the chi-square statistic is no longer seen as a viable goodness-of-fit statistic.

The chi-square test also requires that data demonstrate multivariate normality. If the data are leptokurtic (peaked), a well-fitting model is more likely to be rejected. In contrast, if data are platykurtic (flat), the analysis is likely to fail to reject a poorly fitting model (Anderson and Gerbing 1988). When data demonstrate high or low kurtosis, they should be transformed and/or a search for outliers should be conducted (Tabachnick and Fidell 2012).

Step 2: Reviewing Contemporary Goodness-of-Fit Indices

Today there are numerous alternatives to the chi-square test and it is convention in path analytic research with programs such as PROC CALIS to report at least three. However, no consensus yet exists as to which three should be reported (McDonald and Ho 2002; Sun 2005). We, however, recommend that one absolute index (e.g., Standardized Root Mean Square Residual), a parsimony index (e.g., Root Mean Square Error of Approximation + 90% confidence limit), and an incremental index (e.g., Comparative Fit Index) should be reported. In addition to our suggestions, there are a multitude of other equally acceptable goodness-of-fit statistics reported in the SAS output such as the Adjusted Goodness of Fit Index. These are generated by PROC CALIS and presented again as Output 4.4.

Output 4.4: Output (Page 5), Analysis of Initial Model for Investment Model Study

Covariance Structure Analysis: Maximum Likelihood Estimation

Fit Summary		
Modeling Info	Number of Observations	240
	Number of Variables	6
	Number of Moments	21
	Number of Parameters	17
	Number of Active Constraints	0
	Baseline Model Function Value	2.9937
	Baseline Model Chi-Square	715.4838
	Baseline Model Chi-Square DF	15
	Pr > Baseline Model Chi-Square	<.0001
Absolute Index	Fit Function	0.1564
	Chi-Square	37.3889
	Chi-Square DF	4
	Pr > Chi-Square	<.0001
	Z-Test of Wilson & Hilferty	4.9302
	Hoelter Critical N	61
	Root Mean Square Residual (RMR)	0.2258
	Standardized RMR (SRMR)	0.0645
	Goodness of Fit Index (GFI)	0.9538

Parsimony Index	Adjusted GFI (AGFI)	0.7574
	Parsimonious GFI	0.2543
	RMSEA Estimate	0.1869
	RMSEA Lower 90% Confidence Limit	0.1350
	RMSEA Upper 90% Confidence Limit	0.2438
	Probability of Close Fit	<.0001
	ECVI Estimate	0.3030
	ECVI Lower 90% Confidence Limit	0.2354
	ECVI Upper 90% Confidence Limit	0.4027
	Akaike Information Criterion	71.3889
	Bozdogan CAIC	147.5598
	Schwarz Bayesian Criterion	130.5598
	McDonald Centrality	0.9328
Incremental Index	Bentler Comparative Fit Index	0.9523
	Bentler-Bonett NFI	0.9477
	Bentler-Bonett Non-normed Index	0.8213
	Bollen Normed Index Rho1	0.8040
	Bollen Non-normed Index Delta2	0.9531
	James et al. Parsimonious NFI	0.2527

The Comparative Fit Index (CFI) is likely the most commonly reported **incremental index**, which Bentler (1990) has described as the statistic of choice for analysis of covariance structures. The CFI adjusts for degrees of freedom providing a complete measure of sample covariation (i.e., measures the relative fit between a specified model and a baseline null model; Kline 2005). According to Hu and Bentler (1999), CFI values greater than .94 are suggestive of good fit between data and hypothesized models.

A commonly reported **parsimony index** is the Root Mean Square Error of Approximation (RMSEA), which has also been described as a population-based absolute fit index. As noted by Sun (2005), the RMSEA is one of the most informative goodness-of-fit indices because it considers overall error in the population. In other words, the RMSEA estimates error of approximation in the population (Byrne 2009; Kline 2005). We will discuss the RMSEA again in the next chapter because this statistic can also be used to estimate the statistical power of models with latent variables (Chapters 5 and 6). This ability to estimate fit relative to a reference population allows the RMSEA to be used to estimate sample size requirements when planning to undertake a new study. The latter allows the researcher to determine how many participants need be recruited to attain sufficient statistical power (MacCallum, Browne, and Cai 2006).

Similar to other parsimony indices, the RMSEA adjusts for degrees of freedom, and thus penalizes for model complexity. RMSEA values less than .09 are suggestive of fair or adequate error of approximation, whereas values less than .055 suggest small error (Browne and Cudeck as cited in Byrne, 2009; Hu and Bentler 1999). RMSEA = 0 suggests exact model fit of approximation (MacCallum et al. 1996). Confidence limits are commonly reported for RMSEA values representing the likely range for this statistic within 90% certainty (CL_{90}). In other words, there is a 90% likelihood that the true value within the population for this statistic falls within this range (MacCallum, Browne, and Sugawara 1996).

This feature is another positive aspect of the RMSEA relative to other goodness-of-fit statistics. When the range of confidence limits are within good (i.e., $.09 \geq$ RMSEA $CL_{90} \geq 0$) to ideal parameters (i.e., $.054 \geq$ RMSEA $CL_{90} \geq 0$), the researcher has greater confidence that data fit the model effectively as there is only a 1 in 10 chance that the true RMSEA value within the population falls outside of this range. Strong path models (e.g., large sample size, large numbers of degrees of freedom) will exhibit narrow confidence intervals (MacCallum et al. 1996).

Instead of the chi-square statistic, the Standardized Root Mean Square Residual (SRMR) is an **absolute index** reported more frequently by contemporary social science researchers. The SRMR is calculated as the standardized difference between an observed and predicted correlation. This statistic does not penalize for model complexity in contrast to parsimony indices. A SRMR value of zero indicates perfect fit. Similar to the RMSEA, values less than .09 are suggestive of fair or adequate fit, whereas SRMR values less than .055 suggest good fit (Hu and Bentler 1999). Unlike most other statistics described in this text, both the SRMR and the RMSEA are traditionally reported to three decimal places. This is because small differences often distinguish SRMR and RMSEA values between models.

A final index that can be of use in specific instances is the Expected Cross-Validation Index (ECVI), which is used to compare competing models. Unlike chi-square difference tests, which require that models be nested (i.e., a subset of the other), the ECVI does not have this requirement. (More information on this point regarding *nesting of models* can be found at the end of this chapter.) No specific parameters for model acceptance or rejection exist for ECVI values; instead, this statistic assesses the likelihood that a model cross-validates across similar sized samples from the same population. In other words, the ECVI is used to compare competing models; smaller values are suggestive of greater generalizability (Byrne 1998).

PROC CALIS reports 90% confidence limits for ECVI values similar to the RMSEA. When a model has a lower ECVI value, and when the ECVI value for a competing model is above the upper 90% confidence limit of the first model, you can conclude with greater confidence that the first is the better of the two competing models. For example, assume that for path model A, ECVI = 1.52, and upper and lower 90% confidence limits for this statistic were 1.73 and 1.35, respectively. If the ECVI value for a competing non-nested path model B had an ECVI value greater than 1.73 (e.g., ECVI = 1.84), you would be more confident in your ability to contend that path model A is more likely to be replicated with other samples drawn from this population. The most ideal scenario is when there is no overlap between confidence limits for the ECVI values of the two competing models, but this is not a mandatory criterion to select one non-nested model over another.

Each of these goodness-of-fit statistics is reported by PROC CALIS in Output 4.3. Although the Comparative Fit Index is within optimal parameters (i.e., CFI = .95), neither the Root Mean Square Error of Approximation (RMSEA = .187; $.244 \geq \text{RMSEA CL}_{90} \geq .135$) nor the Standardized Root Mean Residual (SRMR = .065) indicate good model fit to data.

Preliminary results provide mixed support for the model as initially estimated. These indices suggest that there may be merit to this model, but that revisions are likely required to identify relationships among variables more accurately. These results also provide a good example of why it is important to examine and report more than one goodness-of-fit index when computing path analytic models. Had only the CFI been examined, you might have concluded that the model was satisfactory as initially specified.

Step 3: Reviewing R^2 Values for the Endogenous Variables

Remember that path analysis is often conducted because you want to identify the variables that determine variability in the endogenous, or dependent, variables. Although the preceding indices may reflect the goodness of fit between a model and a specific dataset, they do not necessarily reflect the extent to which the independent variables in the model account for variability in the dependent variables.

Fortunately, however, PROC CALIS also reports R^2 values for all endogenous variables included in the model. These R^2 values indicate the percent of variance in the endogenous variables accounted for by their antecedents. As with multiple regression, R^2 values may range from 0 to 1 with larger values indicating a greater percent of explained variance.

In the present analysis, R^2 values appear at the bottom of page 6 of the output, in the table titled "Squared Multiple Correlations." This table is presented as part of Output 4.5.

Output 4.5: Output (Page 6), Analysis of Initial Model for Investment Model Study

Covariance Structure Analysis: Maximum Likelihood Estimation

Linear Equations															
V1	=	0.3888	*	V2	+	0.4828	*	V5	+	-0.5184	*	V6	+	1.0000	E1
Std Err		0.0507		PV1V2		0.0593		PV1V5		0.0492		PV1V6			
t Value		7.6694				8.1486				-10.5450					
V2	=	0.7387	*	V3	+	-0.4308	*	V4	+	1.0000		E2			
Std Err		0.0688		PV2V3		0.0612		PV2V4							
t Value		10.7407				-7.0445									

Estimates for Variances of Exogenous Variables					
Variable Type	Variable	Parameter	Estimate	Standard Error	t Value
Error	E1	VARE1	1.61813	0.14802	10.93161
	E2	VARE2	1.42946	0.13076	10.93161
Observed	V3	VARV3	1.56876	0.14351	10.93161
	V4	VARV4	1.98415	0.18151	10.93161
	V5	VARV5	2.42581	0.22191	10.93161
	V6	VARV6	3.49727	0.31992	10.93161

Covariances Among Exogenous Variables					
Var1	Var2	Parameter	Estimate	Standard Error	t Value
V3	V4	CV3V4	-0.77716	0.12470	-6.23211
V3	V5	CV3V5	1.04288	0.14308	7.28856
V3	V6	CV3V6	-0.95121	0.16353	-5.81680
V4	V5	CV4V5	-0.40675	0.14433	-2.81819
V4	V6	CV4V6	0.92856	0.18067	5.13955
V5	V6	CV5V6	-1.14585	0.20246	-5.65961

Squared Multiple Correlations			
Variable	Error Variance	Total Variance	R-Square
V1	1.61813	4.97275	0.6746
V2	1.42946	3.14850	0.5460

In the present theoretical model, V1 (commitment) was said to be directly determined by V2 (satisfaction), V5 (investments), and V6 (alternatives). The preceding output shows that these three variables account for 67% of the variance in commitment. Most researchers would consider this to be a large percent of the variance (though the figure is probably somewhat inflated since the same method was used to assess both independent and dependent variables). In your model, V2 (satisfaction) was predicted to be directly affected by V3 (rewards) and V4 (costs). These variables account for 55% of the variance in satisfaction, again a substantial proportion of the variance.

Step 4: Reviewing Significance Tests for Path Coefficients and Covariances

As discussed, the CFI, SRMR, and RMSEA are examined to assess overall fit of data to the model. Even if each indicates good fit, it is still necessary to inspect specific features of the model to see if any of the individual features fail to receive support. In other words, goodness-of-fit statistics are examined to assess overall model features; individual elements of the model must also be examined.

For example, the present model predicts that satisfaction will have a significant effect on commitment. To test this prediction, it is necessary to ascertain the significance of the path coefficient that represents the effect of satisfaction on commitment. If this coefficient is significantly different from zero, then prediction receives support.

Significance tests for path coefficients appeared in the present SAS output, under the heading "Standardized Results for Linear Equations." The equation for V2 (satisfaction) in this table is presented again as Output 4.6:

Output 4.6: Standardized Results for Variable V2 (Satisfaction), Analysis of Initial Model for Investment Model Study

V2	=	0.5214	*	V3	+	-0.3420	*	V4	+	1.0000	E2			
Std Err		0.0441		PV2V3		0.0474		PV2V4						
t Value		11.8154				-7.2228								

The first line of this output provides the manifest variable equation itself. V2 (satisfaction) is on the left of the equals sign because it is the endogenous variable being predicted; V3 (rewards) and V4 (costs) are on the right of the equals sign because they are the independent variables.

A given path coefficient appears just before the short name for the predictor variable. The preceding output therefore shows that the path coefficient for V3 is .52, and the coefficient for V4 is -.34. (Notice the minus sign in the equation.) The signs of these coefficients are as you would expect; it makes sense that rewards would be positively associated with satisfaction and that costs would be negatively associated.

Notice that the path coefficient for E2 (the residual term) is 1.00. This is because this coefficient was fixed at 1.00 in the SAS program.

Given the definition of path coefficients, the preceding values indicate that for a one-unit increase in V3, there is an increase of .52 units in V2 while holding constant the effects of the other independent variables. Furthermore, for a one-unit increase in V4, there is a decrease of .34 units in V2, holding constant the effects of the other independent variables.

Just below each path coefficient, the CALIS procedure prints that coefficient's standard error. The standard error for V3 is .04. These standard errors will be valid only if the data are multivariate normal and the sample is adequately large (as discussed earlier in reference to the chi-square test). In addition, the standard error estimates may be incorrect if the analysis is based on a correlation matrix rather than a covariance matrix, even if the sample is relatively large.

The t test for each path coefficient appears below the standard error for that coefficient. These t values are determined by dividing each appropriate path coefficient by its standard error. If the appropriate assumptions are met, these values are used to test the null hypothesis that the corresponding path coefficient is equal to zero in the population. Because these t tests are equivalent to large-sample *z* tests, they are statistically significant at the $p < .05$ level whenever their absolute value exceeds 1.96 (or less than -1.96). In other words, any path coefficient may be viewed as statistically different from zero if the value of the absolute value of its t statistic is greater than 1.96. The path coefficient is significant at the .01 level if $|t|$ exceeds 2.58.

Because the t values for V3 and V4 are 11.82 and -7.22, respectively, it is clear that both of these paths are significant. Nonetheless, even significant path coefficients must be viewed with caution within models where overall fit is questionable as is the case here.

Another set of parameter estimates first appear in the Output 4.5; these are unstandardized estimates. There are two conventions when reporting the results of path analyses: 1) Authors either report unstandardized parameter estimates and their associated standard errors; or 2) they report standardized estimates and associated t values. We recommend the latter. This is because **Standardized path coefficients** are easier to interpret since these coefficients are all on the same metric (i.e., 0 ± 1) and all variables have a standard deviation of 1.00. As a result, it is possible to compare the size of standardized path coefficients of independent variables to determine their relative effect on dependent variables.

The standardized coefficients from the current analysis are again presented here as Output 4.7. You can see that the standardized path coefficient for the effect of V2 on V1 is .31 ($t = 7.66$, $p < .01$). This means that there is an increase of .31 standard deviations in V1 for an increase of 1 standard deviation in V2, while holding constant the effect of the other independent variables.

Output 4.7: Equations with Standardized Path Coefficients, Analysis of Initial Model for Investment Model Study

Standardized Results for Linear Equations													
V1	=	0.3094	*	V2	+	0.3372	*	V5	+	-0.4348	*	V6	+ 1.0000	E1
Std Err		0.0404		PV1V2		0.0412		PV1V5		0.0401		PV1V6		
t Value		7.6574				8.1948				-10.8322				
V2	=	0.5214	*	V3	+	-0.3420	*	V4	+	1.0000		E2		
Std Err		0.0441		PV2V3		0.0474		PV2V4						
t Value		11.8154				-7.2228								

How do I test the significance of the correlations between exogenous variables? Earlier, Output 4.7 reported t values for each covariance specified in the model. These statistics test the null hypothesis that the covariance (or correlation) between a given set of variables is zero in the population. Although the statistical significance of covariances is usually of less interest to researchers than the significance of path coefficients, these tests are still useful for obtaining a more complete understanding of how model variables are related to one another. See the table labeled Standardized Results for Covariances Among Exogenous Variables, p. 7 of the current output

Characteristics of Ideal Fit

As the preceding sections suggest, assessing the adequacy of a path model is no simple matter; there are a number of statistical tests and goodness-of-fit indices that must be consulted. It is possible to simplify things, however, by summarizing the characteristics that would be displayed by a model demonstrating "ideal" fit to data:

- The Comparative Fit Index (CFI) should exceed .94, the Standardized Root Mean Square Residual should be less than .055, and the Root Mean Square Error of Approximation should also be less than .055 and with 90% confidence limits, $.09 \geq \text{RMSEA CL}_{90} \geq 0$ (or ideally $.054 \geq \text{RMSEA CL}_{90} \geq 0$).
- The R2 value for each endogenous variable should be relatively large, compared to what has previously been reported in research examining these variables.
- The absolute value of the t statistics for each path coefficient should exceed 1.96 (or less than -1.96) and the standardized path coefficients should be nontrivial in magnitude

Remember that a model does not necessarily have to display all of these characteristics to be considered acceptable; the literature contains many examples in which models that fail to demonstrate one or more of the preceding traits. For instance, it is not uncommon to obtain an ideal RMSEA value with its confidence limit exceeding the .09 threshhold; this is more likely with smaller samples. Nonetheless, the preceding provides useful standards against which models can be compared. You can have confidence in a path model that does demonstrate these characteristics.

Modifying the Model

When researchers perform path analysis, they usually hope that the results will reveal a good fit between the model and data. If such a fit is obtained, then the model has survived an attempt at disconfirmation (i.e., reject the null hypothesis). The analysis terminates at that point.

But what if the theoretical model does not demonstrate a good fit? In virtually all cases, the researcher needs to modify this initial model to better account for the relationships among observed variables. For example, if the theoretical model predicts no path from variable X to variable Y, but there actually appears to be a rather strong relationship between these variables, the researcher may add the path, revise the PROC CALIS program to reflect this change, recompute the program, and ascertain whether the revised model now demonstrates good fit.

In practice, you typically should rely on a number of modification indices to determine how the model should be adjusted. For example, the CALIS procedure provides a modification index called the Lagrange Multiplier (LM), which estimates the extent to which the model chi-square statistic would decrease (improve) if a given parameter were freed or allowed to be estimated (i.e., added to the model). In most cases, you will modify models by:

- making the single modification that results in the greatest improvement in overall model fit
- analyzing the fit of the revised model
- reviewing the modification indices that are produced by that analysis to identify the next change that would most improve the fit of the revised model

The model is modified in this way, one parameter at a time, until an acceptable fit is obtained.

Problems Associated with Model Modification

Unfortunately, many researchers who modify initial models in the manner described above run the risk of arriving at a final invalid model even though it may effectively fit the data. Such models described as *overfitted* because they will not generalize to other samples from the population of interest (Byrne 1998).

This is because these model modifications very often capitalize on chance characteristics of the sample data. When researchers perform path analysis, they normally hope that their final model will represent the true nature of the relationships among variables in the population, not just in the derived sample. Unfortunately, observed relationships will commonly differ from those existing in the population. Because no sample is perfectly representative of the population from which it is drawn, some differences will exist simply due to sampling error. Therefore, when researchers make many modifications in order to achieve better fit, chances are good that the resulting model will fit data only from that specific sample; it will not generalize to other samples or the overall population.

The modifications described above are known as **data-driven modifications** because they are based on characteristics of sample data, not the overarching theory being tested. MacCallum, Roznowski, and Necowitz (1992) have identified a "worst-case scenario" in which data-driven modifications do not generalize. MacCallum and colleagues (1992) concluded that data-driven model modifications are especially likely to lead to chance findings under the following circumstances:

- **When the sample is small:** MacCallum (1986) has shown that data-driven model modifications based on samples of fewer than 100 observations often lead to poor results. (Another reason samples used for path analysis should exceed 99 as we recommend.)

- **When many modifications are made:** When models are modified, the first few changes typically result in relatively large improvements in fit; successive changes generally result in increasingly smaller improvements. In addition, when the analysis is performed on data from a large sample, the largest discrepancies between the model and data are more likely to be stable (i.e., common to most samples drawn from the population); smaller discrepancies are more likely to be unique to that particular sample and thus unstable. For this reason, only the first few modifications made to the model have a reasonable chance of leading to a model that will generalize. If many modifications are made, the latter changes are more likely to capitalize on chance properties of the sample data.
- **When the modification is not interpretable or justified according to theory or prior research:** Ideally modifications should be made only when they can be justified in light of existing theory or prior research (e.g., Jöreskog and Sörbom 2001). It is bad form to make modifications only because they improve model fit: Revisions should be interpretable and justifiable. In other words, you should be able to explain why adding a given path makes sense in light of what is already known about this phenomenon. Despite frequent warnings, MacCallum and colleagues (1992) contend that very few researchers bother to justify model modifications on substantive grounds in their review of the literature or theory.

Recommendations for Modifying Models

Virtually all path analytic research needs to come to terms with problems associated with model modification. As previously noted, is rare that an initial theoretical model demonstrates good fit. When making model modifications, it is possible to minimize the pitfalls associated with this process by adhering to the following recommendations as suggested by MacCallum and colleagues (1992):

- **Use large samples:** We suggest that path analysis may be performed with 100+ participants. Small samples, in contrast, can lead to unstable solutions particularly if initial analyses are followed by data-driven model modifications. Ideally you should attempt to obtain larger samples especially if many model modifications are anticipated. Ssmaller samples should be acknowledged as possible or probably study limitation
- ... **But not too large:** As previously discussed inordinately large sample sizes can lead to statistically significant - but practically meaningless - results (i.e., over-powered analyses). With larger samples such as 800+ participants, it often makes more sense to randomly split the sample in two halves, derived a model using the first, and then attempt to replicate this model with the second half of the data. (Alternatively, 1 model with men, 1 model with women or any other meaningful comparison). Initial and replicated findings using halves of a large sample are more compelling than single results found with the full sample (1 model only).
- **Make few modifications:** As previously mentioned, only the first few modifications are likely to reflect population relationships; subsequent changes are more likely to be sample specific.
- **Make only changes that can be meaningfully interpreted:** Before making changes, consider whether these modifications can be supported in terms of theory or prior research. These changes should be justified when discussing study findings.
- **Follow a parallel specification search procedure:** A **specification search** is simply the search for changes that will improve model fit. MacCallum and colleagues (1992) recommend that researchers randomly divide large samples in two and undertake separate specification searches. If both lead to the same set of modifications, more confidence can be placed in the stability of the final model. This procedure can also be described as a replication procedure.
- **Compare alternative *a priori* models:** Rather than beginning with a single model and then performing modifications until a final model fits the data, it is often preferable to begin with several competing models, and perform single analyses on each to determine which provides the best fit to data. For example, assume that Theory A states that both attitudes and intentions directly affect behavior, but Theory B states that attitudes affect behavior only indirectly by first influencing intentions (which then directly affect behavior). These alternative models may be tested. This would

involve assessing all variables in a single large sample, then conducting one analysis of the data to test Theory A, and a separate analysis to test Theory B (and Theories C and D, if they exist). The results of these analyses could then be evaluated in terms of overall fit, interpretability, and other criteria to determine which theory obtained the most support. Such an approach is possible both in confirmatory studies based on specific, well-developed theories, as well as in exploratory studies when theory development is nascent.

- **Carefully describe the limitations of your study:** Most path analytic studies in the literature are single-sample studies in which a series of data-driven modifications are made to arrive at a well-fitting model. When this approach is followed, it is important that the research report acknowledge that this approach can result in models that may well not generalize to other samples or the population. The report should state that the model should be considered tentative until replicated with data from new samples.

Modifying the Present Model

Having reviewed the problems associated with model modifications, let's now return to the investment model. Because your sample is not sufficiently large to divide into two samples for the parallel specification procedure, and because we are familiar with no alternative theory specifying the predicted relationships between commitment, satisfaction, rewards, costs, investments and alternatives, neither of these preferred approaches is possible. Instead, you will review the modification indices and other results from the first analysis to see if changes can be made to improve the fit of the first model. To minimize the negative consequences of undertaking this admittedly less desirable approach, you will: 1) make as few revisions as necessary; 2) make only changes that can be justified (theory or existing literature) and; 3) acknowledge and discuss the limitations of this approach in your final report.

In general, there are two types of changes most frequently made when modifying a path analytic model:

- freeing parameters to be estimated (e.g., adding a new directional path or covariance)
- constraining additional parameters to zero (e.g., deleting an existing path or covariance)

Bentler and Chou (1987) argue that, of these two options, the first option (adding new paths or covariances) is more likely to capitalize on chance characteristics of the sample data and lead to inaccuracies. It is more ideal, therefore, to delete nonsignificant associations than to add new associations (especially if these cannot be justified on the basis of theory or existing research). If both are done, nonsignificant pathes are deleted before new paths are added.

The Significance of the Directional Paths

Because it is generally more desirable to drop nonsignificant paths than to add new paths, you next review the output to determine if any path coefficients fail to attain statistical significance. If any are nonsignificant, first re-estimate the model without these paths before adding new parameters.

The significance tests for these equations appear on page 7 of the SAS output and are presented again here as Output 4.8.

Output 4.8: Manifest Variable Equations (Standardized Results), Analysis of Initial Model for Investment Model Study

Standardized Results for Linear Equations															
V1	=	0.3094	*	V2	+	0.3372	*	V5	+	-0.4348	*	V6	+	1.0000	E1
Std Err		0.0404		PV1V2		0.0412		PV1V5		0.0401		PV1V6			
t Value		7.6574				8.1948				-10.8322					
V2	=	0.5214	*	V3	+	-0.3420	*	V4	+	1.0000		E2			
Std Err		0.0441		PV2V3		0.0474		PV2V4							
t Value		11.8154				-7.2228									

Earlier, it was stated that path coefficients are statistically significant at the .05 level if the absolute value of their t statistics exceed 1.96 (or less than -1.96). These results show that all five paths included in the present model are statistically significant and therefore should not be deleted from the model.

For example, consider the first equation, for the endogenous variable V1 (commitment). The path coefficient for independent variable V2 (satisfaction) is .31, and the t statistic for this path coefficient is 7.66. Because this $|t|$ value is larger than the standard cut-off of 1.96, it is significant at $p < .05$. A review of terms in this equation shows that the path coefficients for all independent variables (V2, V5, and V6) hypothesized to predict V1 are statistically significant. In the same way, both independent variables (V3 and V4) assumed to predict V2 (satisfaction) have significant path coefficients. In other words, these results provide no empirical reason to drop any of the paths from this theoretical model.

The Modification Indices

The MODIFICATION option included in the PROC CALIS statement causes two types of modification indices to be produced: The Lagrange Multiplier (LM) test and the Wald test.

First, the **Lagrange Multiplier (LM) test** estimates the decrease in the value of the chi-square statistic that would result from freeing a specific parameter. In practical terms, this means that the test estimates the improvement (or decrease) in chi-square that would result from adding a new path (or a new covariance estimate) to your model. Normally, this test is used to identify new paths (or covariance estimates) that should be added to the model to result in a significant decrease in chi-square. A p value accompanies this statistic; if less than .05, you may assume that adding this path to the model will result in a statistically significant improvement in the model chi-square.

How do I perform a chi-square difference test? The Lagrange Multiplier (LM) provides an approximate test estimating the change in model fit that would result from adding a new path or covariance. A more accurate test can be obtained by performing a **chi-square difference test**. This involves:

- estimating the original model
- estimating the revised model in which the new path has been added
- calculating the difference between the two chi-square values

The resulting chi-square difference also has a chi-square distribution and can be tested for statistical significance. This is done by referring to a table of critical chi-square values. You merely look up the critical value of chi-square for the change in degrees of freedom associated with this test. See Appendix C. The number of degrees of freedom for the chi-square difference test is equal to the difference in degrees of freedom between models. These degrees of freedom are listed to the "Chi-square DF" line of the Fit Summary table (page 5). A table of critical chi-square values appears as Appendix C of this text.

In the present case, your initial model had 4 degrees of freedom, while the revised model with the new path had 3 degrees of freedom. Therefore, the degrees of freedom associated with the chi-square difference test would be 4 – 3; $\Delta df = 1$. A table reporting critical chi-square values shows that the critical value at 1 degree of freedom is 3.84 ($p < .05$). This means that, if the chi-square difference statistic is greater than 3.84, then adding the new path to the original model results in a significant improvement in model fit.

The MODIFICATION option also requests that the **Wald test** be conducted. This test estimates the extent to which chi-square would change if a currently free parameter were fixed at zero. In practical terms, this means that the test estimates the change in chi-square that would result from deleting a path (or covariance) that exists in the current model. Normally, this test is intended to identify unimportant paths or covariances that may be deleted with only a small and nonsignificant increase in chi-square. A p value is also printed for this test.

What if my output doesn't contain the Wald test results? Wald test results are printed only when it is possible to drop paths or covariance estimates from a model without a significant increase in chi-square. For example, the output for the current analysis does not include Wald test results; this means it is probably not possible to drop any of these parameters without negatively affecting model fit. As a result, the following note appears: "All parameters in the model are significant. No parameter can be dropped in the Wald tests."

The Lagrange Multiplier tests requested by the MODIFICATION option are printed on page 9 of the SAS output. These indices are grouped separately for:

- new paths between existing endogenous variables in the current model
- new paths between existing endogenous exogenous variables in the current model predicted by existing exogenous variables
- paths predicting new endogenous variables
- paths predicting covariance between exogenous variables
- paths between error variances and covariances in the current model

PROC CALIS ranks Lagrange Multipliers (LM) by size in descending order. The LM statistics for the current analysis are presented here as Output 4.9.

Output 4.9: Output (Page 9), Analysis of Initial Model for Investment Model Study

The Largest LM Stat for Paths from Endogenous Variables				
To	From	LM Stat	Pr > ChiSq	Parm Change
V2	V1	23.00692	<.0001	0.25212

Rank Order of the 4 Largest LM Stat for Paths from Exogenous Variables				
To	From	LM Stat	Pr > ChiSq	Parm Change
V2	V5	24.31366	<.0001	0.29035
V2	V6	17.50883	<.0001	-0.19375
V1	V4	0.95016	0.3297	0.07125
V1	V3	0.09125	0.7626	0.03013

The Largest LM Stat for Error Variances and Covariances				
Var1	Var2	LM Stat	Pr > ChiSq	Parm Change
E2	E1	0.12752	0.7210	0.05195

Output 4.9: Output (Page 9) Continued, Analysis of Initial Model for Investment Model Study

Rank Order of the 10 Largest LM Stat for Paths with New Endogenous Variables				
To	From	LM Stat	Pr > ChiSq	Parm Change
V3	V2	31.10287	<.0001	-0.54559
V5	V2	16.13929	<.0001	0.27651
V6	V2	9.33446	0.0022	-0.26743
V6	V1	8.32714	0.0039	-0.53205
V5	V1	5.23777	0.0221	0.27293
V4	V2	2.68233	0.1015	0.44922
V4	V1	1.42528	0.2325	0.07844
V3	V1	0.28472	0.5936	-0.02813
V6	V5	0.12752	0.7210	-0.18283
V4	V5	0.12752	0.7210	0.15688

Note: There is no parameter to free in the default LM tests for the covariances of exogenous variables. Ranking is not displayed.

The Largest LM Stat for Error Variances and Covariances				
Var1	Var2	LM Stat	Pr > ChiSq	Parm Change
E2	E1	0.12752	0.7210	0.05195

The interpretation of the modification indices is easiest if you have at hand a figure displaying the model being estimated. Therefore, the initial model being tested in the current analysis is again reproduced here as Figure 4.13.

Figure 4.13: The Initial Model, Investment Model Study

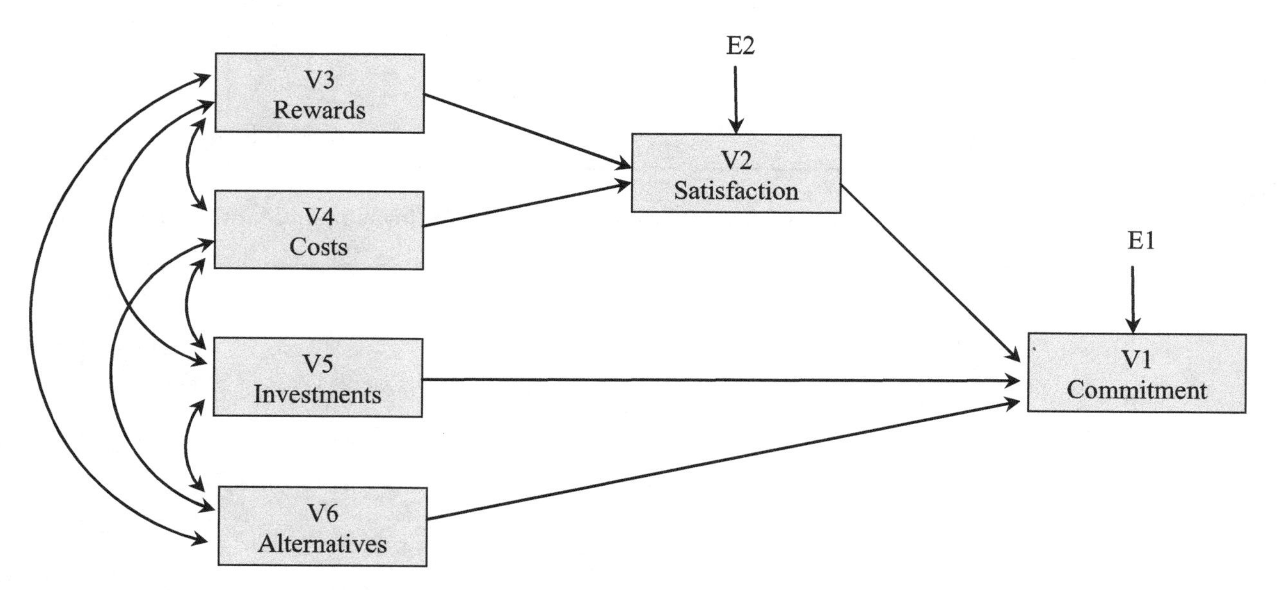

Remember that the LM statistics estimate reduction or improvement in chi-square values that would result from *adding* paths or additional covariance estimates to the model.

In Output 4.9 (middle of page 10), you can see that PROC CALIS has ranked the largest LM statistics for new endogenous variables in the third grouping of statistics under the heading "Rank Order of the 10 Largest LM Stat for Paths with New Exogenous Variables." Here, the largest LM value is 31.10, which means that the model chi-square value would decrease by roughly 31.10 if a path was added from V2 to V3. This would allow rewards (V3) to be predicted by satisfaction (V2). To the right, we see that this LM statistic would result in a statistically significant reduction in the chi-square statistic ($p < .01$). As previously noted, any change greater than 3.84 in chi-square ($\Delta df=1$) is statistically significant.

Such a modification would be questionable, however, because our theory suggests that rewards predict satisfaction —not the other way around. Therefore this revision is not theoretically tenable. In accord with our third recommendation for model revision in the previous section, this change should not be made. (Note, also, that this change would make the model nonrecursive.)

In Output 4.8, the next largest LM statistic appears in the second grouping under the heading "Rank Order of the 4 Largest LM Stat for Paths from Exogenous Variables." The LM statistic here suggests that a reduction in chi-square of roughly 24.31 would occur if a path was added in which investments (V5) was allowed to predict satisfaction (V2). The LM statistic suggests that your model may be significantly improved by adding this path, which again would result in a statistically significant change in the chi-square statistic ($p < .01$). In contrast to the previous change, this would be more consistent with theory (i.e., investment in a relationship predicting satisfaction with that relationship). For the sake of demonstration, let us assume that adding this directional path can be justified on substantive grounds and proceed with the analysis.

In this section of the table, notice also that the LM statistic in which alternatives (V6) is allowed to predict satisfaction (V2) would also be statistically significant ($p < .01$). The change in chi-square in this instance would be associated with an inverse relationship between variables. In other words, this suggests that the fewer one's alternatives, the higher one's satisfaction with a current relationship (negative parameter change estimate).

However, the modification index for this change (LM = 17.51) is smaller than for the previous change (LM = 24.31). Given that these would appear to be equally tenable changes from a theoretical standpoint, ordinarily you would first make the change with the larger LM statistic.

Note also that Output 4.8 also suggests that a statistically significant change would result from including a path from satisfaction (V2) to investments (V5): the opposite of the change we have decided to make. Again, this would suggest a relationship between variables contrary to the way in which our theory was first presented. The LM statistic is also smaller for the V2 to V5 path versus the V5 to V2 path, suggesting that the latter is the more appropriate of the two.

At this point, you might ask: Why not make both revisions at the same time? The reason to make only one revision at a time is that the LM tests conducted in PROC CALIS is a univariate procedure. In other words, the LM statistic for V6 to V2 may no longer be statistically significant after the initial change is made for larger V5 to V2 path estimate. This is because these variables are moderately correlated (r = -.39) as we can see from the correlation matrix in the DATALINES statement of the SAS program. The risk here is overfitting the model when a single change may be sufficient to obtain optimal goodness of fit (i.e., make unnecessary changes which may not generalize to the population). Because you hope to make as few changes as necessary, you will revise your SAS program so that it includes only the new path from V5 to V2, and add the additional paths only if the results of the new model indicate that they are needed. The new model is presented as revised model 1 in Figure 4.14.

Figure 4.14: Revised Model 1, Developed by Adding a Path from V5 (Investments) to V2 (Satisfaction)

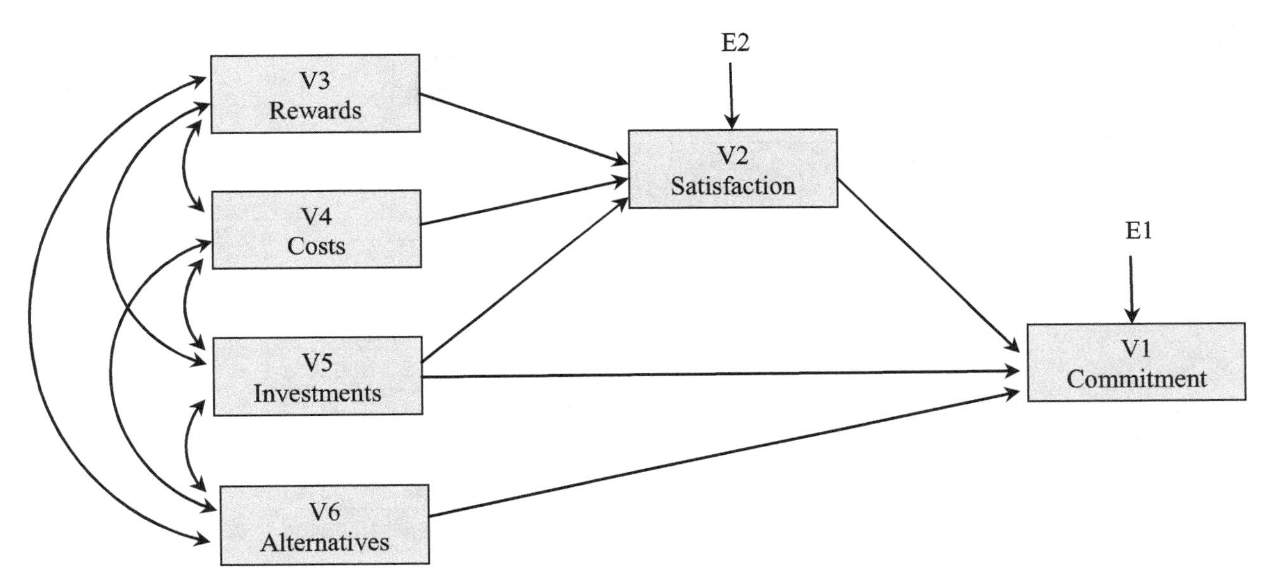

It might also be noted that it is not necessarily best to always choose the path with the largest Lagrange Multiplier as the path that must be added to the model. If a path with a smaller value (also statistically significant) makes better sense in terms of theory or prior research, that path may be added instead.

Creating Revised Model 1

To change the initial model, it is only necessary to revise one line ❶ in the LINEQS statement of the PROC CALIS program. Specifically, the equation for the endogenous variable V2 is modified so that V5 now appears as a predictor of V2. Since the variables that were originally exogenous variables are still exogenous variables, and since all of the previous endogenous variables are still endogenous variables, it was unnecessary to change the VARIANCE or COV statements. Part of the revised program appears below.

```
    proc calis   modification ;
       lineqs
          V1 = PV1V2 V2 + PV1V5 V5 + PV1V6 V6 + E1,
❶         V2 = PV2V3 V3 + PV2V4 V4 + PV2V5 V5 + E2;
       var  V1 V2 V3 V4 V5 V6 ;
    run;
```

Output 4.10: Output (Pages 6 and 7), Analysis of Revised Model 1, Investment Model Study

The CALIS Procedure
Covariance Structure Analysis: Maximum Likelihood Estimation

Linear Equations																	
V1	=		0.3888	*	V2	+	0.4828	*	V5	+	-0.5184	*	V6	+	1.0000		E1
Std Err			0.0507		PV1V2		0.0593		PV1V5		0.0492		PV1V6				
t Value			7.6694				8.1486				-10.5450						
V2	=		0.7387	*	V3	+	-0.4308	*	V4	+	1.0000		E2				
Std Err			0.0688		PV2V3		0.0612		PV2V4								
t Value			10.7407				-7.0445										

Estimates for Variances of Exogenous Variables					
Variable Type	Variable	Parameter	Estimate	Standard Error	t Value
Error	E1	VARE1	1.61813	0.14802	10.93161
	E2	VARE2	1.42946	0.13076	10.93161
Observed	V3	VARV3	1.56876	0.14351	10.93161
	V4	VARV4	1.98415	0.18151	10.93161
	V5	VARV5	2.42581	0.22191	10.93161
	V6	VARV6	3.49727	0.31992	10.93161

Covariances Among Exogenous Variables					
Var1	Var2	Parameter	Estimate	Standard Error	t Value
V3	V4	CV3V4	-0.77716	0.12470	-6.23211
V3	V5	CV3V5	1.04288	0.14308	7.28856
V3	V6	CV3V6	-0.95121	0.16353	-5.81680
V4	V5	CV4V5	-0.40675	0.14433	-2.81819
V4	V6	CV4V6	0.92856	0.18067	5.13955
V5	V6	CV5V6	-1.14585	0.20246	-5.65961

Squared Multiple Correlations			
Variable	Error Variance	Total Variance	R-Square
V1	1.61813	4.97275	0.6746
V2	1.42946	3.14850	0.5460

The CALIS Procedure
Covariance Structure Analysis: Maximum Likelihood Estimation

Standardized Results for Linear Equations																
V1	=	0.3094	*	V2	+	0.3372	*	V5	+	-0.4348	*	V6	+	1.0000		E1
Std Err		0.0404		PV1V2		0.0412		PV1V5		0.0401		PV1V6				
t Value		7.6574				8.1948				-10.8322						
V2	=	0.5214	*	V3	+	-0.3420	*	V4	+	1.0000		E2				
Std Err		0.0441		PV2V3		0.0474		PV2V4								
t Value		11.8154				-7.2228										

Standardized Results for Variances of Exogenous Variables					
Variable Type	Variable	Parameter	Estimate	Standard Error	t Value
Error	E1	VARE1	0.32540	0.03426	9.49928
	E2	VARE2	0.45401	0.04340	10.46110
Observed	V3	VARV3	1.00000		
	V4	VARV4	1.00000		
	V5	VARV5	1.00000		
	V6	VARV6	1.00000		

Standardized Results for Covariances Among Exogenous Variables					
Var1	Var2	Parameter	Estimate	Standard Error	t Value
V3	V4	CV3V4	-0.44050	0.05213	-8.44951
V3	V5	CV3V5	0.53460	0.04620	11.57194
V3	V6	CV3V6	-0.40610	0.05402	-7.51800
V4	V5	CV4V5	-0.18540	0.06246	-2.96824
V4	V6	CV4V6	0.35250	0.05665	6.22273
V5	V6	CV5V6	-0.39340	0.05467	-7.19540

In your analysis of the original model, the Lagrange Multiplier test suggested that adding the path from investments to satisfaction would result in a significant improvement in model chi-square statistic. But remember that the Lagrange Multiplier test merely provides an *estimate* of the improvement in model chi-square. Since the revised model has now been calculated, it is possible to directly determine whether the model improvement was significant by performing the chi-square difference test discussed earlier. This process involves determining the difference between the chi-square values for the two models, and testing this difference for significance.

The chi-square for the initial model was 37.39, while the chi-square for the revised model was 11.75. The difference chi-square is therefore equal to

$$37.39 - 11.75 = 25.64$$

The degrees of freedom for the test are obtained by subtracting the degrees of freedom for the revised model (df = 3) from the degrees of freedom for the original model (df = 4), resulting in 1 degree of freedom for the difference test (Δdf=1). The table of chi-square values (Appendix C) indicates that the critical value for 1 degree of freedom is 3.84 ($p < .05$); since 25.64 is much larger than 3.84, this difference is clearly significant. In fact, this improvement in chi-square is significant at the $p < .01$, since the critical value of chi-square is 6.63 at that level of significance (df = 1).

Notice that the difference in chi-square values is not exactly the same as the LM statistic obtained in Output 4.8. The SAS output for the first model suggested that this chi-square difference would be 24.31; instead, the value above is 25.64. Here, the LM statistic under-estimated the actual change in chi-square values. In other instances, the LM statistic will over-estimate chi-square change. This finding underscores that the LM statistic estimates the actual chi-square change; to be certain that revisions lead to statistically significant improvement, it is prudent to always calculate the resulting difference in chi-square values.

Unfortunately, the goodness-of-fit statistics for this revised model (page 5) still indicate less than ideal fit. These are presented as Output 4.11. The Comparative Fit Index remains in optimal parameters (CFI = .99), and the value of the Standardized Root Mean Square Residual is now optimal as well (SRMR = .030). However, the Root Mean Square Error of Approximation is still large (RMSEA = .111), and almost all of the 90% confidence interval falls outside of ideal parameters ($.180 \geq$ RMSEA $CL_{90} \geq .049$). This result suggests that the path model may benefit from further revision.

Output 4.11: Output (Page 5), Analysis of Revised Model 1 for Investment Model Study

The CALIS Procedure
Covariance Structure Analysis: Maximum Likelihood Estimation

Fit Summary		
Modeling Info	Number of Observations	240
	Number of Variables	6
	Number of Moments	21
	Number of Parameters	17
	Number of Active Constraints	0
	Baseline Model Function Value	2.9937
	Baseline Model Chi-Square	715.4838
	Baseline Model Chi-Square DF	15
	Pr > Baseline Model Chi-Square	<.0001
Absolute Index	Fit Function	0.1564
	Chi-Square	37.3889
	Chi-Square DF	4
	Pr > Chi-Square	<.0001
	Z-Test of Wilson & Hilferty	4.9302
	Hoelter Critical N	61
	Root Mean Square Residual (RMR)	0.2258
	Standardized RMR (SRMR)	0.0645
	Goodness of Fit Index (GFI)	0.9538
Parsimony Index	Adjusted GFI (AGFI)	0.7574
	Parsimonious GFI	0.2543
	RMSEA Estimate	0.1869
	RMSEA Lower 90% Confidence Limit	0.1350
	RMSEA Upper 90% Confidence Limit	0.2438
	Probability of Close Fit	<.0001
	ECVI Estimate	0.3030
	ECVI Lower 90% Confidence Limit	0.2354

	ECVI Upper 90% Confidence Limit	0.4027
	Akaike Information Criterion	71.3889
	Bozdogan CAIC	147.5598
	Schwarz Bayesian Criterion	130.5598
	McDonald Centrality	0.9328
Incremental Index	Bentler Comparative Fit Index	0.9523
	Bentler-Bonett NFI	0.9477
	Bentler-Bonett Non-normed Index	0.8213
	Bollen Normed Index Rho1	0.8040
	Bollen Non-normed Index Delta2	0.9531
	James et al. Parsimonious NFI	0.2527

As before, the LM statistic suggests that adding the path between alternatives (V6) and satisfaction (V2) would significantly improve the model's ability to explain variance within the dataset. You can see that the Lagrange Multiplier for V6 to V2 (LM = 10.39) is statistically significant, $p < .01$, suggesting that adding this path would likely result in significant improved fit. For the sake of demonstration, assume that the addition of such a path is theoretically interpretable, and revise the SAS program to reflect this change.

Output 4.12: Output, Analysis of Revised Model 1, Investment Model Study

The CALIS Procedure
Covariance Structure Analysis: Maximum Likelihood Estimation

The Largest LM Stat for Paths from Endogenous Variables

To	From	LM Stat	Pr > ChiSq	Parm Change
V2	V1	23.00692	<.0001	0.25212

Rank Order of the 4 Largest LM Stat for Paths from Exogenous Variables

To	From	LM Stat	Pr > ChiSq	Parm Change
V2	V5	24.31366	<.0001	0.29035
V2	V6	17.50883	<.0001	-0.19375
V1	V4	0.95016	0.3297	0.07125
V1	V3	0.09125	0.7626	0.03013

Rank Order of the 10 Largest LM Stat for Paths with New Endogenous Variables

To	From	LM Stat	Pr > ChiSq	Parm Change
V3	V2	31.10287	<.0001	-0.54559
V5	V2	16.13929	<.0001	0.27651
V6	V2	9.33446	0.0022	-0.26743
V6	V1	8.32714	0.0039	-0.53205
V5	V1	5.23777	0.0221	0.27293
V4	V2	2.68233	0.1015	0.44922
V4	V1	1.42528	0.2325	0.07844
V3	V1	0.28472	0.5936	-0.02813
V6	V5	0.12752	0.7210	-0.18283
V4	V5	0.12752	0.7210	0.15688

Note: There is no parameter to free in the default LM tests for the covariances of exogenous variables. Ranking is not displayed.

The Largest LM Stat for Error Variances and Covariances

Var1	Var2	LM Stat	Pr > ChiSq	Parm Change
E2	E1	0.12752	0.7210	0.05195

The resulting model with the new path from V6 to V2 appears as revised model 2 in Figure 4.15.

Figure 4.15: Revised Model 2, Developed by Adding a Path from V6 (Alternatives) to V2 (Satisfaction)

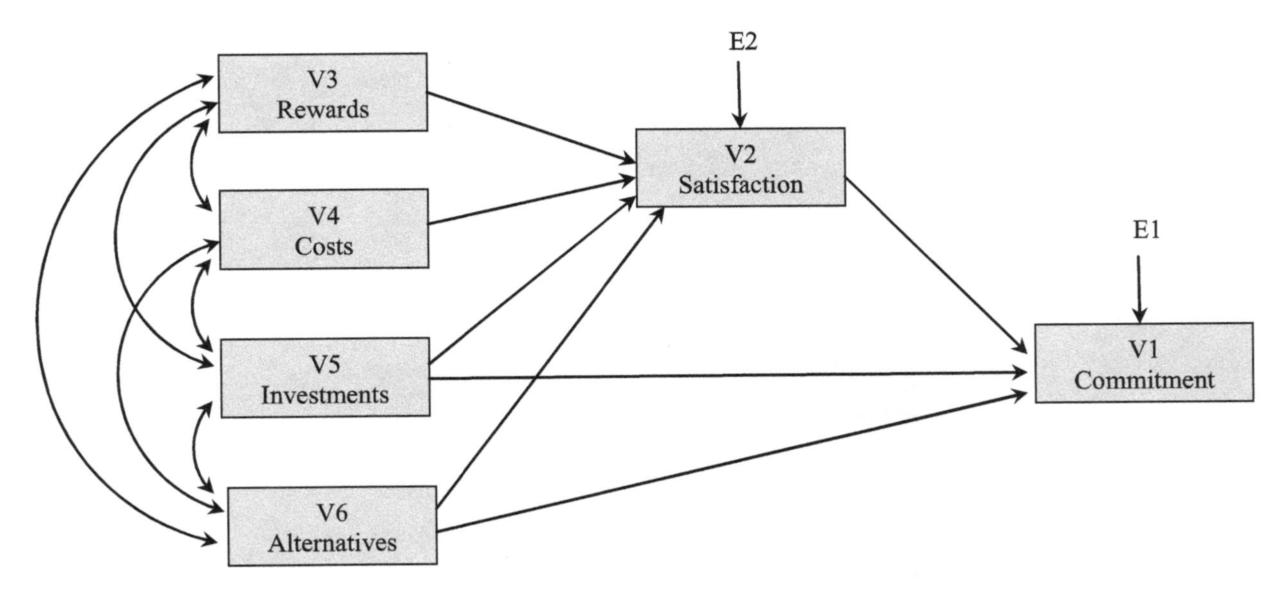

Creating Revised Model 2

Below is the revised SAS program which includes the new path from V6 to V2. It was necessary to change only the LINEQS statement for the endogenous variable V2 on line ❶ to make this modification; all other aspects of the program remain the same.

```
      proc calis   modification ;
         lineqs
          V1 = PV1V2 V2 + PV1V5 V5 + PV1V6 V6          + E1,
❶         V2 = PV2V3 V3 + PV2V4 V4 + PV2V5 V5 + PV2V6 V6 + E2;
         variance
           E1 = VARE1,
           E2 = VARE2,
           V3 = VARV3,
           V4 = VARV4,
           V5 = VARV5,
           V6 = VARV6;
         COV
           V3 V4 = CV3V4,
           V3 V5 = CV3V5,
           V3 V6 = CV3V6,
           V4 V5 = CV4V5,
           V4 V6 = CV4V6,
           V5 V6 = CV5V6;
         VAR  V1 V2 V3 V4 V5 V6 ;
      RUN;
```

The SAS output created by this program was 9 pages long, parts of which are presented here as Output 4.13 and 4.17.

Once again, note that t values for all standardized path coefficients are statistically significant. As presented in Output 4.13 (page 9), we can see that this includes the new path in which (V2) satisfaction is now significantly predicted by (V6) alternatives ($t = -3.30$, $p < .05$). This suggests that inclusion of this path has enhanced the model. We now look to see if there has been significant change in the chi-square statistic followed by the goodness-of-fit indices.

Output 4.13: Analysis of Revised Model 2, Investment Model Study

The CALIS Procedure
Covariance Structure Analysis: Maximum Likelihood Estimation

Standardized Results for Linear Equations																			
V1	=	0.2975	*	V2	+	0.3242	*	V5	+	-0.4180	*	V6	+	1.0000		E1			
Std Err		0.0446		PV1V2		0.0423		PV1V5		0.0408		PV1V6							
t Value		6.6742				7.6742				-10.2460									
V2	=	0.3531	*	V3	+	-0.3222	*	V4	+	0.2144	*	V5	+	-0.1539	*	V6	+	1.0000	E2
Std Err		0.0518		PV2V3		0.0459		PV2V4		0.0494		PV2V5		0.0467		PV2V6			
t Value		6.8111				-7.0225				4.3433				-3.2980					

Standardized Results for Variances of Exogenous Variables					
Variable Type	Variable	Parameter	Estimate	Standard Error	t Value
Error	E1	VARE1	0.30084	0.03254	9.24445
	E2	VARE2	0.39009	0.03941	9.89777
Observed	V3	VARV3	1.00000		
	V4	VARV4	1.00000		
	V5	VARV5	1.00000		
	V6	VARV6	1.00000		

Standardized Results for Covariances Among Exogenous Variables					
Var1	Var2	Parameter	Estimate	Standard Error	t Value
V3	V4	CV3V4	-0.44050	0.05213	-8.44951
V3	V5	CV3V5	0.53460	0.04620	11.57194
V3	V6	CV3V6	-0.40610	0.05402	-7.51800
V4	V5	CV4V5	-0.18540	0.06246	-2.96824
V4	V6	CV4V6	0.35250	0.05665	6.22273
V5	V6	CV5V6	-0.39340	0.05467	-7.19540

The LM statistic test from the analysis of revised model 1 suggested that adding the V2 to V6 path would result in a significant improvement in chi-square. Once again, this should be assessed by performing a chi-square difference test. Revised model 1 resulted in a model chi-square value of 11.75 with 3 degrees of freedom. In the present analysis, the model chi-square for revised model 2 is 1.12 with 2 degrees of freedom. The chi-square difference value is therefore

$$11.75 - 1.12 = 10.63$$

with difference test degrees of freedom equal to

$$3 - 2 = 1$$

The critical value of chi-square with 1 degree of freedom is 3.84 for $p < .05$ (6.63 for $p < .01$), so it is clear that adding the V2 to V6 path did result in a significant improvement in model fit. Once again, this change in chi-square value is slightly larger than estimated by the LM statistic of 10.39 predicted for inclusion of the V6 to V2 path (Output 4.14).

Output 4.14: Output (Page 5), Analysis of Revised Model 2, Investment Model Study

The CALIS Procedure
Covariance Structure Analysis: Maximum Likelihood Estimation

Fit Summary		
Modeling Info	Number of Observations	240
	Number of Variables	6
	Number of Moments	21
	Number of Parameters	19
	Number of Active Constraints	0
	Baseline Model Function Value	2.9937
	Baseline Model Chi-Square	715.4838
	Baseline Model Chi-Square DF	15
	Pr > Baseline Model Chi-Square	<.0001
Absolute Index	Fit Function	0.0047
	Chi-Square	1.1234
	Chi-Square DF	2
	Pr > Chi-Square	0.5702
	Z-Test of Wilson & Hilferty	-0.1914
	Hoelter Critical N	1275
	Root Mean Square Residual (RMR)	0.0202
	Standardized RMR (SRMR)	0.0062
	Goodness of Fit Index (GFI)	0.9984
Parsimony Index	Adjusted GFI (AGFI)	0.9836
	Parsimonious GFI	0.1331
	RMSEA Estimate	0.0000
	RMSEA Lower 90% Confidence Limit	0.0000
	RMSEA Upper 90% Confidence Limit	0.1083
	Probability of Close Fit	0.7256
	ECVI Estimate	0.1685
	ECVI Lower 90% Confidence Limit	0.1724
	ECVI Upper 90% Confidence Limit	0.1961

	Akaike Information Criterion	39.1234
	Bozdogan CAIC	124.2556
	Schwarz Bayesian Criterion	105.2556
	McDonald Centrality	1.0018
Incremental Index	Bentler Comparative Fit Index	1.0000
	Bentler-Bonett NFI	0.9984
	Bentler-Bonett Non-normed Index	1.0094
	Bollen Normed Index Rho1	0.9882
	Bollen Non-normed Index Delta2	1.0012
	James et al. Parsimonious NFI	0.1331

We now turn attention to the goodness-of-fit statistics also on this page (Output 4.14) for this model. Each of the three indices previously examined is now within optimal parameters as: CFI = 1.00; SRMR = .006; and RMSEA = 0. Also of note, the range of RMSEA confidence limit is only marginally higher than the threshold value of .09 ($0 \geq$ RMSEA $CL_{90} \geq .108$), providing greater confidence in this RMSEA statistic.

Ordinarily this would end the process but, for the sake of consistency, let's examine the modification indices to see if any additional revisions might further improve goodness of fit. As presented in Output 4.15 (page 9), the inclusion of no other paths that would further enhance model fit. This is noteworthy when we compare this output with that of the previous model (Output 4.14) in which it appeared as many as seven adjustments might be made to the model to significantly reduce the chi-square statistic. The finding that a single change can result in no further modification indices appear significant underscores the previous point that model revisions should always be made one change at a time.

Output 4.15: Output (Page 9), Analysis of Revised Model 2, Investment Model Study

The CALIS Procedure
Covariance Structure Analysis: Maximum Likelihood Estimation

The Largest LM Stat for Paths from Endogenous Variables

To	From	LM Stat	Pr > ChiSq	Parm Change
V2	V1	0.22161	0.6378	0.04316

Rank Order of the 2 Largest LM Stat for Paths from Exogenous Variables

To	From	LM Stat	Pr > ChiSq	Parm Change
V1	V4	0.94232	0.3317	0.07065
V1	V3	0.07916	0.7784	0.02614

Rank Order of the 10 Largest LM Stat for Paths with New Endogenous Variables

To	From	LM Stat	Pr > ChiSq	Parm Change
V4	V1	1.11832	0.2903	0.07056
V6	V1	0.40593	0.5240	-0.19514
V3	V1	0.38659	0.5341	0.03210
V5	V1	0.28799	0.5915	-0.08223
V3	V5	0.22161	0.6378	-0.18164
V4	V5	0.22161	0.6378	0.22388
V6	V4	0.22161	0.6378	-0.03969
V5	V4	0.22161	0.6378	0.03837
V5	V6	0.22161	0.6378	0.02990
V3	V6	0.22161	0.6378	0.04151

Note: There is no parameter to free in the default LM tests for the covariances of exogenous variables. Ranking is not displayed.

The Largest LM Stat for Error Variances and Covariances

Var1	Var2	LM Stat	Pr > ChiSq	Parm Change
E2	E1	0.22161	0.6378	0.06983

Choosing a Final Model

At this point, three models have been computed for this study:

- the initial investment model (as illustrated in Figure 4.13)
- revised model 1, in which a path from investments to satisfaction had been added to the initial model (as illustrated in Figure 4.14)
- revised model 2, in which a path from alternatives to satisfaction had been added to revised model 1 (as illustrated in Figure 4.15)

In an actual analysis, you would decide at this point which to present as the "final model" in the research report.

The decision is not always straightforward. On empirical grounds, it is difficult to justify accepting the initial modal as a final model given the less than ideal goodness-of-fit statistics (e.g., SRMR = .065, RMSEA = .187; $.135 \leq$ RMSEA $CL_{90} \leq .244$). Of note, only negligible improvement was observed with the first revision in which the path between investments and satisfaction was added even though the t value for this path estimate was statistically significant; the resulting change in the chi-square statistic was also statistically significant ($p < .01$). The third model, however, exhibited ideal goodness-of-fit including the chi-square statistic; also, there were no significant modification indices subsequent to this revision.

To summarize, the positive features of this second revised model are:

- all path coefficients were statistically significant and nontrivial in size
- R^2 values for both predicted variables were large
- CFI > .94, SRMR < .055, and RMSEA < .055 with a 90% confidence interval for the RMSEA largely within acceptable parameters

Furthermore, significant change from the previous chi-square statistic (first revised model) was observed and no remaining modification indices suggested room for additional model enhancements.

On the negative side, however, this second revised model cannot be characterized as parsimonious. A **parsimonious** model is one that accounts for covariation in the data with a minimal number of parameters. In revised model 2, nearly every variable is interconnected by paths or curved double-headed arrows. This lack of parsimony is reflected in the fact that the final model has only two degrees of freedom. This is an instance where it would be ideal to replicate findings with a separate sample or dataset in order to have greater confidence in the generalizability of the model.

The final decision should be based on the preceding considerations along with knowledge of theory and existing research in the topic area. Assuming that prior findings support acceptance of the final model, the following summary provides an example of how to synthesize and describe path analytic research findings.

Preparing a Formal Description of the Analysis and Results for a Paper

The results section of a path analytic study can be much longer than the result sections of studies that use simpler statistical procedures. This is because, in most path analytic studies, the initial model fails to provide adequate fit to the data and is subsequently modified. The research report must describe the rationale for these modifications and how they were made; this is usually done in the results section.

Preparing Figures and Tables

The Figures

It will be much easier for the reader to understand your report if you organize the text around a few figures and tables. In student term papers, it is common to illustrate your initial theoretical model. This figure will appear along with the hypotheses to be tested. It is less common for this initial or hypothesized figure to be depicted these days in scientific journals due to space constraints.

However, it is common for journal articles to depict the "final" version of the path model after all modifications have been made. It is convention to report t values along with standardized estimates (vs. unstandardized estimates and standard errors). Figure 4.16 depicts the final revised model.

Figure 4.16: Revised Model 2

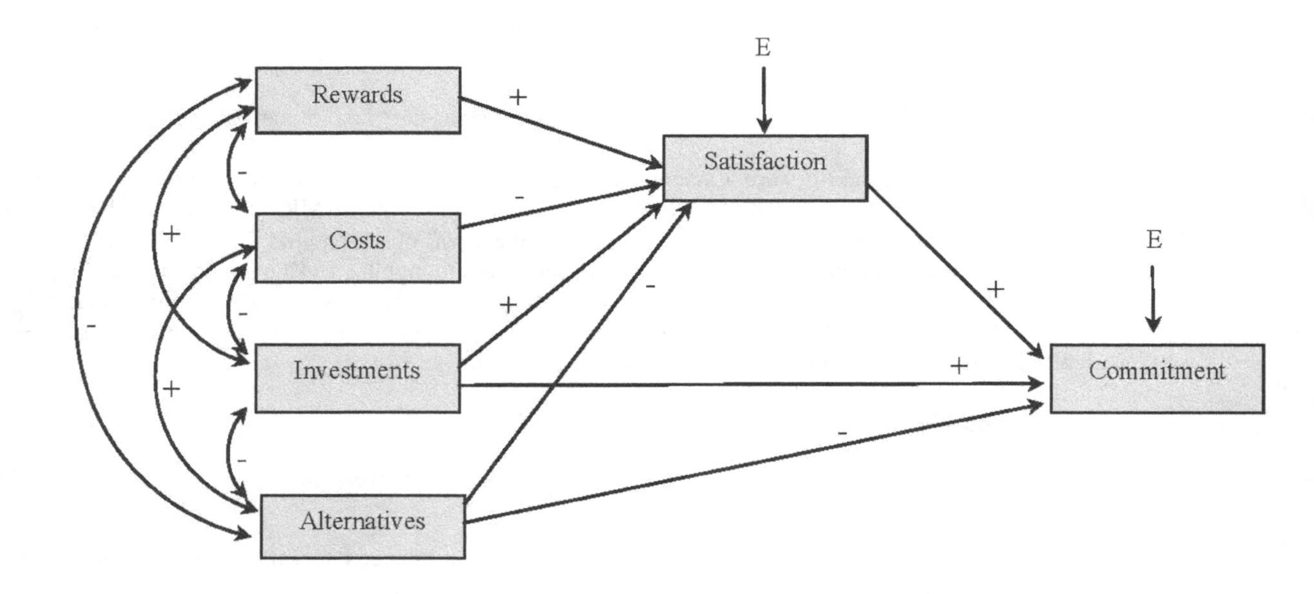

The Tables

In most research papers published in the social sciences (including most path-analytic studies), one of the first tables presented should provide simple descriptive statistics for the study's variables, including the means, standard deviations, and intercorrelations. This enables subsequent researchers to repeat your analyses in accord with the publication guidelines of the American Psychological Association (2009). An example of such a table is presented as Table 4.2.

Table 4.2: Means, Standard Deviations, Internal Consistency Estimates (in Parentheses), and Intercorrelations for the Investment Study Variables

Measure	M	SD	1	2	3	4	5	6
1. Commitment	7.42	2.32	(.81)					
2. Satisfaction	7.88	1.77	.67	(.92)				
3. Rewards	7.92	1.25	.55	.67	(.77)			
4. Costs	3.74	1.41	-.35	-.57	-.44	(.71)		
5. Investment size	6.55	1.50	.64	.52	.53	-.19	(.86)	
6. Alternative value	3.18	1.87	-.69	-.50	-.41	.35	-.39	(.85)

Note: N = 240. An additional table that will prove useful to your readers summarizes the goodness-of-fit indices obtained for the different path models. Such a table is presented as Table 4.3.

Also reported as either part of the final figure itself (cf. Bachner, O'Rourke, and Carmel 2011) or as a separate table, standardized path coefficients and associated t values are reported for the final model. This table is presented here as Table 4.3.

Table 4.3: Standardized Path Coefficients and Associated Significance Values (t values)

Paths	Standardized Path Coefficients	t Values
Satisfaction to commitment	.30	6.67
Investment size to commitment	.32	7.67
Alternative value to commitment	-.42	-10.25
Rewards to satisfaction	.35	6.81
Costs to satisfaction	-.32	-7.02
Investment size to satisfaction	.21	4.34
Alternative value to satisfaction	-.15	3.30

Note: N = 240. Statistically significant t values > |1.96|

* $p < .05$ ** $p < .01$

The columns of Table 4.4 present the goodness-of-fit statistics (and degrees of freedom) for the models computed in this chapter:

- the chi-square statistic, along with degrees of freedom and p values associated with change in the chi-square statistic
- the Comparative Fit Index (CFI)
- the Standardized Root Mean Square Residual (SRMR)
- the Root Mean Square Error of Approximation (RMSEA) and 90% confidence limits

Table 4.4 presents goodness-of-fit indices for four models, though only the last is of primary interest. The first or the "null model" represents a hypothetical path model in which none of the variables are related to any of the other variables. This chi-square value is used as a baseline statistic to compare subsequent models. Here, Table 4.4 reports a null model chi-square value of 715.48 with 15 degrees of freedom. Both subsequent models demonstrate significant reductions in chi-square taking into account change in degrees of freedom. This provides support for model revisions.

Table 4.4: Goodness-of-Fit Indices for Various Models, Investment Model Study (N = 240)

Model	χ^2	df	$\Delta\chi^2$	Δdf	CFI	SRMR	RMSEA	(RMSEA CL_{90})
Baseline model	715.48	15						
Initial model	37.39	4	678.09**	11	.95	.065	.187	(.244-.135)
Revised model 1	11.75	3	26.64**	1	.99	.030	.111	(.180-.049)
Revised model 2	1.12	2	10.63**	1	.98	.006	.000	(.108-.000)

Note: χ^2 = chi-square; df = degrees of freedom; CFI = Comparative Fit Index; SRMR = Standardized Root Mean Square Residual; RMSEA = Root Mean Square Error of Approximation; RMSEA CL_{90} = RMSEA 90% Confidence Limits.

Preparing Text

There is a great deal of variability in the way that the results of path analyses are described. To a large extent, the way the results section is written will depend on factors such as the hypotheses being tested and the number and nature of the models being compared. If you are just learning path analysis, you should review several published studies to see the range of styles (and extent of reported detail) from discipline to discipline. The following represents just one approach that could be used to discuss the results of the present study.

Path analysis was performed to assess the viability of a theoretical model testing elements of Rusbult's (1980) Investment Model. These analyses were conducted using PROC CALIS (maximum likelihood method of parameter estimation) based on the variance-covariance matrix. Our sample size of 240 participants provides sufficient power to detect medium to large effect sizes.

Goodness-of-fit indices for the various models are presented in Table 4.4. The chi-square statistic is reported to enable comparisons between the baseline or null model and subsequent revised models. This table also reports the Standardized Root Mean Square Residual (SRMR), the Comparative Fit Index (CFI), the Root Mean Square Error of Approximation (RMSEA), and 90% confidence limits for the RMSEA statistic (RMSEA CL_{90}). Values for the CFI greater than .94 suggest good fit between data and path models (Hu and Bentler 1999), whereas SRMR and RMSEA values less than .090 suggest acceptable fit, and values less than .055 suggest good model fit (McDonald and Ho 2002). Ideally, the full 90% range for the RMSEA is inacceptable to ideal limits (Byrne 2009; McDonald and Ho 2002).

In accord with theory, the initial theoretical Investment Study model hypothesized that Rewards, Costs, Investment Size, and Alternative Value would each significantly predict Satisfaction with one's current relationship; Satisfaction would, in term, predict Commitment to the relationship (along with Investment Size and Alternative Value). The four exogenous variables (Rewards, Costs, Investment Size, and Alternative Value) were all assumed to be correlated.

Estimated path coefficients for this initial or hypothesized model each differed significantly from zero, χ^2 (4, N=240) = 37.39, $p < .01$. Squared multiple correlation values for both Commitment ($R^2 = .67$) and Satisfaction ($R^2 = .55$) indicate that predictor variables capture large percentages of observed variance for both dependent variables.

Although the CFI for this model suggested good fit to derived (CFI = .95), other indices indicated poor model fit (i.e., SRMR = .065; RMSEA = .187). Of note, the full range of 90% confidence limits for this RMSEA value fell outside of acceptable parameters (.244 ≥ RMSEA CL_{90} ≥ .135). Modification indices were next examined to ascertain if suggested revisions were theoretically tenable.

First, the path coefficients were reviewed to see if any of the paths in the initial model should be deleted. As previously noted, t values for all path coefficients were statistically significant ($p < .05$). Therefore none of the existing paths were eliminated from the initial model as confirmed by examination of the results of Wald tests.

Examination of Lagrange Multipliers (LM) suggested that the model could be significantly improved by adding a path from investment size to satisfaction. Adding this path would be consistent with cognitive dissonance theory (Festinger 1957), which contends that individuals adjust their attitudes in order for these attitudes to be consistent with their decisions and behaviors (i.e., investment of time and effort in a relationship). Because a theoretical basis exists for making this revision, a path from Investment Size to Satisfaction was added to the initial model. The resulting model, called "revised model 1," was then re-estimated.

As expected, the standardized path coefficient for this path proved to be statistically significant, t = 5.23, $p < .05$). A chi-square difference test also confirmed that the addition of this path resulted in a significant improvement in model fit. Comparison of the chi-square statistic for the initial model to the chi-square statistic for revised model 1 indicated that the new path resulted in a significant improvement in model fit, $\Delta\chi^2(\Delta df=1) = 26.64$, $p < .01$. Of further note, the squared multiple correlation value for Satisfaction ($R^2 = .59$) increased with the inclusion of the additional predictor variable; $R^2 = .69$ for Commitment.

However, goodness-of-fit statistics for this revised model remained less than ideal. Both the CFI (.99) and SRMR (.030) are now in ideal parameters, yet the RMSEA (.111) and the full 90% confidence interval for this statistic remain less than ideal (.180 ≥ RMSEA CL_{90} ≥ .049).

Modification indices were again examined to ascertain if another theoretically tenable revision might be made. A statistically significant Lagrange Multiplier indicated that a path from Alternative Value to Satisfaction might be added (LM = 10.39, $p < .01$). Once again, this revision is theoretically tenable as cognitive dissonance theory would support the assertion that relationship satisfaction would be greater when fewer alternatives to that relationship seem to exist (Festinger 1957). We therefore modified revised model 1 by adding a path from Alternatives to Satisfaction. The resulting model was called revised model 2.

The standardized path coefficients for revised model 2 are reported in Table 4.3. Path coefficients, including the new path between Alternatives and Satisfaction, significantly differ from zero (i.e., t values > |1.96|, $p < .05$). As reported in Table 4.4, the chi-square difference test resulted in significant improvement in model fit, $\Delta\chi^2(\Delta df=1) = 10.63$, $p < .01$. Also, squared multiple correlation values have further increased for both Commitment ($R^2 = .70$) and Satisfaction ($R^2 = .61$).

Table 4.4 also indicates that all goodness-of-fit indices are now within ideal parameters as CFI = 1.00, SRMR = .006, and RMSEA = 0 ($.108 \geq$ RMSEA $CL_{90} \geq 0$).

On the basis of these overall findings, revised model 2 appears to best reflect the patterns of association within the derived dataset. Revisions to the initial hypothesized model are theoretically tenable (Festinger 1957) and, in both instances, led to improved model estimation.

We therefore propose the model appearing as Figure 4.16 as the accepted or final model. However, it must be acknowledged that this model is not a parsimonious; with only two remaining degrees of freedom, there is the risk that this model is *overfitted.* Even though revisions to the initial hypothesized model are theoretically tenable, both were data driven. We recommend that future path analytic studies test the validity of this model with larger samples (ideally, randomly recruited), from this and other populations.

Example 2: Path Analysis of a Model Predicting Victim Reactions to Sexual Harassment

The next example comes from the field of gender studies. In recent decades, there has been increased awareness and interest in research dealing with sexual harassment in the work place; some researchers have sought to identify the variables that determine whether women will report instances of sexual harassment (e.g., Kane-Urrabazo 2007).

One approach to studying this phenomenon might involve an analogue study methodology in which women are asked to read a brief scenario in which a fictional woman was being sexually harassed by her supervisor. Participants are then be asked to imagine how they would feel if this were them in this position. They then complete questionnaires to assess a various attitudes, beliefs, and intentions related to the harassment. The purpose of the study would be to identify the attitudes, beliefs, and other variables that influence participants' intention to report the harassment in the workplace.

Figure 4.17 depicts a path model identifying determinants of a participant's intention to report sexual harassment. Although some of these variables were inspired by Brooks and Perot (1991), this model was constructed for purposes of illustration only and should not be regarded as a bona fide test of theory. The model includes the following variables:

- participants' **intention to report** harassment to management, where higher scores indicate greater intention to report
- **expected outcomes** of reporting the harassment, where higher scores indicate stronger belief that reporting the harassment will result in positive results for the victim
- **feminist ideology**, where higher scores indicate egalitarian attitudes about sex roles
- perceived **seriousness of the offense**, where higher scores reveal a stronger belief that the woman in the scenario experienced a serious form of harassment
- **victim marketability**, where higher scores indicate stronger belief that the woman in the scenario could easily find another good job if she were to leave her current position
- **Respondent's age**, in this case, the age of the participant reading the scenario (not the woman described)

- **normative expectations**, where higher scores reflect stronger belief that the victim's family, friends, and coworkers would support her if she reported the harassment

Figure 4.17: Model 1: Determinants of Intent to Report Sexual Harassment Predicting both Direct and Indirect Effects of Feminist Ideology on Participants' Intention to Report

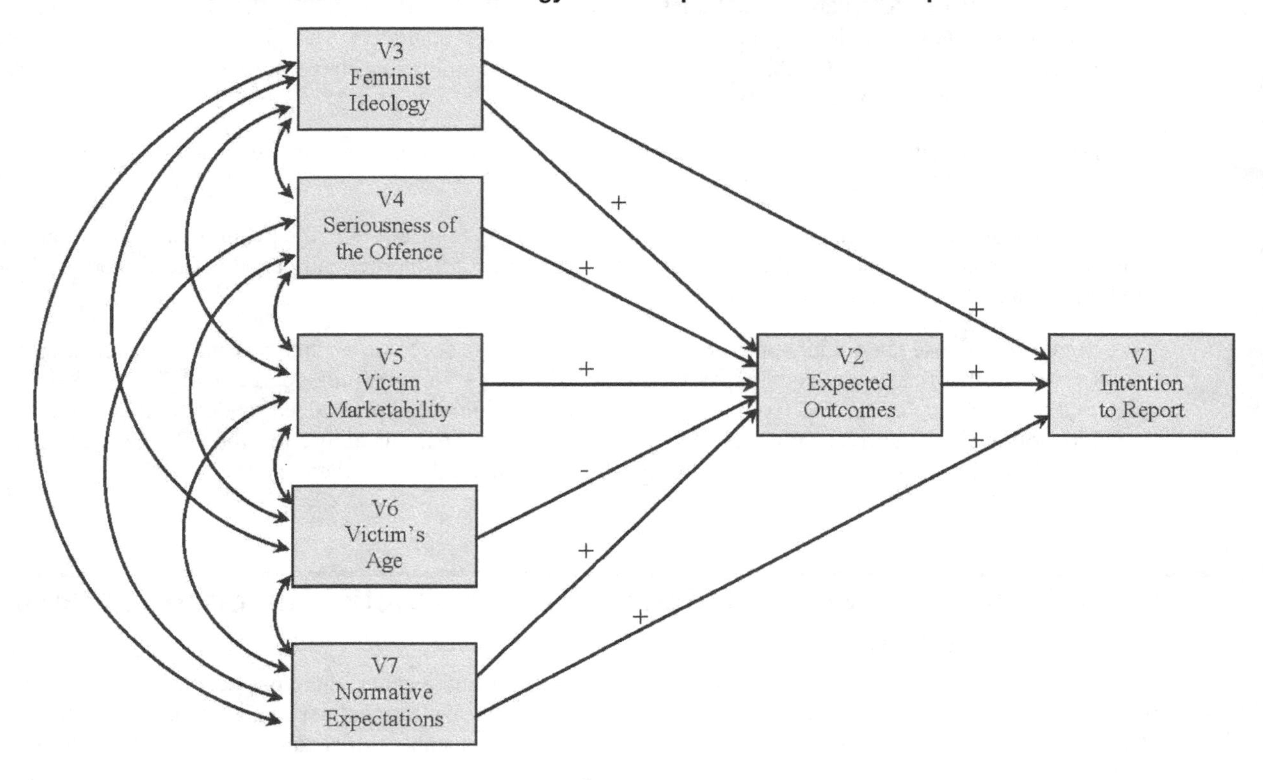

The directional paths hypothesized in this figure are identified by either "+" or "-" signs to indicate whether positive or negative relationships are predicted. The model includes two endogenous variables: Intention to report (V1) and expected outcomes (V2). The model makes the following predictions:

- there will be direct positive paths from expected outcomes, feminist ideology, and normative expectations to intention to report
- there will be direct positive paths from feminist ideology, seriousness of the offense, victim marketability, and normative expectations to expected outcomes
- there will be a direct negative path from victim's age to expected outcomes

Comparing Alternative Models

To make things more interesting, imagine that a controversy exists among scholars studying sexual harassment focused on the relationship between feminist ideology and intention to report; scholars are divided into three schools of thought. The first group believes that feminist ideology is a very important determinant of intent to report. The group believes that feminist ideology has both direct and indirect effects. The direct effect of feminist ideology is represented by the single path that runs from V3 to V1 in Figure 4.17. This first group of scholars believes that women who score high on feminist ideology will be more likely to report harassment because they will view it as a personal obligation as a feminist to take this action. Hence, there is a direct path between these two variables in the figure.

The first group of scholars also believes that feminist ideology has an indirect effect on intentions through its effect on expected outcomes. They argue that, if a woman is a feminist, she will be more likely to believe that reporting harassment will lead to positive outcomes (e.g., punishment for the perpetrator, increased consciousness about harassment in the work place). From this perspective, feminist ideology has a positive effect on expected outcomes (represented by the path from V3 to V2), and expected outcomes, in turn, has a positive effect on intention to report (represented by the path from V2 to V1). In this way feminist ideology has an indirect effect on intention to report. (Of course, any discussion of feminist ideology having an indirect effect on intention to report assumes that the relationship between expected outcomes and intention to report is significant and of meaningful size; if this latter relationship was zero or near-zero, then feminist ideology could not have an indirect effect on intention to report through expected outcomes, no matter how strong the relationship between ideology and expected outcomes.)

The second group of scholars does not believe that feminist ideology is as important as the first group; instead, this second group believes that feminist ideology has only an indirect effect on intention to report. The path model that represents the predictions of this group is presented as Figure 4.18.

Figure 4.18: Model 2, Predicting Only Indirect Effects of Feminist Ideology on Intention to Report

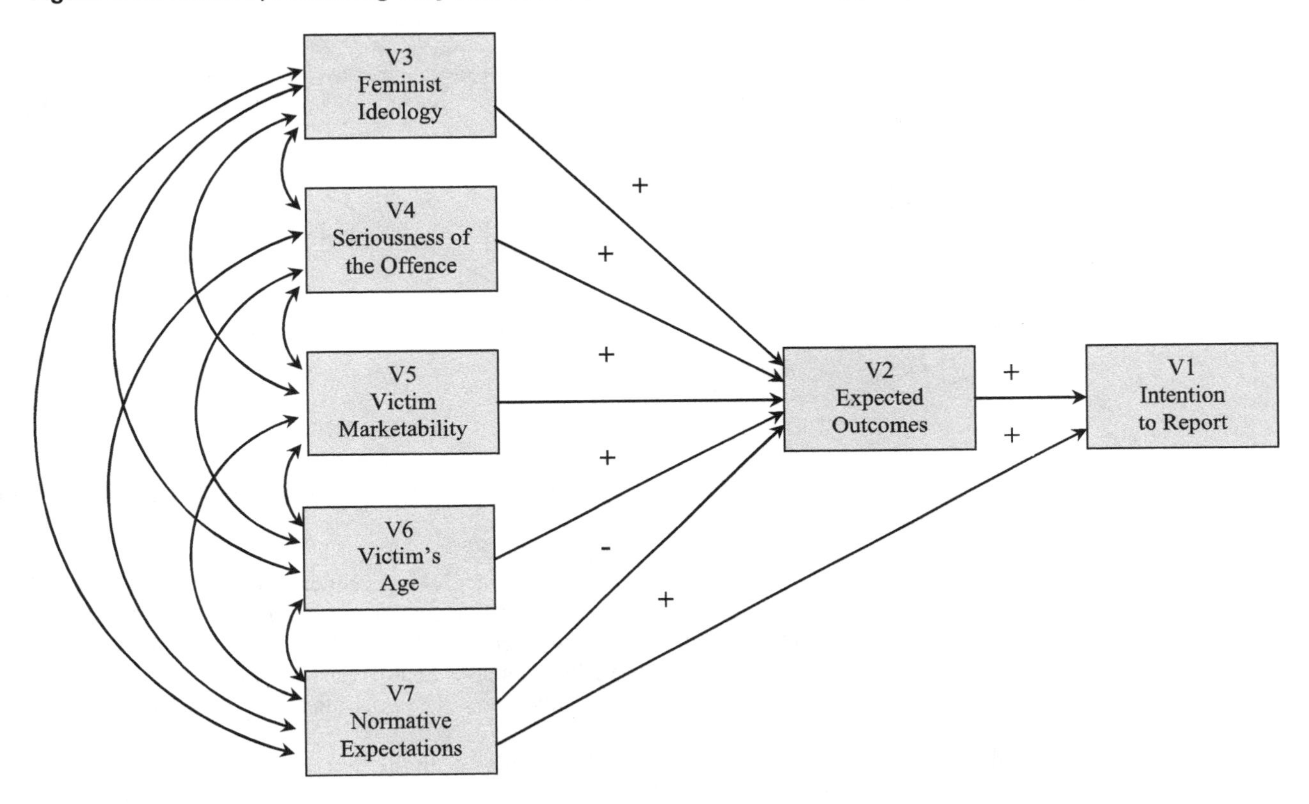

The path model assumed by the second group is identical to that proposed by the first, except that the direct path between feminist ideology and intent to report has been deleted. Feminist ideology is now believed to have only an indirect influence on intention to report via its effect on expected outcomes.

A third and final group of scholars contends that feminist ideology has no effect on the intention to report harassment in the workplace. They believe that the intention to report is directly or indirectly determined by the other variables presented in the preceding model; they assume that feminist ideology does not influence these intentions in any way. The model that represents this last school of thought is presented as Figure 4.19.

Figure 4.19: Model 3, Predicting No Effects of Feminist Ideology on Intention to Report

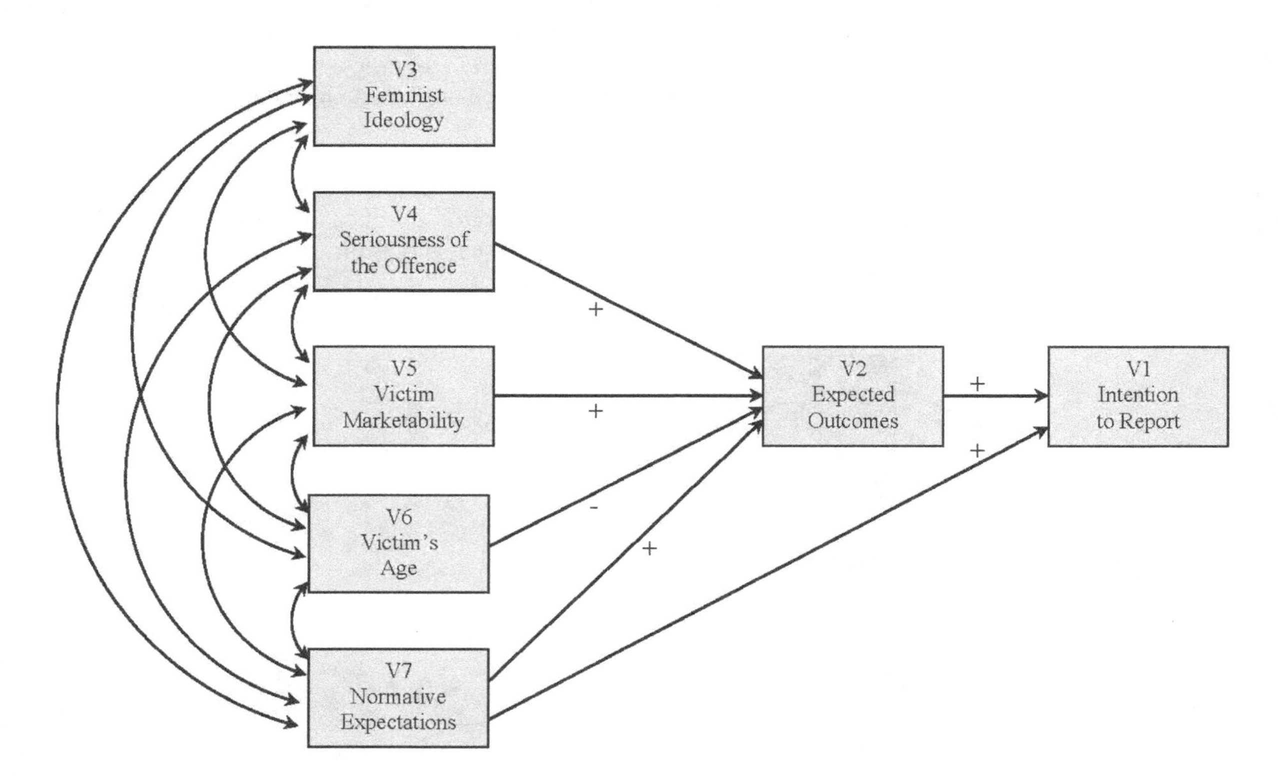

Notice that Figure 4.19 is identical to the Figure 4.18, except that the directional path from feminist ideology to expected outcomes has been deleted.

To our knowledge, there is no such controversy of this nature in the field of gender studies; if there were, this would be an ideal opportunity to test alternative hypotheses (here, presented as *a priori* path models). Earlier, it was noted that it is generally safer to test alternative theory-based models and identify the one with the best fit to the data than it is to begin with a single model and make extensive data-driven modifications to achieve a better fit, as the latter approach is less likely to result in a model that will generalize (MacCallum et al. 1992).

Evaluating these three models will involve conducting three analyses. In the first, a CALIS program will be written to test the first model, referred to here as the **direct and indirect effects model**. The program will then be modified to eliminate the direct path from feminist ideology to intention to report. The resulting model, to be called the **indirect effects model**, will then be re-estimated. If eliminating this path does not result in a significant decrease in fit between the model and data, this will provide support for the "indirect effects" hypothesis. Finally, the program will be modified once more to eliminate the path from feminist ideology to expected outcomes, producing a **no effects model**. If dropping this path does not result in a significant decrease in model fit, this will provide support for the final model.

The SAS Program

The PROC CALIS program that would analyze the direct and indirect effects model is presented below. The data analyzed here are from Murphy, Walters, and Hatcher (1993).

```
data D1(type=corr) ;
  input _type_ $ _name_ $ V1-V7;
  label
     V1 = REPORT
     V2 = EXPECTED_OUTCOMES
     V3 = FEMINIST
     V4 = SERIOUSNESS
     V5 = MARKETABILITY
     V6 = AGE
     V7 = NORMS   ;
datalines;
n      .    202      202      202      202      202      202      202
std    .  2.0355   1.4500   0.4393   2.1873   2.7433   4.0513   1.0552
corr  V1  1.0000    .        .        .        .        .        .
corr  V2   .4815   1.0000    .        .        .        .        .
corr  V3  -.0306    .0014   1.0000    .        .        .        .
corr  V4   .1458    .1683    .1148   1.0000    .        .        .
corr  V5   .0479    .1939    .0128    .0599   1.0000    .        .
corr  V6  -.0302   -.1165    .1479    .1061   -.0998   1.0000    .
corr  V7   .3952    .3700    .0512    .2486    .1275    .0606   1.0000
;
run;
proc calis   modification ;
   lineqs
      V1 = PV1V2 V2 + PV1V3 V3 + PV1V7 V7                         + E1,
      V2 = PV2V3 V3 + PV2V4 V4 + PV2V5 V5 + PV2V6 V6 + PV2V7 V7 + E2;
   variance
      E1 = VARE1,
      E2 = VARE2,
      V3 = VARV3,
      V4 = VARV4,
      V5 = VARV5,
      V6 = VARV6,
      V7 = VARV7;
   cov
      V3 V4 = CV3V4,
      V3 V5 = CV3V5,
      V3 V6 = CV3V6,
      V3 V7 = CV2V7,
      V4 V5 = CV4V5,
      V4 V6 = CV4V6,
      V4 V7 = CV4V7,
      V5 V6 = CV5V6,
      V5 V7 = CV5V7,
      V6 V7 = CV6V7;
   var  V1 V2 V3 V4 V5 V6 V7 ;
run;
```

The indirect effects model (in which the direct path from feminist ideology to intention to report has been deleted) would be identical to the preceding program, except that V3 (feminist ideology) would be removed as an independent variable in the LINEQS statement for V1 (intention to report). The resulting LINEQS statements would be as follows:

```
lineqs
   V1 = PV1V2 V2            + PV1V7 V7                        + E1,
   V2 = PV2V3 V3 + PV2V4 V4 + PV2V5 V5 + PV2V6 V6 + PV2V7 V7 + E2;
```

Similarly, the no effects model (in which the path from feminist ideology to expected outcomes has been deleted) would be identical to the preceding (modified) program except that V3 (feminist ideology) would be dropped as an independent variable in the LINEQS statement for V2 (expected outcomes). This is how the resulting LINEQS statements would appear:

```
lineqs
   V1 = PV1V2 V2             + PV1V7 V7                    + E1,
   V2 =          PV2V4 V4 + PV2V5 V5 + PV2V6 V6 + PV2V7 V7 + E2;
```

All other aspects of the CALIS program remain the same for the three analyses. In all three, V3 (feminist ideology) remains an exogenous variable that is allowed to covary with other exogenous variables. The only aspect of the model that was modified involves its relationship with intention to report and expected outcomes.

Results of the Analysis

Some of the results of the three analyses are presented in Table 4.5. Notice that, in contrast to Table 4.4, Table 4.5 presents the models in ascending order of complexity, with the null (uncorrelated variables) model at the bottom, and the model predicting both direct and indirect effects at the top. Either way of organizing this table is acceptable.

Table 4.5: Goodness-of-Fit Indices for Various Models, Sexual Harassment Study (N = 202)

Model	χ^2	df	$\Delta\chi^2$	Δdf	SRMR	CFI	RMSEA	(RMSEA CL_{90})
Direct/indirect effect	1.20	3	--	--	.012	1.00	.000	.(.082-.000)
Indirect effects	1.75	4	.55	1	.014	1.00	.000	(.070-.000)
No effects	1.77	5	.02	7	.015	1.00	.000	(.048-.000)
Null model	137.89	21	136.12**	16	--	--	--	

Note: χ^2 = chi-square; df = degrees of freedom; SRMR = Standardized Root Mean Square Residual; CFI = Comparative Fit Index; RMSEA = Root Mean Square Error of Approximation; RMSEA CL_{90} = RMSEA 90% Confidence Limits.

** $p < .01$

The goodness-of-fit indices for the direct and indirect effects model seem to suggest a very good fit to the data. The SRMR, CFI, and RMSEA statistics all each within optimal parameters and the confidence limits for the RMSEA are fully within acceptable limits. So is the model acceptable?

Not quite. The t tests for path coefficients (under the heading "Standardized Results for Linear Equations" from Output 4.16) reveal three nonsignificant parameters:

- the path coefficient from the effect of feminist ideology (V3) to intention to report (V1)
- the coefficient for the effect of feminist ideology (V3) on expected outcomes (V2)
- the coefficient for the effect of seriousness of the offense (V4) on expected outcomes (V2)

As you can see, it is quite possible for goodness-of-fit indices to be high even when the model contains nonsignificant paths.

Output 4.16: Output Page 7, Analysis of Direct and Indirect Effects Model, Sexual Harassment Study

Covariance Structure Analysis: Maximum Likelihood Estimation

Linear Equations																				
V1	=	0.5453	*	V2	+	0.4851	*	V7	+	1.0000		E1								
Std Err		0.0900		PV1V2		0.1237		PV1V7												
t Value		6.0569				3.9210														
V2	=	0.0599	*	V4	+	0.0698	*	V5	+	-0.0478	*	V6	+	0.4656	*	V7	+	1.0000		E2
Std Err		0.0439		PV2V4		0.0342		PV2V5		0.0231		PV2V6		0.0912		PV2V7				
t Value		1.3640				2.0377				-2.0652				5.1022						

Estimates for Variances of Exogenous Variables					
Variable Type	Variable	Parameter	Estimate	Standard Error	t Value
Error	E1	VARE1	2.95654	0.29492	10.02497
	E2	VARE2	1.72048	0.17162	10.02497
Observed	V3	VARV3	0.19298	0.01925	10.02497
	V4	VARV4	4.78428	0.47724	10.02497
	V5	VARV5	7.52569	0.75070	10.02497
	V6	VARV6	16.41303	1.63722	10.02497
	V7	VARV7	1.11345	0.11107	10.02497

Covariances Among Exogenous Variables					
Var1	Var2	Parameter	Estimate	Standard Error	t Value
V3	V4	CV3V4	0.11031	0.06822	1.61695
V3	V5	CV3V5	0.01543	0.08501	0.18146
V3	V6	CV3V6	0.26322	0.12690	2.07428
V3	V7	CV2V7	0.02373	0.03274	0.72494
V4	V5	CV4V5	0.35943	0.42400	0.84771
V4	V6	CV4V6	0.94020	0.62854	1.49583
V4	V7	CV4V7	0.57378	0.16775	3.42040
V5	V6	CV5V6	-1.10917	0.78781	-1.40792
V5	V7	CV5V7	0.36908	0.20583	1.79311
V6	V7	CV6V7	0.25906	0.30208	0.85758

Problems with these three paths were also revealed in the Wald tests included in the SAS output. When a path model contains paths or covariances that may be dropped without a significant decrease in fit, these are often identified by the Wald test. The Wald tests for the current analysis appear in Output 4.17.

Output 4.17: Output (Page 9), Wald Test Results, Sexual Harassment Study

Covariance Structure Analysis: Maximum Likelihood Estimation

Stepwise Multivariate Wald Test					
	Cumulative Statistics			Univariate Increment	
Parm	Chi-Square	DF	Pr > ChiSq	Chi-Square	Pr > ChiSq
CV3V5	0.03293	1	0.8560	0.03293	0.8560
CV2V7	0.53353	2	0.7659	0.50060	0.4792
CV6V7	1.10749	3	0.7753	0.57396	0.4487
CV4V5	1.81393	4	0.7699	0.70645	0.4006
PV2V4	3.67448	5	0.5972	1.86055	0.1726
CV4V6	5.64369	6	0.4643	1.96921	0.1605
CV3V4	7.23199	7	0.4051	1.58831	0.2076
CV5V6	10.00987	8	0.2643	2.77787	0.0956
CV5V7	12.99138	9	0.1630	2.98152	0.0842
CV3V6	16.06702	10	0.0977	3.07564	0.0795
PV2V6	19.83429	11	0.0477	3.76727	0.0523

Under the heading "Univariate Increment," the Wald test estimates the change in chi-square that would result from deleting a given parameter from the model. For example, the first line of the output reports the Wald test estimate for deleting the parameter "PV2V3." This is the path from V3 (feminist ideology) to V2 (expected outcomes). The Wald tests estimates that the model chi-square would change by less than .02 if PV2V3 were deleted. This is a very small change suggesting that you may safely delete this path without negatively affecting model fit. In general, the first parameter listed in the Wald test table is the parameter that could be deleted with the least effect upon model fit.

Remember that your study began by comparing the predictions of three groups of scholars who disagreed regarding the importance of feminist ideology in your model. The school proposing a direct and indirect effects model basically predicted that the model must include the path from feminist ideology to intention to report; that dropping this path would seriously harm the model's ability to account for relationships in the data. You can test this prediction directly by performing a chi-square difference test on chi-square values from Table 4.3. The model chi-square for the direct and indirect effects model was 1.20, while the chi-square for the indirect effects model was 1.75. The chi-square difference was therefore

$$1.20 - 1.75 = 0.55$$

The degrees of freedom for this test are equal to the difference between the degrees of freedom for the two models, or

$$4 - 3 = 1$$

The critical value of chi-square ($p < .05$) is 3.84; your obtained difference chi-square is well below that at 0.55. Therefore, you may conclude that there is not a significant difference in fit between the two models. In other words, deleting the path from feminist ideology to intentions did not significantly hurt the model's fit to data. Because the model without the path is the more parsimonious (and therefore the more desirable), you tentatively accept it and reject the "direct and indirect effects" model.

But your analysis is not complete. Is it possible to also drop the path from ideology to expected outcomes without significantly hurting the model's fit? To determine this, you compute a second chi-square difference test, this one comparing the indirect effects model to the no effects model.

Using the appropriate chi-square values from Table 4.5, this chi-square difference is calculated as

$$1.77 - 1.75 = 0.02$$

This obtained value is again well below the critical chi-square value of 3.84 so you know that you may drop this path without causing a significant decrease in model fit. Because the no effects model is the more parsimonious of the two, you tentatively accept it over the indirect effects model.

In short, if you are to believe these results, feminist ideology appears to have no influence on the intention to report harassment: It neither affects these intentions directly nor does it affect them indirectly by first affecting expected outcomes. An interesting feature of these results is the fact that the model still provides a good fit to the data even after deleting these two paths: Table 4.5 shows that the model chi-square for the no effects model is still quite small and nonsignificant; both the RMSEA and the SRMR are very small (with confidence limits fully within acceptable range) and the CFI statistic is very high.

Unfortunately, however, the model is still not quite perfect as the output from the analysis of the no effects model (not reproduced here) shows that the path from seriousness of the offense to expected outcomes remains nonsignificant. If this were an actual investigation, your specification search would continue. (Note, however, that the statistically significant difference in chi-square values between this and the null model in Table 4.3 suggests validity within the general framework of the model.)

IMPORTANT: Compare only nested models using the chi-square. It is hoped that this example has shown how the chi-square difference test can be used to make comparisons between competing *a priori* models. Remember, however, that these difference tests can only be used to compare **nested models**. As previously discussed, a model is said to be nested within another (e.g., fits inside) if it is identical to that model with the exception that one or more paths have been deleted. In this sense, you can see that the indirect effects model was nested within the direct and indirect effects model, and that the no effects model was nested within both of the other models. As previously mentioned, the Expected Cross Validation Index (ECVI, which is reported along with other Modification/ parsimony indices in SAS output) can be used to compare the likelihood of generalizability between competing non-nested models of similar size from the same population (Byrne 1998). Like the RMSEA, PROC CALIS reports 90% confidence limits for ECVI values. When comparing non-nested competing models, when one model has a lower ECVI value than the other, and the upper 90% confidence limit does not contain the ECVI value for the competing model, you can conclude with greater certainty that the this first model has much greater likelihood of replication with other samples from this population.

Conclusion: How to Learn More about Path Analysis

The best way to learn path analysis is to *do it.* Find a good text on path analysis, or perhaps some published research articles reporting path analytic studies, and replicate the analyses that they contain (e.g., Bachner et al. 2011). But be forewarned: Authors do not always indicate whether they analyzed the correlation matrix or the covariance matrix (or raw data), so it may be necessary to perform the analysis both ways in order to obtain results that match the published findings.

Now that you have had a concise introduction to the analysis of simple recursive models, you should be ready for a more in-depth treatment of path analytic procedures. This is particularly important if you want to test more complex models, such as nonrecursive models with reciprocal causation or feedback loops, or time-series designs with repeated measures.

The following two chapters build on the current chapter by introducing models with latent variables. Analyzing models with latent variables has many important advantages over path analysis; the increasing availability of software such as PROC CALIS capable of analyzing these models represents an important advance in applied social science research.

Note

1. The increased risk of developing schizophrenia with these zodiac signs is seen only in Northern Hemisphere. The association is due to the effects of less sunlight in winter months and reduced exposure to vitamin D *in utero* (Pulver et al. 1992). A different set of astrological signs are linked to the risk of developing schizophrenia in the Southern Hemisphere. This is an example of a *spurious association* (i.e., a statistically significant link between phenomena with no causal association). These spurious associations occur due to third variables, here, maternal exposure to less sunlight in winter.

References

American Psychological Association. (2009). *Publication manual of the American Psychological Association* (6th Ed.). Washington, DC: Author.

Anderson, J. C., and Gerbing, D. W. (1988). Structural equation modeling in practice: A review and recommended two-step approach. *Psychological Bulletin, 103,* 411–423.

Bachner, Y. G., O'Rourke, N., and Carmel, S. (2011). Fear of death, mortality communication and psychological distress among secular and religiously observant family caregivers of terminal cancer patients. *Death Studies, 15,* 163–187.

Bentler, P. M. (1990). Comparative fit indexes in structural models. *Psychological Bulletin, 107,* 238–246.

Bentler, P. M. (1989). *EQS structural equations program manual.* Los Angeles, CA: BMDP Statistical Software.

Bentler, P. M., and Bonett, D. G. (1980). Significance tests and goodness-of-fit in the analysis of covariance structures. *Psychological Bulletin, 88,* 588–606.

Bentler, P. M., and Chou, C. P. (1987). Practical issues in structural modeling. *Sociological Methods and Research, 16,* 78–117.

Berry, W. D. (1985). *Nonrecursive causal models.* Beverly Hills, CA: Sage.

Brooks, L., and Perot, A. R. (1991). Reporting sexual harassment: exploring a predictive model. *Psychology of Women Quarterly, 15,* 31–47.

Byrne, B. M. (2009). *Structural equation modeling with AMOS: Basic concepts, applications, and programming* (2nd Ed.). Mahwah, NJ: Lawrence Erlbaum.

Byrne, B. M. (1998). *Structural equation modeling with LISREL, PRELIS, and SIMPLIS: Basic concepts, applications, and programming.* Mahwah, NJ: Lawrence Erlbaum.

Cohen, J. (1992). A power primer. *Psychological Bulletin, 112,* 155–159.

Cohen, J. (1988). *Statistical power analysis for the behavioral sciences* (2nd Ed.). Hillsdale, NJ: Erlbaum.

Festinger, L. (1957). *A theory of cognitive dissonance.* Stanford CA: Stanford University Press.

Giles, H., Ryan, E.B., and Anas, A.P. (2008). Perceptions of intergenerational communication by young, middle-aged, and older Canadians. *Canadian Journal of Behavioural Science, 40,* 121-130.

Jöreskog, K. G., and Sörbom, D. (2001). *LISREL 8: User's reference guide [Computer software manual].* Lincolnwood, IL: Scientific Software.

Hu, L. T., and Bentler, P. M. (1999). Cutoff criteria for fit indices in covariance structure analysis: Conventional criteria versus new alternatives. *Structural Equation Modeling, 6,* 1–55.

Kane-Urrabazo, C. (2007). Sexual harassment in the workplace: It is your problem. *Journal of Nursing Management, 15,* 608–613.

Kenny, D. A. (1979). *Correlation and causality.* New York: John Wiley.

Kline, R. B. (2005). *Principles and practice of structural equation modeling.* (2nd Ed.). New York: Guilford Press.

Le, B., and Agnew, C. R. (2003). Commitment and its theorized determinants: A meta-analysis of the investment model. *Personal Relationships, 10,* 37–57.

Lehmann, E. L. (1999). *Elements of large-sample theory.* New York: Springer.

MacCallum, R. C. (1986). Specification searches in covariance structure modeling. *Psychological Bulletin, 100,* 107–120.

MacCallum, R. C., Roznowski, M., and Necowitz, L. B. (1992). Model modifications in covariance structure analysis: The problem of capitalization on chance. *Psychological Bulletin, 111,* 490–504.

MacCallum, R. C., Browne, M. W., and Cai, L. (2006). Testing differences between nested covariance structure models: Power analysis and null hypotheses. *Psychological Methods, 11,* 19–35.

MacCallum, R. C., Browne, M. W., and Sugawara, H. M. (1996). Power analysis and determination of sample size for covariance structure modeling. *Psychological Methods, 2,* 130–149.

McDonald, R. P., and Ho, M. R. (2002). Principles and practice in reporting structural equation analyses. *Psychological Methods, 7*, 64–82.

Murphy, L., Walters, K., and Hatcher, L. (1993). *Preferred strategies for dealing with sexual harassment: A role-playing experiment.* Paper presented at the Carolinas Psychology Conference, North Carolina State University, Raleigh, NC.

O'Rourke, N. (2000, August). *Dementia, caregiver burden, and marital aggrandizement: A path analytic model.* Paper presented at the annual meeting of the American Psychological Association, Washington, DC.

Pulver, A. E,. Liang, K. Y., Brown, C. H., Wolyniec, P., McGrath, J., Adler, L, Tam, D., Carpenter, W.T., and Childs, B. (1992). Risk factors in schizophrenia: Season of birth, gender, and familial risk. *British Journal of Psychiatry, 160,* 65–71.

Rusbult, C. E. (1980). Commitment and satisfaction in romantic associations: A test of the investment model. *Journal of Experimental Social Psychology, 16,* 172–186.

Sun, J. (2005). Assessing goodness of fit in confirmatory factor analysis. *Measurement and Evaluation in Counseling and Development, 37*, 240–256.

Tabachnick, B., and Fidell, L. (2012). *Using Multivariate Statistics* (6th Ed.). Toronto, ON: Pearson.

Chapter 5: Developing Measurement Models with Confirmatory Factor Analysis

Introduction: A Two-Step Approach to Analyses with Latent Variables

An important advance in social science research has been the development of statistical software for *analyses of covariance structures* (e.g., structural equation modeling). These models are referred to as covariance structure models or latent variable models. For consistency, this text uses the term **structural equation modeling** to describe models in which prediction of latent or unobserved variables is hypothesized. In contrast, path analytic models (Chapter 4) are composed of only observed variables (i.e., no latent variables). As described in the next two chapters, there are significant advantages to the use of structural equation models (where appropriate). This chapter presents confirmatory factor analysis, which is similar to structural equation modeling except that covariance (or correlation), not prediction, between latent variables is assumed (i.e., latent variables are connected by double-headed, not single-headed arrows)

This chapter and Chapter 6, "Structural Equation Modeling," show how PROC CALIS can be used to test these latent variable models. These chapters follow a two-step approach recommended by Anderson and Gerbing (1988). With this approach, the first step involves using confirmatory factor analysis to develop an acceptable measurement model. When testing a measurement model, you look for evidence that indicator variables effectively measure the underlying constructs of interest and that the measurement model demonstrates an acceptable fit to data. As noted above, measurement models do not specify directional relationships between the latent constructs; at this stage of the analysis, latent variables are allowed to correlate freely. The current chapter focuses on how to estimate measurement models, how to assess their psychometric properties, and how to modify them (when necessary) to achieve a better fit.

Chapter 6, "Structural Equation Modeling," builds on this foundation by showing how measurement models can be modified to predict specific relationships between latent variables. Among other things, performing this type of analysis allows you to test hypotheses that certain latent constructs predict other latent constructs.

You should therefore view Chapters 5 and 6 as a two-part introduction to analysis of covariance structures: Chapter 5 shows how to develop measurement models and Chapter 6 shows how to test the (theoretical) models of interest. Even if you are interested only in the topic of "Structural Equation Modeling" (Chapter 6), you will probably still need to read Chapter 5 to obtain the necessary foundational information (unless, of course, you are already familiar with PROC CALIS and latent variable models).

All models discussed in this and the next chapter are **recursive** models: In other words, none of the variables that constitute the structural portion of models are in feedback loops (i.e., reciprocal causation); the structural models discussed will describe unidirectional hypotheses only. A list of references is provided at the end of Chapter 6.

IMPORTANT: This chapter builds on material presented in Chapter 4, "Path Analysis." That chapter introduced basic terminology and concepts, and showed how to write programs for the CALIS procedure. With only a few exceptions, most of that introductory material will not be repeated here. It is assumed that you have completed the preceding chapter before beginning this one. If not, we recommend that you do so.

A Model of the Determinants of Work Performance

We begin with an illustration of a theoretical model that includes only manifest variables (similar to the path analytic models discussed in the preceding chapter). It will then show how models with latent variables differ from the manifest-variable or path models.

Figure 5.1 provides a model predicting the directional relationships among a number of variables related to work performance. The model includes two endogenous variables: (a) work performance, which is predicted to be directly determined by intelligence, motivation and supervisory support; and (b) motivation, which is said to be directly determined by workplace norms and supervisory support. The model includes three exogenous variables (intelligence, workplace norms, and supervisory support), which are expected to covary.

Figure 5.1: Path Model with Manifest Variables

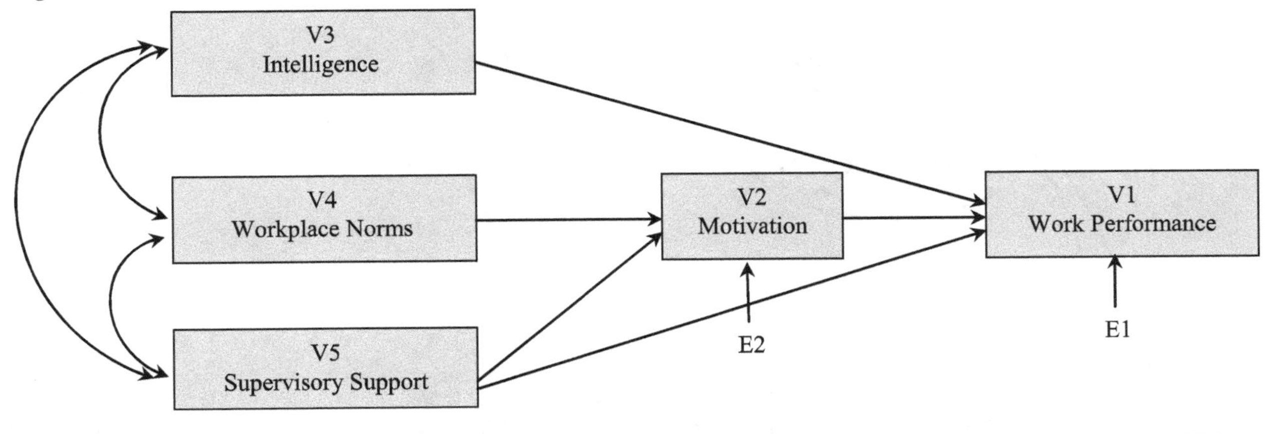

The Manifest Variable Model

This figure displays the manifest variable model presented in the preceding chapter. Recall that that **manifest variables** are variables that are directly measured, such as scores on an intelligence test or motivation scale. Manifest variables are sometimes referred to as **observed variables**, **measured variables**, or **indicator variables**. In path analysis presented in the preceding chapter, all variables in the model were manifest variables.

The Latent Variable Model

Figure 5.2, on the other hand, presents a model with latent variables. In this figure, latent variables are represented by ovals and the manifest variables by rectangles. **Latent variables** are sometimes referred to as **unobserved variables**, **unmeasured variables**, or **latent factors**. Notice that latent variables in the figure are identified by short names such as F1 and F2 (F stands for latent Factor). A latent variable is a **hypothetical construct** that cannot be directly observed. The existence of a latent variable can only be inferred by the way that it influences manifest variables that can be directly measured (more on this shortly).

Figure 5.2: A Structural Equation Model with Latent Variables

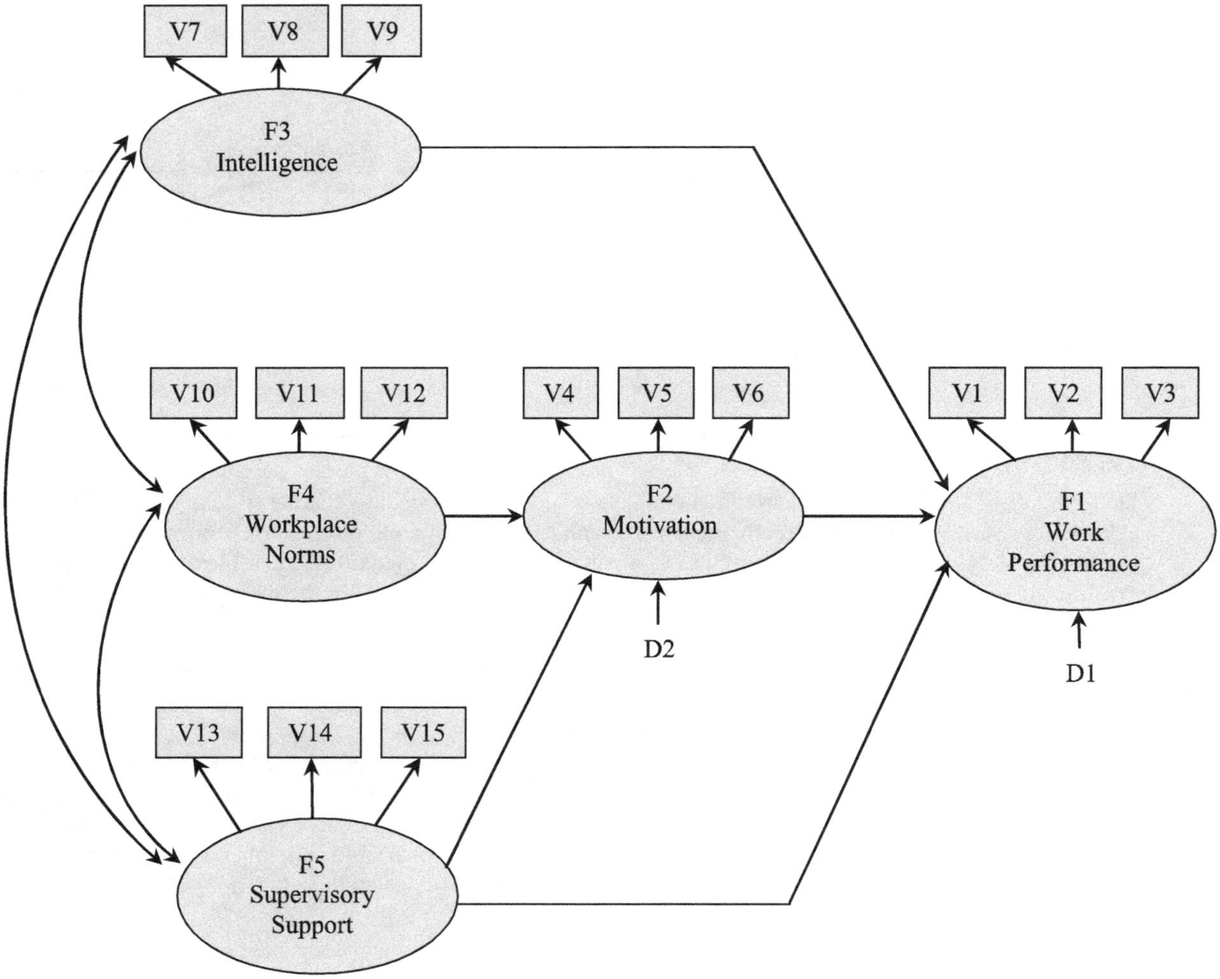

The paths connecting latent variables (ovals) in Figure 5.2 predict the same relationships among latent variables similar to the observed variables in the previous path model. Work performance is predicted by intelligence, motivation, and supervisory support; whereas motivation is predicted by workplace norms and supervisory support.

In this model, work performance is also influenced by a disturbance term, D1 (the "D" stands for **D**isturbance and the "1" corresponds to the "1" in F1). A **disturbance term** for a latent variable is interpreted the same as a residual term, or error term, for a manifest variable (from the preceding chapter). A disturbance term represents the effects on endogenous variables due to such things as omitted variables, measurement error, and misspecification of equations. For example, you would expect a large disturbance term for work performance if other constructs that have important effects on work performance have been omitted from the model. If you leave out variables that have only a minimal effect on work performance, the disturbance term for work performance should be relatively small.

Basic Concepts in Latent Variable Analyses

Latent Variables versus Manifest Variables

The model depicted in Figure 5.2 is referred to as a latent variable model. As previously described, a latent variable is a hypothetical construct that is not directly observed; instead, its existence is inferred from its influence on manifest variables (or other latent variables).

For example, consider the latent variable "intelligence" (F3) in Figure 5.2. The arrows in the figure suggest that this latent factor affects three manifest variables labeled V7, V8, and V9. Assume that these three manifest variables are three different facets of intelligence. For example, V7 may be participants' scores on a measure of analytical problem solving ability, V8 may be scores on measure of creative problem solving, and V9 may be scores on a measure of practical problem solving (Sternberg 1994). If there really is some underlying construct that could reasonably be labeled "intelligence" (and if these three measures are valid), you would expect to see certain results when PROC CALIS is used to analyze your data. For example, the coefficients for the paths going from F3 (intelligence) to V7, V8, and V9 should be relatively large and statistically significant. These findings would provide support for the assumption that there is an underlying intelligence construct which is measured by your manifest variables.

Notice that the model in Figure 5.2 identifies five different latent F variables and indicates which indicator (manifest) variables are expected to contribute to measurement of each. Variables V1 through V3 measure work performance, V4 through V6 measure motivation, and so forth. It is important to remember that the indicator or manifest variables—the variables represented by rectangles—are the variables that you actually gather during the study (e.g., questionnaires completed by participants). Although the hypothetical constructs (represented by ovals) generally represent the variables of greatest interest, they are not directly measured.

Choosing Indicator Variables

A variety of variables may be used as indicators so long as they meet certain conditions discussed below (in the "Necessary Conditions for Confirmatory Factor Analysis and Structural Equation Modeling" section). In the present study, for example, objective measures would be used as indicators of latent variables. In measuring the latent variable "work performance," the indicator V1 could be units produced by participants per hour, V2 could be the quality of units produced, and V3 could be the number of times per month each participant exceeds his or her production quota.

Subjective measures may also be used as indicator variables. In measuring motivation, V4 could be responses to a self-report questionnaire assessing participants' work motivation, V5 could be ratings of the participants' motivation reported by their direct supervisors, and V6 could be ratings of the participants' motivation reported by co-workers.

In designing a study, care should be taken when selecting indicator variables to measure each latent variable. Indicator or manifest variables should be appropriate for use with your population of interest (e.g., supported by theory). Using previously developed instruments is almost always preferred. New measures should only be used if no existing scales to measure that construct currently exist (unless, of course, the primary intent of the study is the development of a new measure; e.g., Chou and O'Rourke 2012).

Regardless of what type of indicator variables you use, remember that you will be able to perform meaningful tests of your directional model only if responses to these indicators show certain psychometric properties. Specifically, it is essential that responses to indicator or manifest variables chosen to measure the same latent construct show **convergent validity**: They must all measure the same underlying construct. In practical terms, this means that the indicators should be moderately correlated with one another. What is more, groups of variables that are intended to measure different latent constructs should display **discriminant validity**. Indicators intended to assess one latent variable (say, F1) should not at the same time measure a different latent variable (say, F2). In practical terms, this means that if V1 through V3 are measuring F1, and V4 through V6 are measuring F2, responses to V1 through V3 should not be strongly correlated with V4 through V6.

Later, you will learn how to assess the convergent and discriminant validity of measurement models in a systematic way. However, these procedures will bring only frustration if you are careless in the initial selection of your indicator variables. Wherever possible, these decisions should be informed by existing theory and/or psychometric research.

The Confirmatory Factor Analytic Approach

The procedures to be discussed in this chapter pertain to confirmatory factor analysis (CFA). These procedures differ in important ways from those discussed in Chapter 2 on exploratory factor analysis. With exploratory factor analysis, you are often unsure of the number of factors being measured; the results of the analysis help to resolve the number-of-factors question. With the procedures described here, however, you not only have a good idea regarding the number of factors being assessed (latent variables), but also which manifest variables load on which factors.

The Measurement Model versus the Structural Model

You should think of the model presented in Figure 5.2 as consisting of two components. The measurement portion of the model, generally referred to as the **measurement model**, describes the relationships among latent factors and their indicator variables. The measurement model is said to provide a good fit to data if V1 through V3 effectively measure work performance (F1), if V4 through V6 effectively measure motivation (F2), and so forth. In this chapter, when manifest variables are used as measures of latent factors, they will be referred to as **indicator variables**.

On the other hand, the structural portion of the model (referred to as the **structural model**) describes the predicted relationships among constructs of central interest. In Figure 5.2, the structural model consists of F1, F2, F3, F4, and F5 as well as the paths that connect them. In this chapter, the variables that constitute this structural model will be called **structural variables**.

Advantages of Covariance Structure Analyses

Covariance structure analyses (e.g., confirmatory factor analyses) have at least two important advantages over path analysis with only manifest variables. First, as was mentioned earlier, the latent variable approach allows researchers to assess the convergent and discriminant validity of their measures. If the proposed measurement model fares well with regard to the tests to be later discussed, support is found for the construct validity of responses to manifest variables. This provides evidence that you really are studying the hypothetical constructs of interest. This is important, because most social science research offer no evidence concerning the construct validity of their variables.

Second, structural equation modeling enables the researcher to estimate measurement error of latent variables. In Chapter 4, it was mentioned that path analysis assumes that all variables are measured without error. This is to say the manifest variables are assumed to be perfectly reliable indices of the constructs they are intended to measure. Needless to say, this assumption is rarely if ever justified with the types of scales used in the social sciences.

For example, assume that you were to perform the work performance study previously described using path analysis. In addition, assume that you use just a measure of analytical problem solving as the sole measure of intelligence. Although responses to your measure of analytical problem solving may be reliable with your population of interest, its reliability is not perfect. Responses will be determined, in large part, by the underlying construct of intelligence.You also assume that responses to your measure will not be a perfect representation of their underlying level of intelligence; some variability in scores will be due to errors of measurement (either random or systematic). This creates a problem because you do not really want to study the relationship between analytical problem solving and work performance; you want to study the relationship between the underlying construct of *intelligence* and work performance. And because analytical problem solving provides an imperfect measure of intelligence, the path coefficients that you obtain from a single-indicator path analysis will likely be biased (Netemeyer, Johnston, and Burton 1990).

Structural equation modeling redresses this limitation by separating error variance from the core measurement of latent variables by estimating error separately. This is illustrated in Figure 5.3 in which residual terms (indicated by the symbol E for **E**rror term) have been added to the model. For illustrative purposes, consider the manifest variables V7 through V9, and assume that these variables are three different aspects of intelligence. The paths leading from the latent construct F3 (intelligence) to V7, V8 and V9 represent the assumptions that variability in responses to these three measures will be determined, in part, by the hypothetical construct of intelligence. Notice also that an arrow points from the residual term E7 to V7. This represents the assumption that variability in V7 is influenced by factors in addition to intelligence such as random and systematic error. This is the component of V7 that is not shared in common with V8 and V9, and so it is modeled separately. In the same way, E8 represents measurement error for V8, and E9 represents measurement error for V9.

As a result, the latent variable F3 consists only of common or shared variance among V7, V8, and V9. Because the error components have been estimated separately, latent variable F3 is more reliable (though not perfectly reliable) as compared to responses to a single manifest variable.

Figure 5.3: Structural Equation Model with Residual Terms for Manifest Variables

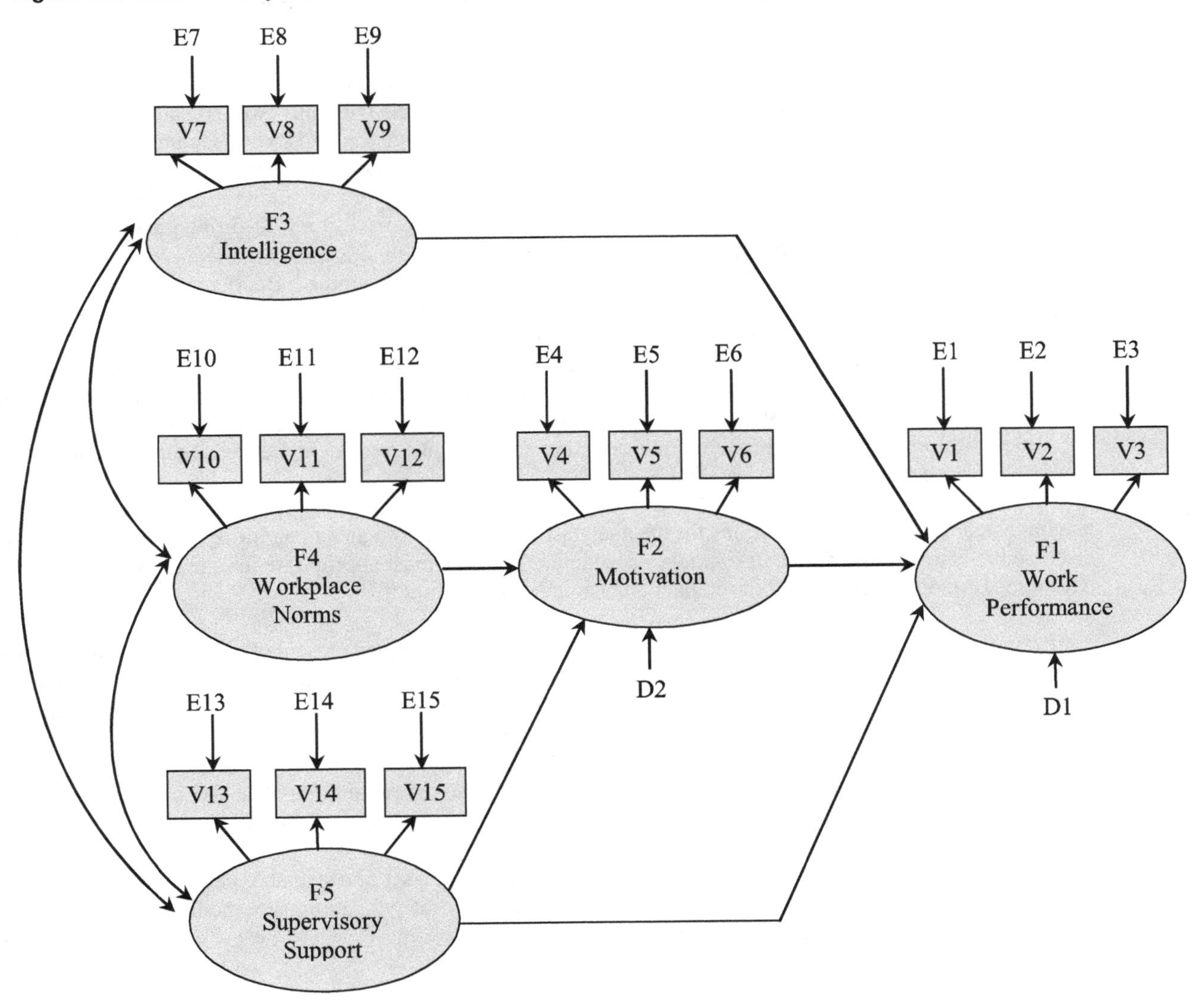

Necessary Conditions for Confirmatory Factor Analysis

The preceding chapter presented a number of requirements that must be met in order to perform path analysis; many of those conditions apply to latent variable analyses as well. These are briefly summarized below. Most of the details explained in the preceding chapter will not be repeated here:

1. **Normally, Dichotomous and Categorical data.**
2. **Linear and additive relationships.** This applies only to the types of analyses discussed in this text. More advanced texts show how to test models with nonlinear relationships; PROC CALIS is capable of testing these models as well.
3. **Absence of multicollinearity** (e.g., r < .80 between observed variables).
4. **Inclusion, within the model, of all nontrivial variables.**
5. **Overidentified model.**

In addition to the preceding, the analysis of latent variable models also requires that the following conditions be met:

6. **At least three indicator variables per latent factor.** Technically, a latent factor may be assessed with just two indicators under certain conditions. However, models with only two indicator variables per factor often exhibit problems with identification and convergence, and so it is recommended that each latent variable be assessed with at least three indicators (Anderson and Gerbing 1988; Byrne 1998). In practice, researchers are well-advised to have at least four or five indicators for each latent factor as it is often necessary to drop some of the indicators in order to arrive at a well-fitting measurement model.
7. **A maximum of 20–30 indicator variables.** One of the limitations of this model involves the maximum number of indicator variables that can be effectively studied. Bentler and Chou (1987) caution that it is easy to become too ambitious when developing structural models, and advise that researchers who lack a great deal of knowledge about the variables of interest should work only with 30 indicator variables or less. Larger numbers of indicator variables often lead to inability to fit your model to the data. Note that this limitation also affects the maximum number of latent factors that may be effectively studied. Researchers who "play it safe" by measuring each latent variable with four indicators each will be limited to investigating a total of just five latent factors (assuming that the study includes just 20 manifest variables).

Having listed the necessary conditions for the analysis, it is important to again emphasize that the procedures discussed in is this chapter are *confirmatory* procedures. They are most appropriate for situations in which you have a fairly good understanding of the phenomena under investigation.

For example, assume that you want to test the model of work performance presented in Figure 5.2. Ideally, your research began with a thorough review of the literature pertaining to all constructs in the model. For instance, you obtained previously developed instruments to measure these constructs appropriate for use with your population of interest. You administered all instruments to participants in a pilot study; and you performed exploratory factor analyses to determine the number of factors assessed, how the factors are associated with one another, and how variables load on their respective factors. In the course of doing this, you may have discovered that responses to some of your instruments were not sufficiently reliable and had to be replaced.

Through this series of exploratory studies, you eventually arrive at a set of acceptable measures and a measurement model that provides a good fit to data. You are now ready to administer the instruments to a new sample and perform confirmatory analyses.

The preceding emphasizes the importance of developing a reasonably good measurement model before obtaining data from additional participants for confirmatory analyses. Many researchers omit this crucial step and learn too late that they cannot test the structural model of interest because their measurement model provides a poor fit to data. In this situation, some researchers may be tempted to perform exploratory factor analyses on their data to discover the factor structure, revise their model to reflect this structure, and then perform confirmatory analyses on the same dataset to test the revised theoretical model. However, performing both exploratory and confirmatory analyses on the same dataset is likely to capitalize on chance characteristics of the sample, and may lead to a "final" model that will not generalize to other samples or to the population. (For more details on this issue, review the "Modifying the Model" section in the preceding chapter.)

Sample Size Requirements for Confirmatory Factor Analysis and Structural Equation Modeling

One subject of considerable confusion pertains to necessary sample sizes required for procedures such as confirmatory factor analysis and structural equation modeling. Because these procedures are based on *large sample theory* (Lehmann 1999), various authors have asserted that minimum sample sizes are required to reliably conduct and report study findings. For instance, a minimum of 200 observations for confirmatory factor analysis has been commonly recommended; however, others have suggested that a threshold of 300 observations is more appropriate (see Floyd and Widaman 1995, for further discussion of this topic). Although this may be a *necessary* condition for confirmatory factor analysis, it would be incorrect to state that this was also a *sufficient* condition. In other words, a minimum number of observations may be required to meet the assumptions of large sample theory; however, factors specific to individual models may necessitate more than 200 observations are required (MacCallum, Widaman, Zhang, and Hong 1999).

What authors commonly neglect to recognize is that the number of model degrees of freedom are also integral when determining sample size requirements. This is because a sufficient number of degrees of freedom is also required to draw correct conclusions about model fit. Failure to reject poor fitting models (Type II errors) occurs when the number of available degrees of freedom (relative to sample size) leads to imprecise goodness-of-fit indices. These measures provide imprecise indicators of fit within the population from which the sample is drawn (MacCallum et al. 1999).

To address this sample size issue, MacCallum, Browne, and Sugawara (1996) recommend that confidence intervals surrounding the Root Mean Square Error of Approximation (RMSEA) be considered when testing the null hypothesis of model non-fit. As you will recall from the previous chapter, PROC CALIS reports both the RMSEA and the upper and lower bounds of the 90% confidence interval for the RMSEA. The RMSEA provides an estimate that a model fits in the population and the interval over which this estimated value is likely to fit the population nine times out of ten. Previously, we noted that RMSEA values less than .055 are suggestive of close model fit, whereas values between .055 and .08 suggest adequate model fit.

Based on these parameter values, MacCallum and colleagues (1996) provided the following SAS program to estimate the *statistical power* of a model (i.e., the ability reject poor fitting models). A full discussion of the topic of statistical power is beyond the scope of this text (Cohen 1988). Suffice to say that a certain number of observations relative to a model's degrees of freedom are required in order to correctly reject poorly fitting latent variable models. It is generally held that statistical power should be equal to, or greater than, .80 (Cohen 1988). As with other statistical procedures, an alpha value of .05 is used as a generally accepted threshold of statistical significance (i.e., $\alpha \leq .05$).

Calculation of Statistical Power

The general form of the SAS program to perform power analysis for analyses of covariance structures is presented below.

```
data D1 ;
   alpha=.05 ;
   rmsea0=.05 ;
   rmseaa=.08 ;
   df = degrees of freedom ;
   n = sample size ;
ncp_null=(n-1)*df*rmsea0**2 ;
ncp_alternate=(n-1)*df*rmseaa**2 ;
if rmsea0 < rmseaa then do ;
   cval=cinv(1 - alpha, df, ncp_null) ;
   power = 1-probchi(cval, df, ncp_alternate) ;
end ;
if rmsea0 > rmseaa then do ;
   cval=cinv(alpha, df, ncp_null) ;
   power=probchi(cval, df, ncp_alternate) ;
end ;
output;
proc print data=D1 ;
   var rmsea0 rmseaa alpha df n power;
run ;
```

You only need to insert the sample size (n) and available degrees of freedom (df) in the above program in order for this SAS program to estimate statistical power for a given model. We appreciate that some of this code has not been previously used in this text and is unfamiliar to most users. Our intent is not to demonstrate the specifics of this programming language but to assist you in determining whether a model has sufficient power to reject poorly fitting models. A specific example will help clarify this procedure.

After working your way through the remainder of this chapter, imagine that you obtained responses from 360 participants and computed a confirmatory factor analytic model with 26 degrees of freedom. This means that there are 26 associations within your covariance matrix for which estimates of association are not calculated (i.e., no associations assumed). After placing these values into the previous SAS program (i.e., n = 360, df = 26), power is estimated at .81. This value suggests good statistical power.

Output 5.1: Estimate of Statistical Power for Hypothetical Confirmatory Factor Analytic Study, n = 360, df = 26

Obs	rmsea0	rmseaa	alpha	df	n	power
1	0.05	0.08	0.05	26	360	0.81070

Of course the best time to estimate whether or not a model will have sufficient statistical power is prior to undertaking the study. In other words, the researcher should estimate the number of participants who need to be recruited in order to reject poorly fitting confirmatory factor analytic models. Even when undertaking analyses of existing data (i.e., secondary data analysis), it is a good idea to first determine if the numbers are sufficient for analyses based on large sample theory. If the sample size is inadequate, it may be necessary to collect new data if confirmatory factor analysis is required to properly answer your research question.

Calculation of Sample Size Requirements

The general form of the SAS program to compute sample size estimates for analyses of covariance structures is presented below (e.g., confirmatory factor analysis); this code is again adapted from MacCallum and colleagues (1996). Upper and lower bounds for RMSEA values provide a range in which a minimal sample size is estimated; this is halved until a close approximation to the desired sample size is determined (MacCallum et al.

1996). Once again, we do not anticipate that this SAS code will necessarily be recognizable to the novice user. It is provided here to provide you with a tool to proactively estimate sample size requirements.

```
data one ;
rmsea0=.05 ; *null hypothesis value ;
rmseaa=.08 ; *alternate hypothesis value ;
df= *degrees of freedom ;
alpha=.05 ; *alpha level ;
powd=.80 ; *desired power ;

*initialize values ;
Powa=0.0 ;
n=0 ;
*begin loop for finding initial level of n ;
do until (powa>powd) ;
n + 100 ;
ncp0=(n-1)*df*rmsea0**2 ;
ncpa=(n-1)*df*rmseaa**2 ;
*compute power ;
   if rmsea0>rmseaa then do ;
   cval=cinv(alpha,df,ncp0) ;
   powa=probchi(cval,df,ncpa) ;
   end ;
   if rmsea0<rmseaa then do ;
   cval=cinv(1-alpha,df,ncp0) ;
   powa=1-probchi(cval,df,ncpa) ;
   end ;
   end ;
* begin loop for interval halving ;
dir=-1 ;
newn=n ;
intv=200 ;
powdiff=powa-powd ;
do until (powdiff<.001) ;
intv=intv*.5 ;
newn + dir*intv*.5 ;
*compute new power ;
ncp0=(newn-1)*df*rmsea0**2 ;
ncpa=(newn-1)*df*rmseaa**2 ;
*compute power ;
   if rmsea0>rmseaa then do ;
   cval=cinv(alpha,df,ncp0) ;
   powa=probchi(cval,df,ncpa) ;
   end ;
   if rmsea0<rmseaa then do ;
   cval=cinv(1-alpha,df,ncp0) ;
   powa=1-probchi(cval,df,ncpa) ;
   end ;
powdiff=abs(powa-powd) ;
if powa<powd then dir=1; else dir=-1 ;
end;
sample=newn ;
sample=int(sample) ;
if sample lt 200 then do sample = 200 ;
end ;
proc print data=one ;
var rmsea0 rmseaa powd alpha df sample powa ;
run ;
```

Using the previous example in which the model had 26 degrees of freedom, by inserting this value in the above program, we see that a sample size of 351 is required in order to have the desired power of .80 (Cohen 1988). See Output 5.2. This should make sense as the previous program with 360 participants was estimated to have a power of .81 (i.e., slightly larger sample size). We should stress that this is the final sample size used to analyze this hypothetical confirmatory factor analytic model. It is good practice to recruit a sample that is at least 10% greater than your final sample size requirement to allow for missing data, outlying cases, etc., which should be removed from datasets before undertaking statistical analyses. In this example, therefore, it would be good research practice to recruit at least 386 participants so that you would be left with a final sample of more than 351 participants.

Output 5.2: Estimate of Sample Size Requirements to Achieve Adequate Power for Hypothetical Confirmatory Factor Analytic Study, df = 26

Obs	rmsea0	rmseaa	powd	alpha	df	sample	powa
1	0.05	0.08	0.80	0.05	26	351	0. 80066

Although the number of degrees of freedom is the only value that needs to be inserted in this SAS program, it is feasible to also adjust both the alpha and power levels; we recommend doing neither. Thorough discussion of statistical power is beyond the scope of this text. The interested reader is encouraged to see Cohen (1988).

Example: The Investment Model

In previous chapters, we made reference to Rusbult's (1980) theory of relationship commitment called the investment model (Le and Agnew 2003). It was noted that one possible interpretation of the investment model predicts that **commitment** to a relationship (e.g., the intention to remain in the relationship) is determined by:

- **satisfaction** with the relationship
- size of personal **investments** (e.g., time, energy) put into the relationship
- **attractiveness of alternatives** to the relationship (e.g., the attractiveness of other potential partners)

Satisfaction, in turn, is expected to be determined by the **rewards** experienced in the relationship (e.g., the good things associated with it) as well as **costs** (e.g., hardships, unpleasant things).

The Theoretical Model

Figure 5.4 presents a latent variable model that illustrates these hypothesized relationships. The structural portion of the model consists of the ovals (latent variables) and the paths that connect them. You can see that commitment (F1) is believed to be directly determined by satisfaction (F2), investments (F5), and alternatives (F6); while satisfaction (F2) is expected to be determined by rewards (F3) and costs (F4).

The measurement portion of the model consists of single-headed arrows from latent variables to the manifest variables that measure them (manifest variables are represented by rectangles). For example, manifest indicators V1, V2, V3, and V4 are predicted to measure F1 (commitment); indicators V5, V6, and V7 are predicted to measure F2 (satisfaction), and so forth.

Research Method and Overview of the Analysis

For purposes of illustration, assume that you have developed a 19-item instrument to assess the six constructs constituting the investment model. All questionnaire items use a 5-point Likert-type response format; items 1 through 4 assessed commitment to their current relationship, items 5 through 7 assessed satisfaction, and so forth. The questionnaire was administered to 247 participants currently involved in romantic relationships; usable responses were obtained from 240 of these. (It is again emphasized that the results presented here are fictitious, and should not be viewed as legitimate tests of the investment model.)

In large part, the analyses reported in this chapter and the next will follow a two-phase procedure (Anderson and Gerbing 1988). With this approach, you begin by developing a measurement model that provides an acceptable fit to the data. In this phase, each latent F variable is allowed to covary with every other latent F variable; the analysis is essentially confirmatory factor analysis. If the initial measurement model is inadequate, variables are reassigned or deleted in order to attain a better fit.

Figure 5.4: The Theoretical Model to Be Tested

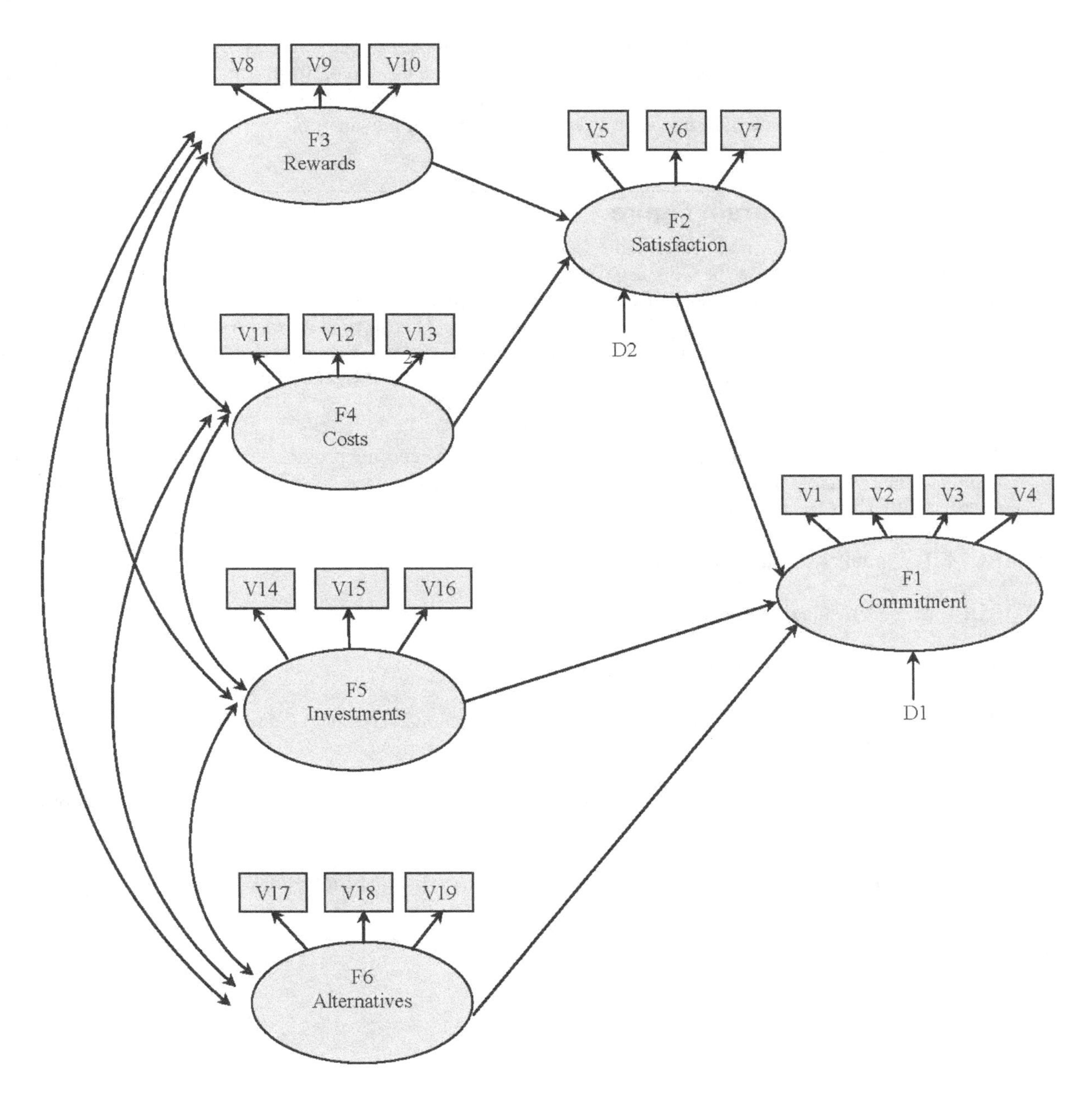

Once an acceptable measurement model has been developed, the analysis moves to a second phase in which the theoretical model itself is tested. This is done by fixing at zero the covariances between some of the F variables in the measurement model, and replacing some other covariance estimates with unidirectional paths so that the relationships between latent variables come to reflect the directional relations to be tested. Testing the resulting theoretical model for goodness of fit allows a simultaneous test of the measurement model developed in phase one, along with the structural model that is of primary substantive interest (e.g., your hypotheses usually pertain to these associations). If the theoretical model survives this test, support for the theory is obtained. If the theoretical model does not survive, it may be modified to attain a better fit.

Testing the Fit of the Measurement Model from the Investment Model Study

Confirmatory factor analysis (CFA) is used to test the fit of the measurement model; PROC CALIS can perform these analyses. In many ways, the program is similar to the one used to perform path analysis as presented in the preceding chapter. Among other things, the program will include one functional equation for each manifest variable. These equations will define each V variable in terms of the latent factor (F variable) that it is believed to measure, as well as its residual term (E term).

Preparing the Program Figure

The typical CALIS program to perform this confirmatory factor analysis is longer than that which performs a path analysis; this is because the CFA usually involves more variables and thus more equations. Writing this program will usually be much easier if you first prepare a **program figure** that identifies latent variables and their indicators, residual terms, and all estimated parameters. This section shows how to prepare a program figure following the same procedure used in the last chapter. Because the general steps followed in preparing a program figure were described in detail in Chapter 4 (in the "Preparing the Program Figure" section), they will be covered more briefly here.

The previous chapter presented a list of rules to guide you in preparing program figures; most of these rules also apply to latent variable models. (There will be a few exceptions to these rules, and they will be discussed where appropriate.) The rules are presented again below for purposes of reference:

RULE 1: Only exogenous variables are supposed to have covariance parameters in the model

RULE 2: A residual term is identified for each endogenous variable in the model

RULE 3: Exogenous variables do not have residual terms.

RULE 4: Variances should be estimated for every exogenous variable in the model, including residual terms.

RULE 5: In most cases, there should be covariance estimates for every pair of manifest or exogenous variables; covariance estimates are not required for endogenous variables.

RULE 6: For simple recursive models, covariance is not estimated for residual terms.

RULE 7: One equation should be created for each endogenous variable with that variable's name to the left of the equals sign.

RULE 8: Variables that have a direct effect on that endogenous variable are listed to the right of the equals sign.

RULE 9: Exogenous variables, including residual terms, are never listed to the left of the equals sign.

RULE 10: To estimate a path coefficient for a given independent variable, we recommend that a unique path coefficient name be created for the path coefficient associated with that independent variable.

RULE 11: The last term in each equation should be the residual (disturbance) term for that endogenous variable; this E (or D) term will have no name for its path coefficient.

RULE 12: To *estimate* a path coefficient , create a name for that parameter.

RULE 13: To *fix* a parameter at a given numerical value, insert that value in the place of the parameter's name.

RULE 14 To *constrain* two or more parameters to be equal, use the same name for those parameters.

Step 1: Drawing the Basic Confirmatory Factor Model

Figure 5.5 presents the basic measurement model to be tested in this phase of the study. As with the theoretical model presented earlier, the model consists of five latent variables or factors: Commitment (F1), satisfaction (F2), rewards (F3), costs (F4), investments (F5), and alternatives (F6). F1 has four manifest indicator variables, which are represented by rectangles. Remember that these indicator variables are responses to questionnaire items. The remaining factors each have three indicators. Each factor is connected to every other factor by a curved two-headed arrow, meaning that every factor is allowed to covary with every other factor.

Figure 5.5: The Initial Measurement Model

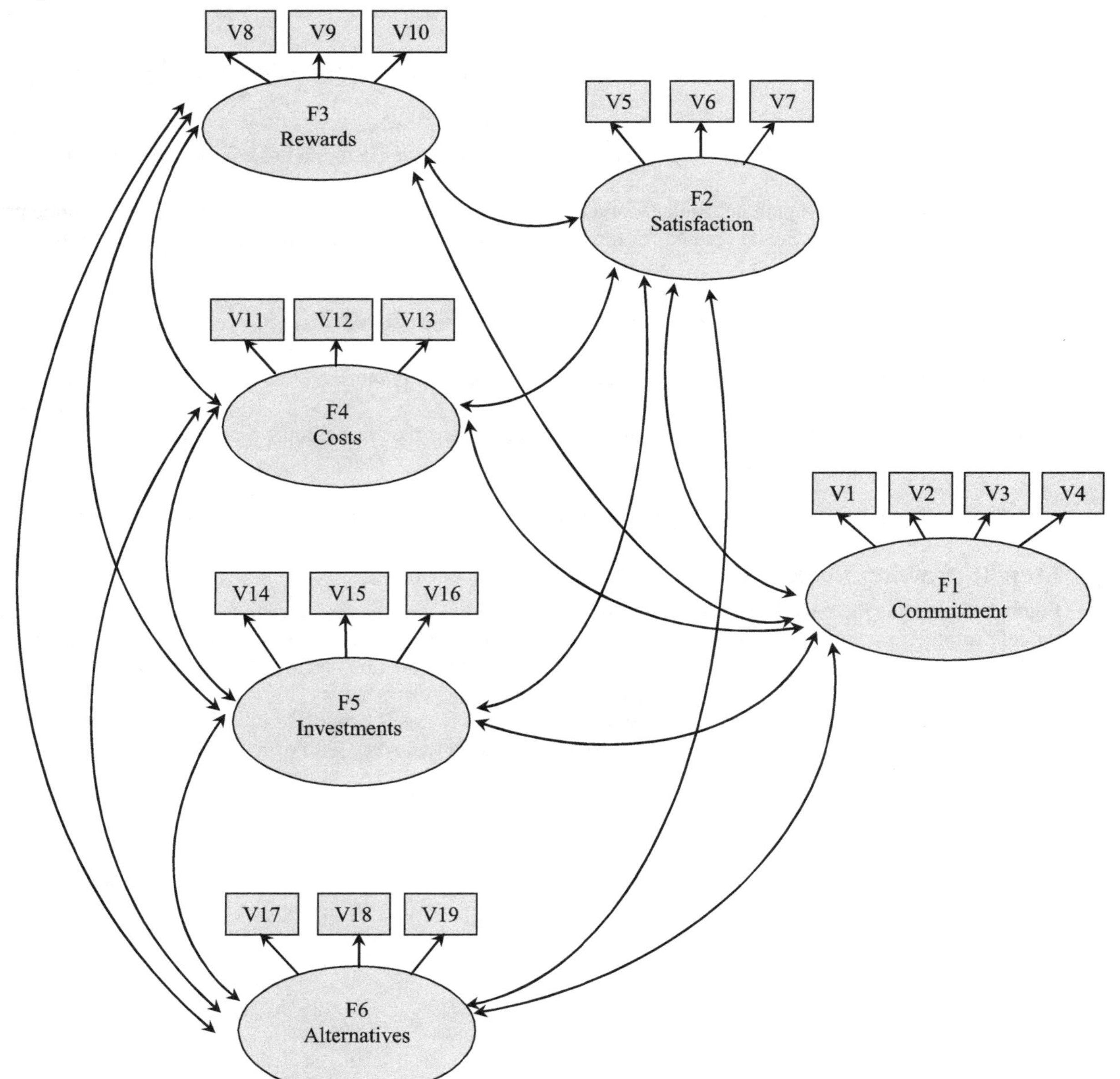

Notice that each indicator variable is assumed to load on only one factor; in other words, there are no **complex variables** (indicators measuring more than one latent variable). Notice also that there are no covariance estimates assumed between any of the indicators. This is because only exogenous variables are supposed to have covariance parameters in the model. (You know that indicator variables are endogenous variables, because a straight, one-headed arrow points to each of them.)

Step 2: Identifying Residual Terms for Endogenous Variables

Rule 2 states that a residual term must be identified for each endogenous variable in the model. You know that all of the indicator variables are endogenous variables because each is affected by F variables. Therefore, a residual term must be created for each indicator. This is illustrated in Figure 5.6.

Figure 5.6: The Initial Measurement Model, Including Residual Terms for Endogenous Variables

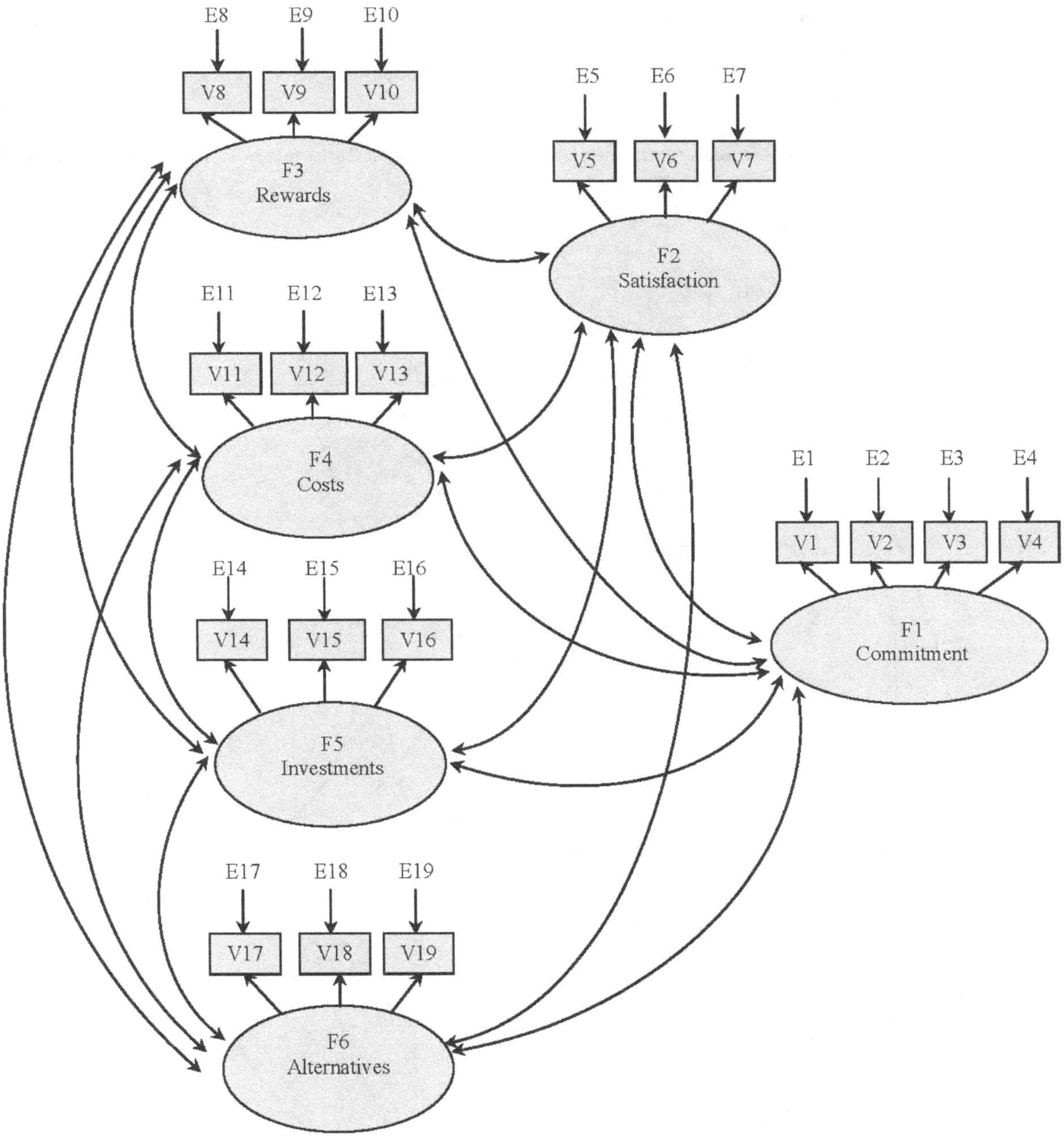

This figure follows the same conventions used in the last chapter; the names for residual terms begin with the letter "E" and end with the same numerical suffix used as the short name for the corresponding indicator. For example, the residual for V1 is E1, the residual for V2 is E2, and so forth.

Figure 5.6 illustrates assumed associations in the confirmatory factor model; you can see that all are relatively simple. Each indicator is affected only by the underlying common factor (F) on which it loads as well as its residual term. That is to say, V1 is affected only by F1 and E1, V5 is affected only by F2 and E5, and so forth.

Step 3: Identifying All Parameters to Be Estimated

Figure 5.7 illustrates all of the parameters to be estimated in the CFA model. Do not be intimidated by the complexity of the figure; the rules presented in the last chapter provided virtually everything you need to know to determine which parameters should be estimated. Only a few new concepts are presented here. Only three types of parameters are estimated in the analysis: Variances for exogenous variables, covariances between latent factors (F variables), and factor loadings.

Figure 5.7: The Initial Measurement Model, After Identifying All Parameters to Be Estimated (Completed Program Figure for Confirmatory Factor Analysis)

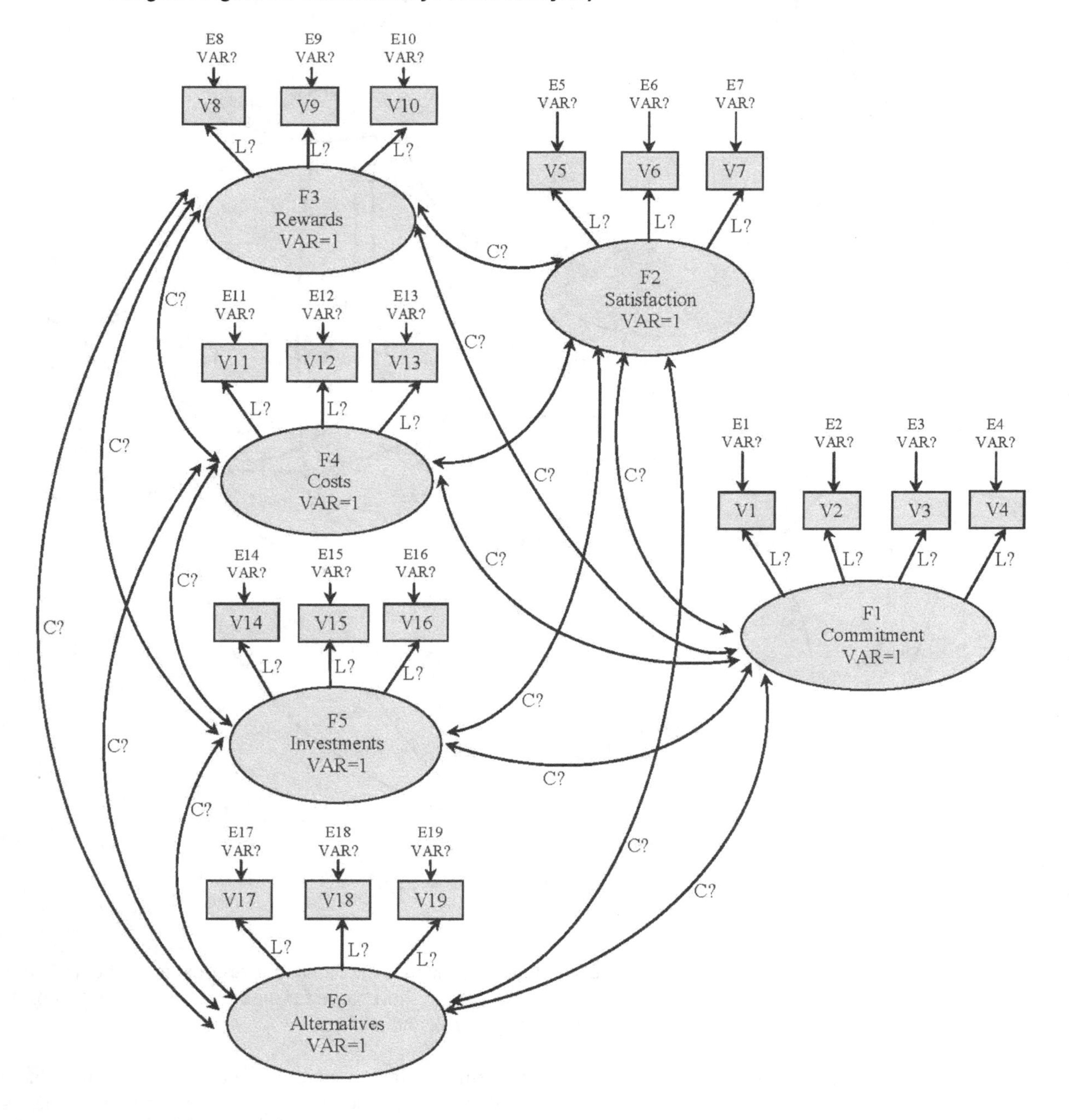

Rule 4 states that variances should be estimated for every exogenous variable in the model, including residual terms. Consistent with this, notice that the symbol "VAR?" appears below all of the E residual terms in the model. (Here, any symbol ending with a "?" indicates a parameter to be estimated; in this case "VAR?" represents a variance parameter to be estimated.) This means that variance will be estimated for each E term in the model. Notice that there are no VAR? symbols under the V terms, because all manifest variables in this CFA model are endogenous variables andvariances are not estimated for endogenous variables.

This figure introduces a new concept, and an exception to one of last chapter's rules. Notice that all of the latent factors (the F variables) are exogenous variables. You can tell this because an exogenous variable is a variable that is not affected by a straight single-headed arrow. Rule 4 states that variance should be estimated for every exogenous variable in the model, but this will not apply to the F variables in confirmatory factor analysis. This is because there exists a basic indeterminacy problem involving the variance of F variables and the factor loadings for the manifest indicators that measure those F variables. Because it is a hypothetical construct (rather than a manifest variable), an F variable has no established metric or scale. This is known as the **scale indeterminacy problem**.

To solve this problem, you can give all factors unit variances by fixing their variances at 1. This establishes a scale or metric for the F variables and helps ensure that the model is identified. This is illustrated in the program figure (Figure 5.7) by indicating "VAR=1" below the long name of each F variable. For example, "VAR=1" is listed below "Commitment" in the oval for F1, below "Satisfaction" in the oval for F2, and so forth. This convention for dealing with the scale indeterminacy problem is summarized in Rule 15:

RULE 15:
In confirmatory factor analysis, variances of the latent F variables are usually fixed at 1.

Having identified the variances to be estimated or fixed, you now turn to the covariance estimates. Rule 5 states that covariance estimates are usually estimated for all possible pairs of manifest exogenous variables. The counterpart to this rule in CFA is that covariances should be estimated for every possible pair of latent factors. In Figure 5.7, this is illustrated by the curved or two-headed arrows that connect all sets of F variables. The "C?" symbol on each curved arrow represents the covariance estimate.

Notice that covariance estimates are specified only for the latent F variables; V variables do not covary because these are endogenous variables. Nor is covariation estimated between residual terms in most hypothesized models. (Certain instances when these terms are allowed to covary include time-series or nonrecursive models.)

Finally, it is necessary to estimate factor loadings for the model; these are represented by the symbol "L?" in Figure 5.7. In this model, factor loadings are just path coefficients for the paths leading from a factor to an indicator variable. For example, the L? symbol appears on the arrow from F1 (commitment) to V1, one of its indicator variables. If the path coefficient (or factor loading) for this path is relatively large and significantly different from zero, it means that V1 effectively measures F1. You can see that factor loadings are estimated for every path (single-headed arrow) leading from a factor to an indicator variable.

Step 4: Verifying That the Model Is Overidentified

The preceding chapter discussed a number of necessary but not sufficient procedures that can be used to determine whether a model is identified. These include:

- verifying that the number of data points in the analysis is larger than the number of parameters to be estimated
- checking the SAS log and output carefully to see if PROC CALIS specified an identification problem
- repeating the analysis several times using different starting values and verifying that the same parameter estimates are obtained with each run

Along with these reminders, it is again noted that researchers are more likely to run into identification problems if they:

- measure a latent variable with less than three indicators
- fail to establish the metric or scale of a factor (e.g., failing to fix its variance at 1)
- analyze nonrecursive structural models

The last chapter stated that recursive path models with manifest variables will always be either just-identified or overidentified. In contrast, models with latent variables may be just-identified, overidentified, or underidentified. Remember that the procedures listed above cannot conclusively prove that a model is identified; worse, if PROC CALIS fails to detect an underidentified model, it will still produce seemingly interpretable, though meaningless, results. Kline (2005) provides a useful discussion of the identification problem and approaches for identification.

Preparing the SAS Program

The DATA Step

Since the DATA step was given extensive coverage in the preceding chapter, only a few key points will be repeated here. First, whenever possible, perform initial analyses on raw data as this allows for use of the KURTOSIS option; this option prints various descriptive statistics to help identify observations with kurtotic and skewed distributions. In this way, you may identify outliers that bias parameter estimates if not deleted.

It is also possible to input the data as a correlation or covariance matrix. If inputting a correlation matrix, standard deviations must be included so that PROC CALIS can transform the correlation matrix into a covariance matrix. The procedures discussed in this chapter should generally be performed on the covariance matrix.

If you have missing data, it is possible to use full information maximum likelihood (FIML) estimation, which uses all available data to perform the analysis. To use FIML, raw data (not a correlation or covariance matrix) must be used as input.

Table 5.1 presents (fictitious) standard deviation values and correlation coefficients between manifest variables analyzed in the present study. Readers who want to replicate the analyses reported here may analyze this dataset by inputting it as a type=CORR dataset. Because the last chapter showed how to input a type=CORR dataset, instructions for managing the DATA step of the SAS program will not be repeated.

Table 5.1: Standard Deviations and Correlation Coefficients between Manifest Variables, Investment Model Study

Intercorrelations

	V1	V2	V3	V4	V5	V6	V7	V8	V9	V10	V11	V12	V13	V14	V15	V16	V17	V18	V19
N	240	240	240	240	240	240	240	240	240	240	240	240	240	240	240	240	240	240	240
VARIANCE	2.486	2.909	2.724	2.926	1.929	2.113	2.056	1.417	1.408	1.724	2.595	2.691	2.360	2.102	2.219	1.874	2.001	1.966	2.185
CORR V1	1.000	.	.	.	.	.	.	.	.	.	.	.	.	.	.	.	.	.	.
CORR V2	.734	1.000	.	.	.	.	.	.	.	.	.	.	.	.	.	.	.	.	.
CORR V3	.819	.786	1.000	.	.	.	.	.	.	.	.	.	.	.	.	.	.	.	.
CORR V4	.672	.732	.751	1.000	.	.	.	.	.	.	.	.	.	.	.	.	.	.	.
CORR V5	.514	.362	.496	.471	1.000	.	.	.	.	.	.	.	.	.	.	.	.	.	.
CORR V6	.534	.346	.452	.452	.713	1.000	.	.	.	.	.	.	.	.	.	.	.	.	.
CORR V7	.522	.345	.507	.546	.720	.764	1.000	.	.	.	.	.	.	.	.	.	.	.	.
CORR V8	.346	.293	.341	.294	.337	.375	.285	1.000	.	.	.	.	.	.	.	.	.	.	.
CORR V9	.209	.147	.167	.214	.251	.306	.357	.390	1.000	.	.	.	.	.	.	.	.	.	.
CORR V10	.349	.241	.287	.236	.282	.351	.304	.506	.492	1.000	.	.	.	.	.	.	.	.	.
CORR V11	.051	.082	.005	-.038	-.161	-.166	-.117	-.091	-.055	-.063	1.000	.	.	.	.	.	.	.	.
CORR V12	-.040	.013	-.057	-.090	-.189	-.150	-.190	-.102	-.036	-.018	.714	1.000	.	.	.	.	.	.	.
CORR V13	-.029	-.012	-.066	-.023	-.110	-.101	-.083	-.043	.055	-.003	.379	.403	1.000	.	.	.	.	.	.
CORR V14	.559	.428	.581	.485	.433	.451	.470	.305	.345	.333	.003	-.005	.043	1.000	.	.	.	.	.
CORR V15	.434	.322	.454	.424	.472	.418	.415	.205	.231	.241	-.037	.007	-.062	.595	1.000	.	.	.	.
CORR V16	.375	.326	.431	.311	.253	.256	.225	.157	.151	.140	.093	.022	.054	.457	.410	1.000	.	.	.
CORR V17	-.141	-.075	-.145	-.275	-.300	-.282	-.305	-.198	-.200	-.188	.226	.216	.051	-.220	-.256	-.065	1.000	.	.
CORR V18	-.135	-.184	-.154	-.266	-.192	-.184	-.204	-.158	-.133	-.243	.132	.119	.107	-.144	-.149	-.074	.529	1.000	.
CORR V19	-.167	-.142	-.145	-.327	-.234	-.209	-.251	-.288	-.298	-.220	.147	.142	.076	-.179	-.202	-.083	.460	.572	1.000

Overview of the PROC CALIS Program

The PROC CALIS program that performs confirmatory factor analysis is very similar to the program that performs path analysis. It is somewhat longer (especially the LINEQS statement) because there are more parameters to estimate; the basic principles for writing the statements remain the same, however. Below is the complete program (minus the DATA step) for performing CFA on the model presented in Figure 5.7:

```
❶   proc calis  modification ;
❷      lineqs
           V1  = LV1F1  F1 + E1,
           V2  = LV2F1  F1 + E2,
           V3  = LV3F1  F1 + E3,
           V4  = LV4F1  F1 + E4,
           V5  = LV5F2  F2 + E5,
           V6  = LV6F2  F2 + E6,
❸          V7  = LV7F2  F2 + E7,
           V8  = LV8F3  F3 + E8,
           V9  = LV9F3  F3 + E9,
           V10 = LV10F3 F3 + E10,
           V11 = LV11F4 F4 + E11,
           V12 = LV12F4 F4 + E12,
           V13 = LV13F4 F4 + E13,
           V14 = LV14F5 F5 + E14,
           V15 = LV15F5 F5 + E15,
           V16 = LV16F5 F5 + E16,
           V17 = LV17F6 F6 + E17,
           V18 = LV18F6 F6 + E18,
           V19 = LV19F6 F6 + E19;
        variance
❹          F1 = 1,
           F2 = 1,
           F3 = 1,
           F4 = 1,
           F5 = 1,
❺          F6 = 1,
❻          E1-E19 = VARE1-VARE19;
        cov
❼          F1 F2 = CF1F2,
```

```
            F1 F3 = CF1F3,
            F1 F4 = CF1F4,
            F1 F5 = CF1F5,
❽           F1 F6 = CF1F6,
❾           F2 F3 = CF2F3,
            F2 F4 = CF2F4,
            F2 F5 = CF2F5,
❿           F2 F6 = CF2F6,
⓫           F3 F4 = CF3F4,
            F3 F5 = CF3F5,
⓬           F3 F6 = CF3F6,
            F4 F5 = CF4F5,
            F4 F6 = CF4F6,
            F5 F6 = CF5F6;
⓭       var  V1-V19 ;
     run;
```

The PROC CALIS Statement

The PROC CALIS statement, which requests the CALIS procedure, appears on line ❶ of the preceding program. Any options desired for the analysis are listed in this statement, separated by at least one space. The PROC CALIS statement in the preceding program includes the modification option which requests Lagrange Multiplier and Wald test modification indices. The MODIFICATION option is useful when determining how the model might be modified if it does not demonstrate an adequate fit.

The LINEQS Statement

The LINEQS statement is used to indicate which manifest variables load on which latent factors. This is done with a series of equations; there is a separate equation for each manifest variable. Below is the general form for the LINEQS statements that appear in a confirmatory factor analysis:

```
lineqs
   V = L  F + E,
   V = L  F + E,
   V = L  F + E;
```

where

V = manifest variable (indicator variable)

L = coefficient name for the factor loading

F = factor that the manifest variable loads on

E = residual term for corresponding manifest variable

Although the preceding provides the general form for three equations, any number of equations is actually possible. The program provided earlier includes 19 equations: One for each indicator variable.

For a concrete illustration, consider the first seven equations and the last three equations from your program:

```
❷       lineqs
            V1  = LV1F1  F1 + E1,
            V2  = LV2F1  F1 + E2,
            V3  = LV3F1  F1 + E3,
            V4  = LV4F1  F1 + E4,
            V5  = LV5F2  F2 + E5,
            V6  = LV6F2  F2 + E6,
❸           V7  = LV7F2  F2 + E7,
            .
            .
            .
            V17 = LV17F6 F6 + E17,
            V18 = LV18F6 F6 + E18,
            V19 = LV19F6 F6 + E19;
```

The LINEQS statement begins with the word LINEQS and ends with a semicolon. Each equation in the statement is separated by a comma. These equations are prepared following the same conventions discussed in the previous chapter.

The preceding LINEQS equations reflect the hypothesized relationships depicted in Figure 5.7. Figure 5.7 shows that only indicator variables are endogenous in the confirmatory factor analysis model; only these indicators appear to the left of the equals sign in the LINEQS equations (e.g., V1, V2, and V3). Figure 5.7 also shows that each V variable is affected by a single latent factor (an F variable) and a single residual term (an E variable). This is reflected in the LINEQS equations as well: V1 is affected by F1 and E1, V2 is affected by F1 and E2, and so forth.

Figure 5.7 shows which variables load on which factors. If an arrow goes from a factor to an indicator variable, then that indicator loads on that factor. The loadings illustrated in Figure 5.7 are also reflected in the LINEQS equations. If an equation for a V variable includes a certain F variable, it means that the V variable loads on that factor. It is in this way that the following statements show that V1 through V4 load on F1, while V5 through V7 load on F2:

```
      V1   = LV1F1   F1 + E1,
      V2   = LV2F1   F1 + E2,
      V3   = LV3F1   F1 + E3,
      V4   = LV4F1   F1 + E4,
      V5   = LV5F2   F2 + E5,
      V6   = LV6F2   F2 + E6,
❸     V7   = LV7F2   F2 + E7,
```

Although a group of variables may share a common factor, each still has its own unique residual term. In other words, V1 has the residual term E1, V2 has the residual term E2, and so forth.

The only parameters estimated by these LINEQS equations are factor loadings; these are estimated as coefficients for paths leading from F variables to V variables. For example, PROC CALIS may estimate that the factor loading for the path from F1 to V1 is .89, or that the coefficient for the path from F1 to V2 is .65.

Rule 12 from the preceding chapter said that, in order to estimate a parameter, you must provide a name for that parameter. This means that you must create names for the parameter estimates you want to estimate in this analysis. Technically, you may select any name for these coefficients so as long as they comply with the usual SAS rules for variable names. You may also let CALIS name them for you by specifying an asterisk * in place of the path name. To create more meaningful names, however, we recommend the system introduced in Chapter 4.

Specifically, we recommend that each factor loading be given a name that begins with the letter "L" (for **L**oading), next the short name for the indicator variable being affected, and last the short name for the underlying factor involved. For example, LV1F1 is the name for the factor loading of V1 on F1, LV17F6 is the name for the factor loading of V17 on F6, and so forth. Review the names of the factor loadings included in the preceding program to verify that these names accurately reflect the relationships between factors and variables portrayed in Figure 5.7. First creating a diagram such as Figure 5.7 reduces the possibility of incorrectly specifying these labels and path estimates.

One advantage of this system is that it provides each parameter estimate with a name that is both unique and meaningful. For example, if you saw the parameter name LV14F5 among the parameter names listed in the Wald test, you would immediately know that this is the factor loading for manifest variable V14 on latent factor F5.

Again consider the LINEQS statements for indicators V1 through V7:

```
V1   = LV1F1   F1 + E1,
V2   = LV2F1   F1 + E2,
V3   = LV3F1   F1 + E3,
```

```
         V4  = LV4F1  F1 + E4,
         V5  = LV5F2  F2 + E5,
         V6  = LV6F2  F2 + E6,
❸        V7  = LV7F2  F2 + E7,
```

Note, dependent variables are always listed to the left of the equals sign, and independent variables with direct effects on these variables are listed to the right of the sign. Also note that the parameter name appears to the immediate left of an independent variable in order to estimate the parameter for that independent variable. The preceding equations reflect these conventions; placing the name for a parameter estimate to the left of an F variable means that it will be estimated to describe the relationship between the V variable and that F variable.

But notice that each of the preceding statements also includes a second independent variable: The indicator variable's E term (residual term). Why not, however, create a name for the path coefficient that represents the relationship between each E term and its corresponding V variable? We do not recommend this. As with path analysis, the paths for these E terms are fixed at 1. You accomplish this by simply leaving the names for these path coefficients out of the equation; this automatically fixes path estimates from the E term to the V variable at 1.

The VARIANCE Statement

The VARIANCE statement is used to specify which parameters are to be estimated and which are to be fixed. In the last chapter, you learned that variance is estimated or fixed for all exogenous variables in a model, but never estimated for endogenous variables. A glance at Figure 5.7 reveals two types of exogenous variables in your confirmatory path model: The F variables and the E variables. This means that F and E variables are listed in the VARIANCE statement. Since all manifest variables are endogenous variable, none of the V terms appear here.

Below is the VARIANCE statement for the current program:

```
        variance
❹          F1 = 1,
           F2 = 1,
           F3 = 1,
           F4 = 1,
           F5 = 1,
❺          F6 = 1,
❻          E1-E19 = VARE1-VARE19;
```

The VARIANCE statement begins with the word VARIANCE and ends with a semicolon. Equations that constitute the statement are each separated by commas.

In Figure 5.7, you wrote "VAR=1" below the long name of each of the latent factors. This was to indicate that the variance of each factor was to be fixed at 1 to solve the scale indeterminacy problem. In the PROC CALIS program, these variances are actually fixed at 1 by listing the short name of the F variable to the left of the equals sign and the number "1" to the right of the sign. This is presented between lines ❹ to ❺ of the program.

Figure 5.7 shows that the symbol "VAR?" appears below the short name of each of the E terms indicating that the variance for each of these terms is to be estimated. Remember that in order to estimate any parameter it is necessary to include a name for that parameter in the PROC CALIS program. This is done on line ❻ of the program. The short names for the E terms (E1–E19) appear to the left of the equals sign and the names for the corresponding variance parameter estimates (VARE1–VARE19) appear to the right of the equals sign. This single equation will estimate variance for all 19 of the specified residual terms.

The COV Statement

The COV statement is used to identify pairs of variables that are expected to covary. The statement begins with the letters COV and ends with a semicolon; each equation is separated by a comma.

If two variables are expected to covary, the short names for these variables are listed to the left of the equals sign; the name for the covariance estimate appears to the right of the equals sign. Below is the COV statement for the present program:

```
        cov
❼          F1 F2 = CF1F2,
           F1 F3 = CF1F3,
           F1 F4 = CF1F4,
           F1 F5 = CF1F5,
❽          F1 F6 = CF1F6,
❾          F2 F3 = CF2F3,
           F2 F4 = CF2F4,
           F2 F5 = CF2F5,
❿          F2 F6 = CF2F6,
⓫          F3 F4 = CF3F4,
           F3 F5 = CF3F5,
⓬          F3 F6 = CF3F6,
           F4 F5 = CF4F5,
           F4 F6 = CF4F6,
           F5 F6 = CF5F6;
```

In confirmatory factor analysis, all latent factors generally are allowed to covary (at least initially). This means that covariance estimates will be calculated for every pairing of F variables; the preceding equations achieve this.

To ensure that you do not miss any covariances, we advise that you follow the general format presented above: Lines ❼ to ❽ specify pairings between variable F1 and variables F2 to F6; lines ❾ to ❿ specify pairings between F2 and variables F3 to F6; lines ⓫ to ⓬ specify pairings between F3 and variables F4 to F6, and so forth. Following this systematic approach assures that all pairings of the F variables are listed.

The names for covariance estimates appear to the right of the equals sign. We recommend the convention presented in the last chapter in which each estimate begins the letter "C" (for **C**ovariance) followed by the short names for the two F variables for which covariation is estimated. Covariance estimates are not (initially) calculated between residual error terms.

The VAR Statement

The VAR statement appears as line ⓭ of the program. It is reproduced below:

```
⓭       var  V1-V19 ;
```

The VAR statement begins the letters VAR and ends with a semicolon; it specifies the variables to be analyzed by PROC CALIS.

Making Sure That the SAS Log and Output Files Look Right

When the analysis is completed, SAS creates two new files. The log file contains the lines of the original SAS program along with notes, warnings, and error messages; the output file ontains the results of the analysis. Both files should be reviewed to verify that the analysis was performed in the desired manner.

Specifically, the log file should be inspected for any notes, warnings, or error messages that indicate a problem in the analysis. Toward the end of the log, look for the statement "Convergence criterion satisfied."

Below is a brief description of some of the information contained in this output:

- Page 1 lists the endogenous variables and exogenous variables specified in the LINEQS statement, which is the general form of the structural equations specified in the LINEQS statement.
- Page 2 reports the simple statistics for observed variables.
- Page 3 reports initial parameter estimates and their respective gradients.
- Page 4 provides optimization results including the iteration history.
- Page 5 reports goodness-of-fit statistics, which will be discussed later.
- Page 6 reports unstandardized results and square multiple correlations.
- Page 7 reports standardized variance and covariance estimates for exogenous variables.
- Page 8 reports the modification indices: Lagrange Multipliers (LM).

- Page 9 reports the modification indices: Wald test results.

Before reviewing the substantive results of the analysis (e.g., the goodness-of-fit indices, the factor loadings), you should routinely review the first pages of the output to verify that the program was executed as expected. The first seven pages of the current output are reproduced here as Output 5.3.

Output 5.3: PROC CALIS Output Pages 1 to 4, Analysis of Initial Measurement Model, Investment Model Study

The CALIS Procedure
Covariance Structure Analysis: Model and Initial Values

Modeling Information	
Maximum Likelihood Estimation	
Data Set	WORK.D1
N Obs	240
Model Type	LINEQS
Analysis	Covariances

Variables in the Model		
Endogenous	Manifest	V1 V10 V11 V12 V13 V14 V15 V16 V17 V18 V19 V2 V3 V4 V5 V6 V7 V8 V9
	Latent	
Exogenous	Manifest	
	Latent	F1 F2 F3 F4 F5 F6
	Error	E1 E10 E11 E12 E13 E14 E15 E16 E17 E18 E19 E2 E3 E4 E5 E6 E7 E8 E9
Number of Endogenous Variables = 19 Number of Exogenous Variables = 25		

Initial Estimates for Linear Equations								
V1	=		.	*	F1	+	1.0000	E1
					LV1F1			
V2	=		.	*	F1	+	1.0000	E2
					LV2F1			
V3	=		.	*	F1	+	1.0000	E3
					LV3F1			
V4	=		.	*	F1	+	1.0000	E4
					LV4F1			
V5	=		.	*	F2	+	1.0000	E5
					LV5F2			
V6	=		.	*	F2	+	1.0000	E6
					LV6F2			
V7	=		.	*	F2	+	1.0000	E7
					LV7F2			
V8	=		.	*	F3	+	1.0000	E8
					LV8F3			
V9	=		.	*	F3	+	1.0000	E9
					LV9F3			
V10	=		.	*	F3	+	1.0000	E10
					LV10F3			
V11	=		.	*	F4	+	1.0000	E11
					LV11F4			
V12	=		.	*	F4	+	1.0000	E12
					LV12F4			
V13	=		.	*	F4	+	1.0000	E13
					LV13F4			
V14	=		.	*	F5	+	1.0000	E14
					LV14F5			
V15	=		.	*	F5	+	1.0000	E15

V16	=		.	*	F5	+	1.0000	E16
					LV16F5			
V17	=		.	*	F6	+	1.0000	E17
					LV17F6			
V18	=		.	*	F6	+	1.0000	E18
					LV18F6			
V19	=		.	*	F6	+	1.0000	E19
					LV19F6			

Initial Estimates for Variances of Exogenous Variables			
Variable Type	Variable	Parameter	Estimate
Latent	F1		1.00000
	F2		1.00000
	F3		1.00000
	F4		1.00000
	F5		1.00000
	F6		1.00000
Error	E1	VARE1	.
	E2	VARE2	.
	E3	VARE3	.
	E4	VARE4	.
	E5	VARE5	.
	E6	VARE6	.
	E7	VARE7	.
	E8	VARE8	.
	E9	VARE9	.
	E10	VARE10	.
	E11	VARE11	.
	E12	VARE12	.
	E13	VARE13	.
	E14	VARE14	.
	E15	VARE15	.
	E16	VARE16	.
	E17	VARE17	.
	E18	VARE18	.
	E19	VARE19	.

Initial Estimates for Covariances Among Exogenous Variables			
Var1	Var2	Parameter	Estimate
F1	F2	CF1F2	.
F1	F3	CF1F3	.
F1	F4	CF1F4	.
F1	F5	CF1F5	.
F1	F6	CF1F6	.
F2	F3	CF2F3	.
F2	F4	CF2F4	.
F2	F5	CF2F5	.
F2	F6	CF2F6	.
F3	F4	CF3F4	.
F3	F5	CF3F5	.
F3	F6	CF3F6	.
F4	F5	CF4F5	.
F4	F6	CF4F6	.
F5	F6	CF5F6	.

Output 5.3 (Page 2)

The CALIS Procedure
Covariance Structure Analysis: Descriptive Statistics

Simple Statistics

Variable	Mean	Std Dev
V1	0	2.48600
V2	0	2.90900
V3	0	2.72400
V4	0	2.92600
V5	0	1.92900
V6	0	2.11300
V7	0	2.05600
V8	0	1.41700
V9	0	1.40800
V10	0	1.72400
V11	0	2.59500
V12	0	2.69100
V13	0	2.36000
V14	0	2.10200
V15	0	2.21900
V16	0	1.87400
V17	0	2.00100
V18	0	1.96600
V19	0	2.18500

Output 5.3 (Page 3)

The CALIS Procedure
Covariance Structure Analysis: Optimization

Initial Estimation Methods	
1	Instrumental Variables Method
2	McDonald Method

Optimization Start Parameter Estimates

N	Parameter	Estimate	Gradient
1	LV1F1	2.25837	0.05797
2	LV2F1	2.27508	-0.08232
3	LV3F1	2.51331	-0.02411
4	LV4F1	2.49990	0.07426
5	LV5F2	1.64639	0.04049
6	LV6F2	1.76717	-0.04187
7	LV7F2	1.81048	0.00902
8	LV8F3	1.03948	0.05279
9	LV9F3	0.83597	-0.03769
10	LV10F3	1.24764	-0.00981
11	LV11F4	2.23401	0.00286
12	LV12F4	2.18994	-0.01539
13	LV13F4	1.14346	0.02096
14	LV14F5	1.79425	0.00750
15	LV15F5	1.54589	-0.01170
16	LV16F5	1.04648	0.01381
17	LV17F6	1.58106	0.06120
18	LV18F6	1.20957	-0.10600
19	LV19F6	1.60319	0.03672
20	VARE1	1.07995	-0.16346
21	VARE2	3.28627	0.05236
22	VARE3	1.10347	0.00482
23	VARE4	2.31196	-0.09158

24	VARE5	1.01045	-0.06994
25	VARE6	1.34188	0.05255
26	VARE7	0.94928	-0.02264
27	VARE8	0.92737	-0.06540
28	VARE9	1.28363	0.02245
29	VARE10	1.41557	0.00480
30	VARE11	1.74322	0.0001174
31	VARE12	2.44563	0.01563
32	VARE13	4.26210	-0.00546
33	VARE14	1.19908	-0.00924
34	VARE15	2.53417	0.00747
35	VARE16	2.41677	-0.00581
36	VARE17	1.50427	-0.07676
37	VARE18	2.40209	0.05052
38	VARE19	2.20401	-0.03267
39	CF1F2	0.62146	-0.12105
40	CF1F3	0.44124	-0.04690
41	CF1F4	-0.02340	-0.01554
42	CF1F5	0.69530	-0.11532
43	CF1F6	-0.29380	-0.10907
44	CF2F3	0.53391	0.05401
45	CF2F4	-0.22638	-0.03640
46	CF2F5	0.63886	0.09986
47	CF2F6	-0.39374	0.08333
48	CF3F4	-0.08957	0.04064
49	CF3F5	0.49481	-0.03725
50	CF3F6	-0.43534	-0.01748
51	CF4F5	0.01104	0.01748
52	CF4F6	0.26790	-0.00860
53	CF5F6	-0.32012	0.02420
Value of Objective Function = 1.2296108485			

Output 5.3 (Page 4)

The CALIS Procedure
Covariance Structure Analysis: Optimization
Levenberg-Marquardt Optimization
Scaling Update of More (1978)

Parameter Estimates	53
Functions (Observations)	190

Optimization Start			
Active Constraints	0	Objective Function	1.2296108485
Max Abs Gradient Element	0.1634612729	Radius	1

Iteration	Restarts	Function Calls	Active Constraints	Objective Function	Objective Function Change	Max Abs Gradient Element	Lambda	Ratio Between Actual and Predicted Change
1	0	4	0	1.04697	0.1826	0.1561	0	0.846
2	0	6	0	1.03685	0.0101	0.0151	0	0.891
3	0	8	0	1.03633	0.000515	0.00626	0	0.841
4	0	10	0	1.03629	0.000040	0.000850	0	0.796
5	0	12	0	1.03629	3.776E-6	0.000567	0	0.757
6	0	14	0	1.03629	3.863E-7	0.000100	0	0.725
7	0	16	0	1.03629	4.186E-8	0.000061	0	0.703
8	0	18	0	1.03629	4.682E-9	0.000013	0	0.687
9	0	20	0	1.03629	5.35E-10	6.798E-6	0	0.677

Optimization Results			
Iterations	9	Function Calls	23
Jacobian Calls	11	Active Constraints	0
Objective Function	1.0362903002	Max Abs Gradient Element	6.7983308E-6
Lambda	0	Actual Over Pred Change	0.6766855591
Radius	0.0001183677		

Convergence criterion (GCONV=1E-8) satisfied.

Page 1 of Output 5.3 identifies the endogenous and exogenous variables of the analysis and the general form of structural equations; these should be reviewed to verify that PROC CALIS analyzed the model as intended by you. Page 4 provides information about the number of parameter estimates, functions (or observations), and iteration history. This page also indicates whether the convergence criterion was satisfied. If the information presented on these pages appears to be in order, you may proceed to assess the fit between model and data, along with other results of the analysis.

Assessing the Fit between Model and Data

When conducting confirmatory factor analysis, you begin with a model that predicts the existence of a specific number of latent factors and predicts which indicator variables load on each factor. You then test the model by measuring these variables in a sample of participants drawn from the population of interest. If the model provides a reasonably good approximation, it should do a good job of accounting for the observed relationships in the dataset. In other words, the model should provide a good fit to data.

The procedures for determining whether path models fit the data were presented in the last chapter; each of these procedures can also be used to assess the fit of confirmatory factor analytic models. A few modifications will be necessary (since CFA models tend to be somewhat more complex than path analytic models), but the basic strategy for assessing fit remains the same. The process begins by reviewing significance tests for factor loadings, overall goodness-of-fit indices (e.g., SRMR, CFI, and RMSEA), and then proceeds to other indices such as R^2 values and modification indices.

Step 1: Assessing the Statistical Power of the Model

As previously discussed, confirmatory factor analytic models need to have a sufficient number of observations relative to degrees of freedom to minimize the likelihood of Type II errors. In other words, models need to possess sufficient statistical power to reject poorly fitting models (i.e., to have confidence in reported goodness-of-fit indices). With 240 observations, we know that our sample size exceeds the minimum threshold of 200 (though below the ideal level of 300). The number of observations appears as the first line on page 5. This table is presented here as Output 5.4.

Output 5.4: Number of Observations and Goodness-of Fit-Statistics, Analysis of Initial Measurement Model, Investment Model Study

Fit Summary		
Modeling Info	N Observations	240
	N Variables	19
	N Moments	190
	N Parameters	53
	N Active Constraints	0
	Baseline Model Function Value	10.2915
	Baseline Model Chi-Square	2459.6733
	Baseline Model Chi-Square DF	171
	Pr > Baseline Model Chi-Square	<.0001
Absolute Index	Fit Function	1.0363
	Chi-Square	247.6734
	Chi-Square DF	137
	Pr > Chi-Square	<.0001
	Z-Test of Wilson & Hilferty	5.4581
	Hoelter Critical N	160
	Root Mean Square Residual (RMSR)	0.2373
	Standardized RMSR (SRMSR)	0.0465
	Goodness of Fit Index (GFI)	0.9064
Parsimony Index	Adjusted GFI (AGFI)	0.8702
	Parsimonious GFI	0.7262
	RMSEA Estimate	0.0581
	RMSEA Lower 90% Confidence Limit	0.0464
	RMSEA Upper 90% Confidence Limit	0.0696
	Probability of Close Fit	0.1221
	ECVI Estimate	1.5203
	ECVI Lower 90% Confidence Limit	1.3490
	ECVI Upper 90% Confidence Limit	1.7277
	Akaike Information Criterion	353.6734
	Bozdogan CAIC	591.1472
	Schwarz Bayesian Criterion	538.1472
	McDonald Centrality	0.7941
Incremental Index	Bentler Comparative Fit Index	0.9516
	Bentler-Bonett NFI	0.8993
	Bentler-Bonett Non-normed Index	0.9396
	Bollen Normed Index Rho1	0.8743
	Bollen Non-normed Index Delta2	0.9524
	James et al. Parsimonious NFI	0.7205

Within the "Absolute Index" section of the table, we can see that our initial model has 137 degrees of freedom. This appears to the right of the line "Chi-Square DF." We can now proceed to estimate statistical power using the formula from MacCallum and colleagues (1996), which was previously presented. With n = 240 and df = 137, this model is estimated to have power greater than .99 (i.e., power ≥ .80; Cohen 1988). On this basis, we can proceed to interpret goodness-of-fit indices with greater confidence.

Step 2: Reviewing Goodness-of-Fit Indices

As noted in Chapter 4, it remains customary to first examine the chi-square statistic to test of the null hypothesis that models fit exactly in the population (i.e., $p > .05$). As previously discussed, however, this statistic is often significant even when models provide good fit to data (Byrne 1998; MacCallum, Browne, and Cai 2006). This is particularly true with CFA (and structural equation models) which tend to be more complex than path models. Therefore the chi-square statistic should not be seen as a bona fide goodness-of-fit index; instead, chi-square values are useful primarily when modifying models to ensure that changes are statistically significant (i.e., viable model revisions).

Instead, indices such as the Standardized Root Mean Square Residual (SRMR), the Comparative Fit Index (CFI), and the Root Mean Square Error of Approximation (RMSEA & RMSEA CL_{90}) provide more accurate information regarding goodness of model fit. There statistics are reported on page 5 among a multitude of statistics. As mentioned in the last chapter, no universal consensus exists as to which provide the best reflection of model fit; instead, it is common practice to report at least three goodness-of-fit indices, at least one absolute index (e.g., SRMR), one parsimony index (e.g., RMSEA), and one incremental index (e.g., CFI). Goodness-of-fit indices are also presented as Output 5.4.

In the previous chapter, we recommended that the SRMR, CFI, and RMSEA values be examined and reported as well as 90% confidence limits for the RMSEA. SRMR values less than .055 are ideal. In contrast, CFI values between .90 and .94 suggest adequate fit, but values greater than .94 are ideal. Similar to the SRMR, smaller RMSEA values reflect good model fit. A RMSEA value above .09 is deemed to be poor; values between .08 and .10 are deemed to be mediocre, and values between .055 and .08 suggest fair model fit; whereas values less than .055 are viewed as most ideal (Hu and Bentler 1999; MacCallum et al. 1996). In addition, the range of RMSEA confidence limits should be relatively narrow; 90% confidence limits between $.090 \geq \text{RMSEA } CL_{90} \geq 0$ are adequate whereas limits between $.054 \geq \text{RMSEA } CL_{90} \geq 0$ are ideal. In these instances, the researcher has greater confidence that data fit the model effectively as there is only a 1 in 10 chance that the true RMSEA value within the population falls outside of these ranges. (As mentioned in the previous chapter, it is customary to report SRMR and RMSEA values to three decimal places unlike the CFI and most other statistics.)

For our current model, our goodness-of-fit indices can be found within the table of values. These are presented below.

Standardized RMSR (SRMSR)	0.0465
RMSEA Estimate	0.0581
RMSEA Lower 90% Confidence Limit	0.0464
RMSEA Upper 90% Confidence Limit	0.0696
Bentler Comparative Fit Index	0.9516

At first glance, goodness-of-fit statistics appear to be within good parameters (CFI = .95, SRMR = .047, RMSEA = .058). The comparatively narrow range of the RMSEA confidence limits provides further confidence in this model. Overall, these values provide a good first indication that the overall structure of the model fits the data. Before accepting this model, however, we need to examine parameter estimates to see if associations between variables are statistically significant as initially hypothesized. We turn now to looking within the model itself.

Step 3: Reviewing Significance Tests for Factor Loadings

Remember that a factor loading is equivalent to a path coefficient from a latent factor to an indicator variable. A nonsignificant parameter estimate therefore indicates that the observed variable does not significantly contribute to measurement of the underlying factor and should be deleted from the model. As discussed in Chapter 4, it is generally ideal to interpret and report standardized path coefficients and covariance estimates along with their respective t values instead of unstandardized estimates and standard errors. The former are presented here as Output 5.5.

Output 5.5

Procedure
Covariance Structure Analysis: Maximum Likelihood Estimation

Standardized Results for Linear Equations									
V1	=		0.8743	*	F1	+	1.0000		E1
Std Err			0.0183		LV1F1				
t Value			47.6959						
V2	=		0.8389	*	F1	+	1.0000		E2
Std Err			0.0219		LV2F1				
t Value			38.3140						
V3	=		0.9343	*	F1	+	1.0000		E3
Std Err			0.0129		LV3F1				
t Value			72.5754						
V4	=		0.8114	*	F1	+	1.0000		E4
Std Err			0.0247		LV4F1				
t Value			32.8794						
V5	=		0.8270	*	F2	+	1.0000		E5
Std Err			0.0254		LV5F2				
t Value			32.5806						
V6	=		0.8649	*	F2	+	1.0000		E6
Std Err			0.0222		LV6F2				
t Value			38.9535						
V7	=		0.8769	*	F2	+	1.0000		E7
Std Err			0.0213		LV7F2				
t Value			41.2164						
V8	=		0.6646	*	F3	+	1.0000		E8
Std Err			0.0500		LV8F3				
t Value			13.3023						
V9	=		0.6369	*	F3	+	1.0000		E9
Std Err			0.0514		LV9F3				
t Value			12.3931						
V10	=		0.7496	*	F3	+	1.0000		E10
Std Err			0.0464		LV10F3				
t Value			16.1718						

Standardized Results for Linear Equations									
V11	=		0.8254	*	F4	+	1.0000		E11
Std Err			0.0450		LV11F4				
t Value			18.3574						
V12	=		0.8649	*	F4	+	1.0000		E12
Std Err			0.0449		LV12F4				
t Value			19.2725						
V13	=		0.4634	*	F4	+	1.0000		E13
Std Err			0.0570		LV13F4				
t Value			8.1234						
V14	=		0.8428	*	F5	+	1.0000		E14
Std Err			0.0335		LV14F5				
t Value			25.1792						
V15	=		0.7085	*	F5	+	1.0000		E15
Std Err			0.0409		LV15F5				
t Value			17.3359						
V16	=		0.5496	*	F5	+	1.0000		E16
Std Err			0.0520		LV16F5				
t Value			10.5672						
V17	=		0.6809	*	F6	+	1.0000		E17
Std Err			0.0473		LV17F6				
t Value			14.3948						
V18	=		0.7622	*	F6	+	1.0000		E18
Std Err			0.0439		LV18F6				
t Value			17.3575						
V19	=		0.7284	*	F6	+	1.0000		E19
Std Err			0.0452		LV19F6				
t Value			16.1192						

Standardized Results for Variances of Exogenous Variables					
Variable Type	Variable	Parameter	Estimate	Standard Error	t Value
Latent	F1		1.00000		
	F2		1.00000		
	F3		1.00000		
	F4		1.00000		
	F5		1.00000		
	F6		1.00000		

Error	E1	VARE1	0.23552	0.03206	7.34710
	E2	VARE2	0.29627	0.03673	8.06512
	E3	VARE3	0.12703	0.02406	5.28026
	E4	VARE4	0.34170	0.04004	8.53312
	E5	VARE5	0.31606	0.04198	7.52815
	E6	VARE6	0.25187	0.03841	6.55721
	E7	VARE7	0.23108	0.03731	6.19319
	E8	VARE8	0.55828	0.06641	8.40619
	E9	VARE9	0.59433	0.06547	9.07838
	E10	VARE10	0.43804	0.06950	6.30283
	E11	VARE11	0.31869	0.07423	4.29352
	E12	VARE12	0.25194	0.07763	3.24534
	E13	VARE13	0.78527	0.05287	14.85376
	E14	VARE14	0.28966	0.05642	5.13376
	E15	VARE15	0.49798	0.05792	8.59836
	E16	VARE16	0.69798	0.05716	12.21082
	E17	VARE17	0.53640	0.06441	8.32768
	E18	VARE18	0.41909	0.06694	6.26112
	E19	VARE19	0.46948	0.06582	7.13241

Standardized Results for Covariances Among Exogenous Variables					
Var1	Var2	Parameter	Estimate	Standard Error	t Value
F1	F2	CF1F2	0.61919	0.04618	13.40693
F1	F3	CF1F3	0.43845	0.06624	6.61906
F1	F4	CF1F4	-0.02585	0.07280	-0.35504
F1	F5	CF1F5	0.71293	0.04412	16.15965
F1	F6	CF1F6	-0.25908	0.07192	-3.60222
F2	F3	CF2F3	0.53381	0.06246	8.54576
F2	F4	CF2F4	-0.22446	0.07131	-3.14773
F2	F5	CF2F5	0.63499	0.05210	12.18736
F2	F6	CF2F6	-0.37430	0.06911	-5.41600
F3	F4	CF3F4	-0.09210	0.08171	-1.12707
F3	F5	CF3F5	0.51620	0.06899	7.48177
F3	F6	CF3F6	-0.42374	0.07524	-5.63209
F4	F5	CF4F5	0.00795	0.07897	0.10065
F4	F6	CF4F6	0.25393	0.07621	3.33202
F5	F6	CF5F6	-0.30005	0.07710	-3.89156

These t values represent large-sample t tests of the null hypothesis that factor loading are equal to zero in the population. Remember that t values greater than 1.96 (or less than -1.96) are significant at $p < .05$ and those greater than 2.58 are significant at $p < .01$ (or less than -2.58). The obtained t values in the output show that all factor loadings are significant at $p < .01$.

Standardized path coefficients appear in the "Standardized Results for Linear Equations" section. A given factor loading appears next to its name (e.g., the standardized factor loading for V1 on F1 is .87). This output shows that the standardized loadings range in size from .46 to .93, and that only two are below .60. These large parameter estimates are not unexpected given their respective t values (i.e., $p < .01$ for each). Also note that t values for all error estimates also differ significantly from zero. These are reported under the heading "Standardized Results for Variances of Exogenous Variables."

Modifying the Measurement Model

As discussed in Chapter 4, the most justifiable model revisions are deletion of nonsignificant paths. These revisions make models more parsimonious and unlike the decision to add paths not initially hypothesized, deleting paths does not risk *overfitting* models (i.e., arriving at improved goodness-of-fit indices with models unlikely to be generalizable to the population). In contrast, deleting paths originally estimated increases the number of available degrees of freedom.

As we can see from page 7, t values for three covariance estimates are not statistically significant. These appear in Output 5.5 under the heading "Standardized Results for Covariances Among Exogenous Variables." We see that covariance estimates between F1 and F4 (commitment and costs), between F3 and F4 (rewards and costs), and between F5 and F4 (investments and costs) do not differ significantly from zero.

The Wald Test

If we examine the results of the **Wald Tests**, we also see that these three covariance estimates are each listed. As presented in Output 5.6, Wald tests report that these three estimated parameters can be deleted from the model without negatively affecting goodness-of-fit indices.

Despite this, these parameters will not be dropped from the model. With confirmatory factor analysis, all factors are normally allowed to covary during this stage of the analyses; at this juncture, you will not remove nonsignificant covariance estimates. (For illustrative purposes we will retain these estimates.)

When computing measurement models, you primarily look to the results of Wald tests to indicate where *factor loadings* could be dropped from the model without significantly affecting the chi-square statistic. Such findings indicate that indicator variables do not effectively contribute to measurement of their respective latent factors to which they are initially assigned. In this example, it should come as no surprise that no factor loadings were reported in the Wald test (under the heading "Stepwise Multivariate Wald Test").

Output 5.6: Wald Test Results for Initial Measurement Model, Investment Model Study

The CALIS Procedure
Covariance Structure Analysis: Maximum Likelihood Estimation

Stepwise Multivariate Wald Test					
	Cumulative Statistics			Univariate Increment	
Parm	Chi-Square	DF	Pr > ChiSq	Chi-Square	Pr > ChiSq
CF4F5	0.01013	1	0.9198	0.01013	0.9198
CF1F4	0.28973	2	0.8651	0.27960	0.5970
CF3F4	1.72700	3	0.6309	1.43727	0.2306

Strictly speaking, models should not be revised solely on the basis of modification indices to achieve acceptable goodness of fit. These decisions should, instead, be based on theory or existing research as opposed to these statistical criteria alone. In general, goodness of fit should first be consulted; if they provide evidence of a poor or questionable fit, you may then turn to the modification indices to determine if specific modifications might improve model fit.

The modification option included in the PROC CALIS statement requests that two modification indices be computed. (See the "Overview of the PROC CALIS program"section.) As discussed above, the Wald test identifies parameters that can be dropped from the model. Conversely, the **Lagrange Multiplier or LM statistic** identifies parameters that, if added to the model, would improve indices of fit.

The Lagrange Multiplier Test

The LM statistic estimates reduction in the model chi-square statistic that would result from freeing a fixed parameter and allowing it to be estimated. In other words, the LM statistic approximates the degree to which chi-square would improve if a new factor loading or covariance estimate were added to your model. The LM statistic appears in Output 5.7 on page 9.

Output 5.7: LM Statistics, Initial Model, Investment Model Study

The CALIS Procedure
Covariance Structure Analysis: Maximum Likelihood Estimation

Rank Order of the 10 Largest LM Stat for Paths from Endogenous Variables				
To	From	LM Stat	Pr > ChiSq	Parm Change
V4	V19	23.08136	<.0001	-0.26583
V2	V7	19.19557	<.0001	-0.29071
V4	V17	14.31861	0.0002	-0.22785
V1	V6	13.99959	0.0002	0.19068
V4	V7	11.98741	0.0005	0.24315
V4	V18	10.65562	0.0011	-0.20125
V2	V6	10.11926	0.0015	-0.20396
V2	V14	9.88472	0.0017	-0.21644
V2	V4	9.76054	0.0018	0.23420
V4	V2	9.76051	0.0018	0.27328

Rank Order of the 10 Largest LM Stat for Paths from Exogenous Variables				
To	From	LM Stat	Pr > ChiSq	Parm Change
V4	F6	22.57525	<.0001	-0.66957
V2	F2	18.91095	<.0001	-0.69783
V2	F5	14.90731	0.0001	-0.78694
V1	F2	8.92517	0.0028	0.38258
V4	F2	6.40940	0.0114	0.42934
V1	F3	5.35242	0.0207	0.27589
V17	F2	5.05799	0.0245	-0.29736
V8	F1	4.55541	0.0328	0.21967
V18	F2	4.20707	0.0403	0.27137
V2	F4	4.15274	0.0416	0.24841

Rank Order of the 10 Largest LM Stat for Paths with New Endogenous Variables				
To	From	LM Stat	Pr > ChiSq	Parm Change
F1	V2	21.88319	<.0001	0.22464
F6	V4	20.39608	<.0001	-0.19091
F1	V9	9.41017	0.0022	-0.16027
F2	V2	9.18722	0.0024	-0.10672
F1	V1	8.10737	0.0044	-0.18929
F2	V1	7.36459	0.0067	0.13064
F5	V2	7.10657	0.0077	-0.09998
F6	V18	7.08181	0.0078	0.38733
F1	V8	5.74101	0.0166	0.13190
F5	V3	5.12155	0.0236	0.17840

Note: No LM statistic in the default test set for the covariances of exogenous variables is nonsingular. Ranking is not displayed.

Rank Order of the 10 Largest LM Stat for Error Variances and Covariances				
Var1	Var2	LM Stat	Pr > ChiSq	Parm Change
E7	E4	10.46578	0.0012	0.45668
E4	E2	9.76053	0.0018	0.68514
E4	E19	9.01301	0.0027	-0.60904
E6	E1	8.98764	0.0027	0.32831
E9	E7	8.68713	0.0032	0.27228
E4	E1	7.64382	0.0057	-0.50050
E2	E18	7.42459	0.0064	-0.45929
E19	E17	7.08019	0.0078	-0.90779
E5	E15	7.06181	0.0079	0.36042

Rank Order of the 10 Largest LM Stat for Error Variances and Covariances				
Var1	Var2	LM Stat	Pr > ChiSq	Parm Change
E8	E7	6.95186	0.0084	-0.24164

LM statistics are presented in five sections; in each, a maximum of 10 values per category are listed in descending order of size. Within each grouping, the fourth column indicates how much the chi-square statistic is likely to be reduced with the inclusion of this additional path or covariance estimate (along with the estimated statistical significance of this reduction). In theory, any change resulting in a chi-square reduction in which $p < .05$ can be considered (shown as "Pr > ChiSq").

These five groupings of LM statistics are:

- paths from endogenous variables
- paths from exogenous variables
- paths with new endogenous variables
- paths with new exogenous variables
- paths between error variances and covariances estimates

It makes little sense to add path estimates between observed variables (first grouping, from endogenous variables). Our model assumes that these observed variables each contribute to measurement of their respective latent variables; observed variables are not assumed to predict one another. These modification indices will not be considered. Nor are we interested in LM statistics appearing in the third grouping (Paths with New Endogenous Variables). Our model assumes that each of the latent constructs or factors predicts responses to their respective observed variables; these LM statistics reflect the opposite associations. The modification indices in this third grouping are not theoretically tenable.

The fourth grouping lists additional covariance estimates between exogenous variables which would result in reduction of the chi-square statistic. In the current example, none are listed. This is because each of our six latent variables was already assumed to covary with all others. In other words, there are no additional covariance estimates between factors to include in the model.

The final grouping (error variances and covariances) will not be considered here. (As previously noted, however, there are circumstances when it is viable to allow for correlation between error estimates particularly with time series models.)

Instead, we are most interested in LM statistics in the second grouping that indicate if new paths might be added from latent factors (F variables) to observed variables (V variable). Here we see that inclusion of an additional path from F6 (alternatives) to V4 would mathematically improve model fit. The value of this LM statistic is 22.57 ($p < .01$), meaning that the chi-square statistic would be reduced by approximately 22.57 if this path were included added to the model.

Modifying the Model

In a situation such as this, you have a number of options. One alternative is to simply add the path from F6 to V4 and leave the remainder of the model untouched. This means that V4 would become a complex variable, contributing to measurement of more than one factor.

This alternative, however, is generally undesirable when developing a measurement model. This is because the theoretical model (to be assessed later) is more easily interpreted if all of the indicators are unifactorial (i.e., each indicator loads on only one factor). In most cases, it is preferable to reassign or completely drop an indicator from a model rather than assign it to two factors simultaneously.

Should you reassign V4 so that it loads on F6 but not on F1 (the factor to which it was originally assigned)? This option is also unsatisfactory as the output has already shown that V4 loads significantly on factor F1 as initially hypothesized ($p < .01$). In other words, this is not an instance in which a unifactorial variable was initially assigned to the wrong factor.

In this case, the best alternative is to drop V4 from the analysis entirely. This should not create any identification problems for F1, because this factor will still be measured by three indicators even after V4 is deleted. (This is why it is important to have at least four or five indicators for each factor in the initial measurement model.)

The revised measurement model appears as Figure 5.8. This model is identical to that presented in Figure 5.7 except that the path from F1 (commitment) to V4 has been deleted. As a result, the variable V4 itself has been deleted from the model.

Figure 5.8: The Revised Measurement Model

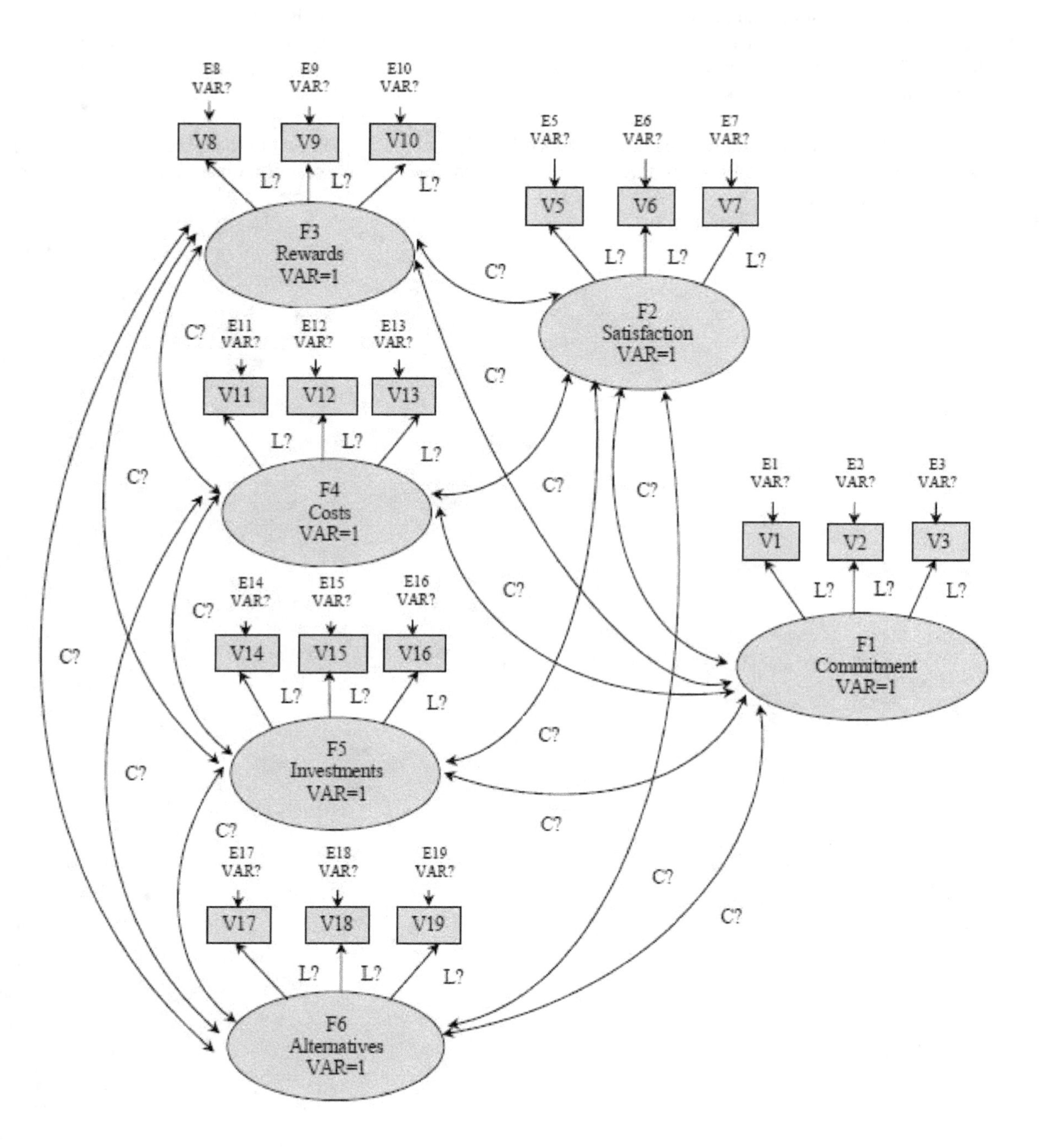

Estimating the Revised Measurement Model

The revised model must now be re-estimated to see if it provides an acceptable fit to the data. Below is a portion of the revised PROC CALIS program that will compute the new model:

```
      proc calis  modification ;
         lineqs
            V1  = LV1F1  F1 + E1,
            V2  = LV2F1  F1 + E2,
            V3  = LV3F1  F1 + E3,
❶
            V5  = LV5F2  F2 + E5,
            V6  = LV6F2  F2 + E6,
            V7  = LV7F2  F2 + E7,
            V8  = LV8F3  F3 + E8,
            V9  = LV9F3  F3 + E9,
            V10 = LV10F3 F3 + E10,
            V11 = LV11F4 F4 + E11,
            V12 = LV12F4 F4 + E12,
            V13 = LV13F4 F4 + E13,
            V14 = LV14F5 F5 + E14,
            V15 = LV15F5 F5 + E15,
            V16 = LV16F5 F5 + E16,
            V17 = LV17F6 F6 + E17,
            V18 = LV18F6 F6 + E18,
            V19 = LV19F6 F6 + E19;
         variance
            F1 = 1,
            F2 = 1,
            F3 = 1,
            F4 = 1,
            F5 = 1,
            F6 = 1,
❷           E1-E3   = VARE1-VARE3  ,
❸           E5-E19  = VARE5-VARE19;
         cov
            F1  F2 = CF1F2,
            F1  F3 = CF1F3,
            F1  F4 = CF1F4,
            F1  F5 = CF1F5,
            F1  F6 = CF1F6,
            F2  F3 = CF2F3,
            F2  F4 = CF2F4,
            F2  F5 = CF2F5,
            F2  F6 = CF2F6,
            F3  F4 = CF3F4,
            F3  F5 = CF3F5,
            F3  F6 = CF3F6,
            F4  F5 = CF4F5,
            F4  F6 = CF4F6,
            F5  F6 = CF5F6;
❹       var  V1-V3 V5-V19 ;
      run;
```

Three changes have been made to the initial program for it to estimate the revised model. Because the path from F1 to V4 had been dropped from the initial model, the equation for V4 was deleted from the program. Initially, this equation appeared on line ❶; you can see that this line is now blank in the revised program. A second change was made to the VARIANCE statement. In the initial program, an error term was attached to V4 because it was a predicted or endogenous variable. The error term E4 no longer needs to be estimated because the observed variable V4 has been dropped from the model. The revised program now needs only to estimate error terms for the remaining endogenous variables. This can be seen in the final two lines of the VARIANCE statement, which now estimates error terms E1 through E3 ❷ and E5 through E19 ❸.

The final change appears on line ❹, where V4 is no longer specified as a variable within the model. As the correlation matrix for this program continues to specify associations between V4 and all other observed variables, LM statistics will consider V4 for re-inclusion in the model if it is not removed from the list of variables. If this final revision is not made, LM statistics will consider V4 a prospective new endogenous variable to add to the model. (**Note:** Removing V4 from the VAR statement is considerably easier than recalculating the correlation matrix!)

It was not necessary to revise the COV statement in the program, because only the covariances between latent factors were specified in this statement. In the revised program, all factors are allowed to covary as before.

After reviewing the SAS log and output files to verify that the revised program ran properly, the same fit indices discussed earlier should be consulted to see if the revised model provides a better fit to the data.

Page 5 of the output for this program provides goodness-of-fit indices for the revised measurement model (these indices appear in Output 5.8).You can see that the model chi-square value for the revised model is 180.87, with 120 degrees of freedom. Of note, this chi-square value for the revised model has declined substantively from the initial measurement model, where chi-square was 247.67, with 137 degrees of freedom. By eliminating V4 from the analysis, model chi-square decreased by 66.80, with a reduction of 17 degrees of freedom. If we consult Appendix C, "Critical Values of the Chi-Square Distribution," and find the intersection point between column .05 and row 17 (degrees of freedom), we find a critical value of 27.59, which is considerable smaller than the observed change of 66.80 in the chi-square statistic between models. This change is also significant at .01 where the critical value is 33.41. So far, it appears that deleting V4 improved model fit.

Output 5.8: Goodness-of-Fit Indices for Revised Measurement Model, Investment Model Study

Covariance Structure Analysis: Maximum Likelihood Estimation

Fit Summary		
Modeling Info	Number of Observations	240
	Number of Variables	18
	Number of Moments	171
	Number of Parameters	51
	Number of Active Constraints	0
	Baseline Model Function Value	9.0702
	Baseline Model Chi-Square	2167.7711
	Baseline Model Chi-Square DF	153
	Pr > Baseline Model Chi-Square	<.0001
Absolute Index	Fit Function	0.7568
	Chi-Square	180.8717
	Chi-Square DF	120
	Pr > Chi-Square	0.0003
	Z-Test of Wilson & Hilferty	3.4488
	Hoelter Critical N	194
	Root Mean Square Residual (RMR)	0.1970
	Standardized RMR (SRMR)	0.0422
	Goodness of Fit Index (GFI)	0.9251

Parsimony Index	Adjusted GFI (AGFI)	0.8933
	Parsimonious GFI	0.7256
	RMSEA Estimate	0.0461
	RMSEA Lower 90% Confidence Limit	0.0316
	RMSEA Upper 90% Confidence Limit	0.0594
	Probability of Close Fit	0.6711
	ECVI Estimate	1.2204
	ECVI Lower 90% Confidence Limit	1.0841
	ECVI Upper 90% Confidence Limit	1.3933
	Akaike Information Criterion	282.8717
	Bozdogan CAIC	511.3843
	Schwarz Bayesian Criterion	460.3843
	McDonald Centrality	0.8809
Incremental Index	Bentler Comparative Fit Index	0.9698
	Bentler-Bonett NFI	0.9166
	Bentler-Bonett Non-normed Index	0.9615
	Bollen Normed Index Rho1	0.8936
	Bollen Non-normed Index Delta2	0.9703
	James et al. Parsimonious NFI	0.7189

Because change in chi-square values also follow a chi-square distribution, we can look at this difference between values (relative to the associated change in degrees of freedom) to assess change between nested models.

Also in Output 5.8, we see that the CFI value has increased (CFI = .97), the SRMR = .042, RMSEA = .046 and the 90% confidence limits for the RMSEA has narrowed and is within acceptable parameters (i.e., $.059 \geq \text{RMSEA CL}_{90} \geq .032$). This range of values suggests that the RMSEA would be less than .06 for this model 9 times out of 10 with other samples derived from this population. These goodness-of-fit indices provide general support for this revised model.

Statistical power for this revised model remains above .99 using the SAS code by MacCallum and colleagues (1996) presented earlier in this chapter (n = 240, df = 120). This finding increases confidence in the accuracy of findings. Despite the reduction in degrees of freedom, the remaining number (relative to sample size) is sufficient to accurately detect a poor fitting model.

The standardized factor loadings also appear in this section (Output 5.9). For example, see that the first standardized coefficient for LV1F1 is 0.89. These results indicate that all standardized loadings except two are greater than .60.t tests and their associated p values for standardized parameter and error estimates appear on page 7 of Output 5.9. Remember that the t test for parameter estimates appear below that loading, and to the right of the heading "t Value" for indicator variables. For example, you can see that the t value for the first loading (LV1F1) is 48.11 ($p < .01$). Remember that these tests are significant if the observed t value is greater than 1.96 (or less than –1.96). With this in mind, you can see that the t tests on page 16 of Output 5.9 show that the factor loadings for all 18 indicator variables remain significantly different from zero.

Output 5.9: PROC CALIS Output page 7 for Revised Measurement Model, Investment Model Study

Standardized Results for Linear Equations										
V1	=		0.8851	*	F1	+	1.0000		E1
Std Err			0.0184		LV1F1				
t Value			48.1094						
V2	=		0.8246	*	F1	+	1.0000		E2
Std Err			0.0239		LV2F1				
t Value			34.5236						
V3	=		0.9366	*	F1	+	1.0000		E3
Std Err			0.0146		LV3F1				
t Value			64.1463						
V5	=		0.8275	*	F2	+	1.0000		E5
Std Err			0.0254		LV5F2				
t Value			32.6325						
V6	=		0.8659	*	F2	+	1.0000		E6
Std Err			0.0222		LV6F2				
t Value			39.0827						
V7	=		0.8755	*	F2	+	1.0000		E7
Std Err			0.0214		LV7F2				
t Value			40.8893						
V8	=		0.6660	*	F3	+	1.0000		E8
Std Err			0.0498		LV8F3				
t Value			13.3610						
V9	=		0.6346	*	F3	+	1.0000		E9
Std Err			0.0515		LV9F3				
t Value			12.3261						
V10	=		0.7505	*	F3	+	1.0000		E10
Std Err			0.0463		LV10F3				
t Value			16.2198						
V11	=		0.8258	*	F4	+	1.0000		E11
Std Err			0.0450		LV11F4				
t Value			18.3700						
V12	=		0.8646	*	F4	+	1.0000		E12
Std Err			0.0449		LV12F4				
t Value			19.2692						
V13	=		0.4633	*	F4	+	1.0000		E13
Std Err			0.0571		LV13F4				
t Value			8.1201						
V14	=		0.8434	*	F5	+	1.0000		E14
Std Err			0.0334		LV14F5				
t Value			25.2655						

V15	=		0.7071	*	F5	+	1.0000		E15
Std Err			0.0409		LV15F5				
t Value			17.2800						
V16	=		0.5507	*	F5	+	1.0000		E16
Std Err			0.0519		LV16F5				
t Value			10.6113						
V17	=		0.6833	*	F6	+	1.0000		E17
Std Err			0.0472		LV17F6				
t Value			14.4816						
V18	=		0.7599	*	F6	+	1.0000		E18
Std Err			0.0440		LV18F6				
t Value			17.2750						
V19	=		0.7282	*	F6	+	1.0000		E19
Std Err			0.0452		LV19F6				
t Value			16.1122						

Standardized Results for Variances of Exogenous Variables					
Variable Type	Variable	Parameter	Estimate	Standard Error	t Value
Latent	F1		1.00000		
	F2		1.00000		
	F3		1.00000		
	F4		1.00000		
	F5		1.00000		
	F6		1.00000		

Error	E1	VARE1	0.21654	0.03257	6.64841
	E2	VARE2	0.32011	0.03939	8.12735
	E3	VARE3	0.12273	0.02735	4.48717
	E5	VARE5	0.31522	0.04197	7.51059
	E6	VARE6	0.25019	0.03837	6.52055
	E7	VARE7	0.23351	0.03749	6.22827
	E8	VARE8	0.55640	0.06640	8.37921
	E9	VARE9	0.59734	0.06534	9.14265
	E10	VARE10	0.43681	0.06944	6.29001
	E11	VARE11	0.31808	0.07424	4.28428
	E12	VARE12	0.25251	0.07758	3.25475
	E13	VARE13	0.78539	0.05286	14.85848
	E14	VARE14	0.28870	0.05631	5.12736
	E15	VARE15	0.49996	0.05787	8.63876
	E16	VARE16	0.69670	0.05717	12.18729
	E17	VARE17	0.53306	0.06449	8.26611
	E18	VARE18	0.42262	0.06685	6.32239
	E19	VARE19	0.46979	0.06581	7.13811

Standardized Results for Covariances Among Exogenous Variables					
Var1	Var2	Parameter	Estimate	Standard Error	t Value
F1	F2	CF1F2	0.60873	0.04733	12.86262
F1	F3	CF1F3	0.43995	0.06647	6.61925
F1	F4	CF1F4	-0.01632	0.07311	-0.22319
F1	F5	CF1F5	0.71440	0.04435	16.10878
F1	F6	CF1F6	-0.22331	0.07333	-3.04531
F2	F3	CF2F3	0.53383	0.06245	8.54758
F2	F4	CF2F4	-0.22450	0.07131	-3.14807
F2	F5	CF2F5	0.63479	0.05211	12.18058
F2	F6	CF2F6	-0.37506	0.06911	-5.42735
F3	F4	CF3F4	-0.09224	0.08170	-1.12901
F3	F5	CF3F5	0.51592	0.06898	7.47897
F3	F6	CF3F6	-0.42402	0.07524	-5.63599
F4	F5	CF4F5	0.00805	0.07896	0.10192
F4	F6	CF4F6	0.25464	0.07622	3.34103
F5	F6	CF5F6	-0.30048	0.07710	-3.89728

For this revised model, we again requested modification indices. In Output 5.10, we see that LM statistics indicate that additional revisions might further improve fit of the model to derived data. More specifically, in the second grouping of statistics we see that additional observed variables appear to load across factors (i.e., complex variables). Under the heading, "LM Stats for Paths from Exogenous Variables," it is apparent that V1 would also contribute significantly to measurement of F2 (satisfaction) as well as the commitment factor to which it was initially assigned.

Recall from Chapter 4 that the Lagrange Multiplier (LM) is a univariate statistic and only one revision should be considered at a time. Usually this will be the largest theoretically viable LM statistic. In other words, it is not appropriate to identify all statistically significant LM statistics and add several new path estimates in one step. As we saw in the last chapter, one change can make any subsequent revisions unnecessary. Adding more than one new parameter estimate at a time risks overfitting models.

Output 5.10: LM Statistics for the Revised Measurement Model, Investment Model Study

Rank Order of the 10 Largest LM Stat for Paths from Exogenous Variables				
To	From	LM Stat	Pr > ChiSq	Parm Change
V2	F2	12.34147	0.0004	-0.58375
V1	F2	11.89215	0.0006	0.44947
V2	F5	11.17253	0.0008	-0.72042
V1	F3	5.89356	0.0152	0.29318
V17	F2	4.96039	0.0259	-0.29453
V8	F1	4.45493	0.0348	0.21839
V9	F1	4.14497	0.0418	-0.20781
V18	F2	4.04611	0.0443	0.26581
V17	F4	3.62608	0.0569	0.24098
V15	F2	3.19261	0.0740	0.35651

In this instance, should we also delete V2 from the model and re-estimate the measurement model once again? In this case, the answer is no. Keeping with Rule #9 (necessary conditions for confirmatory factor analysis) as presented earlier in this chapter, there should be at least three observed variables for each latent factor; deleting V2 from the model would violate this rule. In some instances, an argument can be made for having only two observed variables contributing to measurement of a latent variable when the overall average of observed variables to latent factors in a model is three or more (i.e., other latent variables measured by four or more observed variables). That exception cannot be applied here either as all factors in our revised measurement model as presented in Figure 5.8 each have only three observed variables. This scenario provides a good example as to why it is good practice to have more than three observed variables per factor when computing measurement models. Had that been the case in this example, we would have had the flexibility to make further refinements to the measurement model.

In summary, the preceding output suggests that the revised measurement model:

- provides good fit to data as measured by the SRMR, CFI, and RMSEA, and narrow 90% confidence limits for the RMSEA
- retains sufficient power to reduce the likelihood of Type II errors
- displays no nonsignificant factor loadings

Combined, these findings provide general support for the revised measurement model. Before accepting this as your final model, however, you should first perform a few additional tests to assess the reliability and validity of measurement.

Assessing Reliability and Validity of Constructs and Indicators

One of the primary advantages of analyses with latent variables is the opportunity to assess the reliability and validity of responses to study variables. Broadly speaking, **reliability** refers to consistency of measurement (Devellis 2012). Responses to a test are reliable if, for example, they provide essentially the same set of scores upon repeated testing of the same participants. Reliability of responses can be assessed in a variety of ways (e.g., test-retest reliability, alternate-forms reliability, split-half reliability, internal consistency).

Validity, on the other hand, refers to the extent to which an instrument measures what it is intended to measure. If, for example, you develop a scale designed to measure locus of control, and scores on the scale do in fact reflect participants' underlying levels of perceived control, then the responses to the scale are valid. As with reliability, there are several different ways that validity can be assessed (e.g., construct validity, criterion-related validity).

Reliability is not an all-or-nothing phenomenon; rather, it is assessed along a continuum. Responses to an instrument may reflect a relatively high level of reliability, a relatively low level of reliability, or any amount in between. Indices of reliability may also differ across populations. As noted in Chapter 3, reliability and validity are properties of responses to scales, not scales themselves.

This section shows how the results of a CFA (confirmatory factor analysis) using PROC CALIS can be used to assess item reliability, composite reliability, variance extraction estimates, convergent validity, and discriminant validity. Combined, these procedures provide evidence concerning the extent to which responses to indicators measure what they are intended to measure.

Indicator Reliability

The reliability of an indicator variable is defined as the square of the correlation between a latent factor and that indicator. In other words, reliability is estimated by the percent of variation in the indicator that is explained by the factor that it is supposed to measure.

The reliability of an indicator can be computed in a very straightforward manner by squaring the standardized factor loadings obtained in the analysis. For example, the loading that represents the path from F1 to V1 was given the name LV1F1. The standardized loadings for all indicators are provided on page 7 of Output 5.11; you can see that the parameter estimate for LV1F1 is .8851. The square of this loading is .78, meaning that the reliability of responses to V1 is .78.

Fortunately, it is not necessary to actually perform this calculation, as these squared multiple correlation values are provided on page 6, presented below as of Output 5.11. The last column of this table (titled "R-square") indicates the percent of variance in each indicator that is accounted for by the common factor to which it was assigned. These R^2 values are indices of item reliability.

When assessing the contribution to measurement by scale items upon their respective factors, R^2 values greater than .39 are considered ideal. Only V13 and V16 fall short of this mark as shown in Output 5.11. These are the same two items with standardized path coefficient values below .60; this should not be surprising as R^2 values are derived from standardized path coefficients.

Output 5.11: PROC CALIS Output, Squared Multiple Correlation Values for Observed Variables, Revised Measurement Model, Investment Model Study

Squared Multiple Correlations			
Variable	Error Variance	Total Variance	R-Square
V1	1.33825	6.18020	0.7835
V2	2.70887	8.46228	0.6799
V3	0.91070	7.42018	0.8773
V5	1.17293	3.72104	0.6848
V6	1.11706	4.46477	0.7498
V7	0.98706	4.22714	0.7665
V8	1.11719	2.00789	0.4436
V9	1.18420	1.98246	0.4027
V10	1.29827	2.97218	0.5632
V11	2.14194	6.73403	0.6819
V12	1.82858	7.24148	0.7475
V13	4.37432	5.56960	0.2146
V14	1.27560	4.41840	0.7113
V15	2.46180	4.92396	0.5000
V16	2.44671	3.51188	0.3033
V17	2.13437	4.00400	0.4669
V18	1.63351	3.86516	0.5774
V19	2.24289	4.77423	0.5302

You can see that reliability estimates for indicators vary from a low of .21 for V13 to a high of .88 for V3. Some factors are measured by indicators which all display relatively high reliability estimates. For example, F1 (commitment) is measured by V1, V2, and V3, and the reliability estimates for these indicators are .78, .68, and .88, respectively. Other factors are assessed by indicators with relatively low reliability estimates. For example, F3 (rewards) is assessed by V8, V9, and V10, and the reliability estimates for these indicators are only .44, .40, and .56, respectively. It will be interesting to see whether the composite reliability for F3 is low. This topic is covered next.

Composite Reliability

When conducting research with multiple-item scales, it is appropriate to compute coefficient alpha values to estimate the reliability of scale responses (Cronbach 1951). This coefficient (represented by the Greek letter α) is an index of internal consistency. Alpha values should be greater than .69; however, values between .80 and .90 viewed as ideal. As discussed in Chapter 3 of this text, alpha values greater than .90 may suggest item redundancy.

Similarly, when performing confirmatory factor analysis, it is possible to compute a **composite reliability** index for each latent factor included in the model. This index is analogous to the coefficient alpha, and reflects the internal consistency of indicators measuring a given factor. The formula for this composite reliability index is presented below:

$$\text{Composite reliability} = \frac{(\Sigma L_i)^2}{(\Sigma L_i)^2 + \Sigma \quad \text{Var}(E_i)}$$

where

L_i = standardized factor loadings for that factor,

and

$\text{Var}(E_i)$ = error variance associated with the individual indicator variables.

Computing composite reliability for each scale will be easier if you first prepare a table that summarizes the necessary information. Table 5.2 provides the information needed to compute composite reliability estimates for the present measurement model. You can see that Table 5.2 includes the standardized loading for each indicator along with reliability estimates for each indicator (defined earlier as the square of the standardized loading). The last column in the table provides the error variance associated with each indicator.

Table 5.2: Information Needed to Compute Composite Reliability and Variance Extracted Estimates

Construct and Indicators	Standarized Loading	Indicator Reliability[a]	Error Variance[b]
Commitment (F1)			
V1	.89	.78	.22
V2	.82	.68	.32
V3	.94	.88	.12
Satisfaction (F2)			
V5	.83	.68	.32
V6	.87	.75	.25
V7	.88	.77	.23
Rewards (F3)			
V8	.67	.44	.56
V9	.63	.40	.60
V10	.75	.56	.44
Costs (F4)			
V11	.83	.68	.32
V12	.86	.75	.25
V13	.46	.21	.79
Investment size (F5)			
V14	.84	.71	.29
V15	.71	.50	.50
V16	.55	.30	.70
Alternative value (F6)			
V17	.68	.47	.53
V18	.76	.58	.42
V19	.73	.53	.47

[a] Calculated as the square of the standardized factor loading.
[b] Calculated as 1 minus the indicator reliability.

The error variance is calculated as $1 - L_i^2$, or 1 minus the square of the standardized factor loading for that variable. Because reliability is estimated by the square of the factor loading, you can calculate error variances by simply subtracting the reliability estimates from 1. Thus, for indicator V1, 1 - .78 = .22, for indicator V2, 1 - .68 = .32, and so forth. Readers are encouraged to prepare a table with the columns similar to those in Table 5.2 before computing composite reliabilities.

With error variances computed, you may now insert the values in Table 5.2 into the appropriate parts of the formula for composite reliability. The numerator of the formula appears below:

$$(\sum L_i)^2$$

Remember that operations that appear within parentheses are performed prior to performing operations outside of parentheses. The "$\sum$" symbol indicates that you are to add all factor loadings (which are symbolized by L_i). After this is done, square the resulting sum (as is indicated by the superscript "2" that appears outside of parentheses).

$$\Sigma L_i = .885 + .824 + .937$$
$$= 2.646$$

Now square of this sum is calculated:

$$(\Sigma L_i)^2 = (2.646)^2$$
$$= 7.001$$

This quantity is now inserted in the appropriate sections of the equation:

$$\frac{(\Sigma L_i)^2}{(\Sigma L_i)^2 + \Sigma \text{ Var}(E_i)} = \frac{7.001}{7.001 + \Sigma \text{ Var}(E_i)}$$

To calculate $\sum$ Var(Ei), you will simply sum the error variances associated with V1, V2, and V3 (from Table 5.2):

$$\Sigma \text{ Var}(E_i) = .217 + .321 + .122$$
$$= .660$$

This sum is inserted in the appropriate location in the formula, and you may now calculate the composite reliability for F1:

$$\frac{(\Sigma L_i)^2}{(\Sigma SL_i)^2 + \Sigma \text{ Var}(E_i)} = \frac{7.001}{7.001 + .660}$$
$$= \frac{7.001}{7.661}$$
$$= .914$$

So the composite reliability for F1 (the commitment construct) is .91. Similar to Cronbach's alpha, you should generally think of .70 as being the minimally acceptable level of reliability for instruments used in research whereas $.80 \leq \alpha \leq .90$ is preferred. Clearly, composite reliability for the commitment construct exceeds this requirement.

For purposes of contrast, the composite reliability for the F3 construct (rewards) is computed below:

$$\frac{\Sigma L_i^{\ 2}}{\Sigma L_i^{\ 2} + \Sigma \text{ Var}(E_i)} = \frac{(.67 + .63 + .75)^2}{(.67 + .63 + .75)^2 + (.56 + .60 + .44)}$$
$$= \frac{(2.05)^2}{(2.05)^2 + (1.60)}$$
$$= \frac{4.20}{4.20 + 1.60}$$
$$= \frac{4.20}{5.80}$$
$$= .72$$

And so the composite reliability for the rewards construct is .72; not as high as the reliability for commitment, but above the minimally acceptable level.

Table 5.3 reports reliability estimates for all variables included in the final measurement model. The third column of figures provides reliability data. Composite reliability estimates for latent factors are flagged with a superscript [b] symbol; individual indicator estimates are indented two spaces. You may choose to use a format similar to this when summarizing properties of a measurement model in research reports.

Table 5.3: Properties of the Revised Measurement Model

Construct and Indicators	Standardized Loading	t[a]	Reliability	Variance Extracted Estimate
Commitment (F1)			.91[b]	.78
V1	.89	48.11	.78	
V2	.82	34.52	.68	
V3	.94	64.15	.88	
Satisfaction (F2)			.89[b]	.73
V5	.83	32.63	.68	
V6	.87	39.08	.75	
V7	.88	40.89	.77	
Rewards (F3)			.72[b]	.47
V8	.67	13.36	.44	
V9	.63	12.32	.40	
V10	.75	16.22	.56	
Costs (F4)			.77[b]	.55
V11	.83	18.37	.68	
V12	.86	19.27	.75	
V13	.46	8.12	.21	
Investment size (F5)			.75[b]	.50
V14	.84	25.27	.71	
V15	.71	17.28	.50	
V16	.55	10.61	.30	
Alternative value (F6)			.77[b]	.53
V17	.68	14.48	.47	
V18	.76	17.27	.58	
V19	.73	16.11	.53	

[a] All t tests were significant at $p < .01$.
[b] Denotes composite reliability.

Variance Extracted Estimates

Variance extracted estimates are next calculated to assess the amount of variance captured by factors in relation to variance attributable to measurement error. The formula appears below:

$$\text{Variance extracted} = \frac{\Sigma L_i^2}{\Sigma L_i^2 + \Sigma \operatorname{var}(E_i)}$$

Notice that the preceding differs from the formula for composite reliability in that the $\sum L_i$ term is no longer within parentheses. This means that each factor loading is squared first, and then these squared factor loadings are summed. Because a squared factor loading for an indicator is an estimate of that indicator's reliability, this is equivalent to simply summing the reliability estimates for a given factor's indicators. To illustrate, the variance extracted estimate for F1 is calculated below by summing the reliabilities and error variance terms from Table 5.2:

$$\begin{aligned}\frac{\Sigma L_i^2}{\Sigma L_i^2 + \Sigma \operatorname{Var}(E_i)} &= \frac{(.783 + .679 + .878)}{(.783 + .679 + .878) + (.217 + .321 + .122)} \\ &= \frac{2.340}{2.340 + .660} \\ &= \frac{2.340}{3.000} \\ &= .780\end{aligned}$$

So the variance extracted estimate for the commitment factor was .78, meaning that 78% of variance is captured by your commitment construct; only 22% (or 1 - .78 = .22) is error. Fornell and Larcker (1981) suggest that constructs should have variance extracted estimates greater than .49; estimates less than .50 indicate that measurement error is larger than variance captured by the factor. This may call into question the validity of the latent construct as well as its indicators. This test is quite conservative, however; very often variance extracted estimates will be below .50 even when reliability estimates are acceptable.

The last column of Table 5.3 provides variance extracted estimates for the six study constructs. Note that all exceed the .50 criterion except for rewards (F3), for which the variance extracted estimate was .47. On the whole, however, the constructs in the model fared fairly well. (The average variance estimate is .59 across the six factors.)

Convergent Validity

Convergent validity and discriminant validity are commonly associated with use of the multi-trait, multi-method (MTMM) approach to validation in which multiple constructs are each assessed using more than one assessment method. This MTMM approach is believed to provide a stronger test of convergent (and discriminant) validity than the following procedures. Nonetheless, they provide a measure of convergent and discriminant validity of constructs within measurement models; these are useful in situations when it is not possible to follow the MTMM approach.

Stated simply, **convergent validity** is demonstrated when scores from different instruments used to measure the same construct are strongly correlated. For example, imagine that tests using different methods (e.g., a written test and an oral test) are both used to measure some technical skill in a sample of participants. Further imagine that the correlation coefficient between tests is measured as $r = .70$. This finding would provide initial evidence of convergent validity; the strong correlation suggests that both instruments are measuring the same construct even though they used different methods.

Convergent validity for the current study is, instead, assessed by reviewing the t tests for the factor loadings. If all factor loadings for the indicators measuring the same construct are statistically significant (greater than twice their standard errors), this suggests convergent validity of those indicators. The finding that t values are significant for all path coefficients suggests that indicators effectively measure the same construct.

For example, the standardized factor loadings from the current analysis and the t tests for these loadings are presented in Table 5.3. Consider the convergent validity for V1, V2, and V3 which are each assumed to measure the commitment construct. Results show that the t values for these three indicators range from 15.32 to 18.78.These t values are all significantly different from zero as $p < .01$ (i.e., the three t values exceed the critical t threshold value of 2.58 for $p < .01$). These results support the convergent validity of V1, V2, and V3 as measures of commitment. A quick review of the remaining constructs shows that $p < .01$ for all t values.

Discriminant Validity

Conversely, **discriminant validity** is demonstrated when different instruments are used to measure unrelated or divergent constructs and the correlation coefficients between the measures are weak or strongly negative. A test displays discriminant validity when it is demonstrated that the test does *not* measure a construct that it was not designed to measure.

As with convergent validity, discriminant validity is often studied using the multi-trait, multi-method procedure (MTMM). For example, assume that you are studying psychological needs and have developed an instrument to measure the need for power. You not only want to demonstrate that your scale successfully measures the need for power but also that it does not measure a similar psychological construct such as the need for achievement. You can obtain this evidence using the MTMM approach. Assume that you administer the following four instruments to a sample of participants:

Test A: Your new measure of the need for power (a self-report scale).

Test B: A previously validated test of the need for power (an observer-rated test).

Test C: A previously validated test of the need for achievement (a self-report scale).

Test D: A second previously validated test of the need for achievement (an observer-rated test).

Note that you are assessing multiple traits (need for power versus need for achievement) and are measuring each via multiple methods (a self-report scale versus an observer-rated test).

When the data are analyzed, you will hope for a number of results. First, you hope that Test A will show a relatively strong correlation with Test B. This would mean that your new need for power scale is strongly related to another measure of the need for power. This outcome would demonstrate convergent validity of responses to the new scale.

To support the discriminant validity of the scale, you also hope that Test A (a test of the need for power) will be weakly correlated with both Tests C and D (two tests of the need for achievement). At the very least, you will hope that Test A will show a weaker correlation with Tests C and D than it shows with Test B. This outcome will provide evidence supporting the discriminant validity of responses to Test A. It will show that Test A is apparently not measuring the need for achievement.

The MTMM approach provides a relatively strong test of discriminant validity. Unfortunately, these tests cannot be utilized for the present investment model study as multiple methods were not used to assess the different constructs. Nonetheless, evidence regarding discriminant validity may still be obtained from the present analysis through use of three procedures:

- the chi-square difference test

- the confidence interval test
- the variance extracted test

Remember that the procedures discussed here do not necessarily have to be performed as a matter of course each time a latent variable model is analyzed but are recommended in cases where discriminant validity is in doubt.

With the **chi-square difference test**, you assess the discriminant validity of two constructs by:

- estimating the standard measurement model in which all factors are allowed to covary
- creating a new measurement model identical to the previous one except that the correlation between the two factors of interest is fixed at 1
- computing the chi-square difference statistic for the two models

Discriminant validity is demonstrated if chi-square is significantly lower for the first model suggesting that the better model was the one in which the two constructs were viewed as distinct (but correlated) factors.

To illustrate, this procedure will be used to assess the discriminant validity of commitment (F1) and investment size (F5). Output 5.12 presents "Covariances Among Exogenous Variables" from your analysis of the revised measurement model in which all factors were allowed to covary. Notice that this table shows that the correlation between F1 and F5 is .71 (this is the 4th entry from the top under the "Estimate" heading). In one respect this is encouraging as the investment model predicts that investments have a positive effect on commitment; this correlation is consistent with that prediction. What is disconcerting, however, is the size of the coefficient. These two constructs are strongly correlated so it is reasonable to question whether you are, in fact, measuring two different constructs. It is possible that items V1 to V3 and items V14 to V16 are measuring the same underlying construct. If true, then responses to these items lack discriminant validity.

Output 5.12: Correlations Coefficients between Exogenous Variables for Revised Measurement Model, Investment Model Study

Covariances Among Exogenous Variables					
Var1	Var2	Parameter	Estimate	Standard Error	t Value
F1	F2	CF1F2	0.60873	0.04733	12.86262
F1	F3	CF1F3	0.43995	0.06647	6.61925
F1	F4	CF1F4	-0.01632	0.07311	-0.22319
F1	F5	CF1F5	0.71440	0.04435	16.10878
F1	F6	CF1F6	-0.22331	0.07333	-3.04531
F2	F3	CF2F3	0.53383	0.06245	8.54758
F2	F4	CF2F4	-0.22450	0.07131	-3.14807
F2	F5	CF2F5	0.63479	0.05211	12.18058
F2	F6	CF2F6	-0.37506	0.06911	-5.42735
F3	F4	CF3F4	-0.09224	0.08170	-1.12901
F3	F5	CF3F5	0.51592	0.06898	7.47897
F3	F6	CF3F6	-0.42402	0.07524	-5.63599
F4	F5	CF4F5	0.00805	0.07896	0.10192
F4	F6	CF4F6	0.25464	0.07622	3.34103
F5	F6	CF5F6	-0.30048	0.07710	-3.89728

To assess the discriminant validity of F1 and F5, you will modify your CALIS program so that covariance between the two factors is fixed at 1. This will require that you change only one equation in the COV statement of your program (The COV statement had appeared in the earlier section of this chapter titled "Overview of the PROC CALIS program"). Below is the revised cov statement:

```
        COV
           F1 F2 = CF1F2,
           F1 F3 = CF1F3,
           F1 F4 = CF1F4,
❶          F1 F5 = 1,
           F1 F6 = CF1F6,
           F2 F3 = CF2F3,
           F2 F4 = CF2F4,
           F2 F5 = CF2F5,
           F2 F6 = CF2F6,
           F3 F4 = CF3F4,
           F3 F5 = CF3F5,
           F3 F6 = CF3F6,
           F4 F5 = CF4F5,
           F4 F6 = CF4F6,
           F5 F6 = CF5F6;
```

Notice that this statement is identical to the original COV statement with one exception; line ❶ of the new statement fixes covariance between F1 and F5 at 1. The model created as a result of this modification will be referred to as the **unidimensional model**, and the model in which covariance between F1 and F5 is free to be estimated will be referred to as the **standard measurement model.**

When estimated, the unidimensional model produced a model chi-square value of 251.13 251.13ith 121 degrees of freedom. (The output from this analysis is not reproduced here.) The summary table for your standard measurement model (from Output 5.11) has already shown that the chi-square for that model was 180.87 with 120 degrees of freedom. You may now calculate the difference in chi-square between the two models:

$$\begin{array}{r} 251.13 \\ -\ 180.87 \\ \hline 70.26 \end{array}$$

So the difference in chi-square values is 70.26. To determine whether this value is statistically significant, find the critical chi-square value for the degrees of freedom associated with the test. The df for the test is found by subtracting the df for the two models:

$$\begin{array}{r} 121 \\ -\ 120 \\ \hline 1 \end{array}$$

Since there is 1 degree of freedom associated with this chi-square difference test, you turn to a table of chi-square in Appendix C of this text and find that, with 1 df, the critical values of chi-square are 3.84 at $p < .05$, 6.64 at $p < .01$. With a chi-square difference value of 70.26 the difference between the two models was clearly significant at $p < .01$. In other words, the standard measurement model in which the factors were viewed as distinct but correlated constructs provided a fit that was significantly better than the fit provided by the unidimensional model. In short, this test supports the discriminant validity of F1 and F5.

In some cases, you may want to test the discriminant validity for every possible pair of F factors. This would require a series of tests, in which covariance between just two factors is fixed at 1, the model is estimated, and you compute the resulting chi-square difference from the chi-square value for the standard measurement model. In the present case, this would result in 15 different models (and consequently 15 difference tests) as there are 15 separate covariances between the six factors in the model.

Performing such a large number of tests, however, creates problems involving the overall significance level for the family of tests. If you perform just one difference test and use the critical value of chi-square associated with $p = .05$, it is clear that the significance level for that test is .05. However, if you perform a series of tests, the overall significance level for that series of tests will be larger. The overall significance level for a family of tests can be computed with this formula:

$$a_0 = 1 - (1 - a_i)^t$$

where

a_0 = the overall significance level for the family of tests

a_i = the significance level used for each individual difference test

t = the number of tests performed

For example, imagine that you perform just two tests, and use the significance level of .05 for both individual tests. (This means that you used a critical value of 3.84 for both tests.) What is the actual overall significance level for the family of tests? That is, what is the probability that you will incorrectly reject a true null hypothesis for at least one of the tests? You may find this by inserting the appropriate figures in the preceding formula:

$$a_0 = 1 - (1 - a_i)^t$$

$$= 1 - (1 - .05)^2$$

$$= 1 - (.95)^2$$

$$= 1 - .9025$$

$$= .0975$$

So the actual significance level for the series of tests is almost. 10. This means that there is close to a 10% chance that you would incorrectly reject a true null hypothesis for at least one of the two tests. This actual significance level is higher than the standard level of .05, making this an unacceptable possibility.

The formula shows that the overall significance level quickly reaches an unacceptable level when many individual tests are performed. For example, if you set p at .01 for individual tests and perform all 15 of the comparisons, the overall significance level is actually .14 for the series of tests.

You can do two things to address this problem. First, you can perform what is known as a **Bonferroni correction** whereby the alpha value is adjusted according to the number of tests performed. In this case for example, the standard alpha of .05 would be divided by 15 to arrive at a modified significance threshold of .003 (i.e., .05/15 = .003). This procedure, however, is generally regarded as overly conservative.

Second, you should perform as few individual tests as are necessary, conducting only those of substantive importance. For example, if factor F1 is being measured by new, unvalidated indicators, and factors F2 to F6 are measured by older tests and scales with established validity of responses, then you may want to perform just those tests that specifically assess the discriminant validity of F1.

The chi-square difference test is not without its limitations (MacCallum et al. 2006). We therefore recommend that you may also perform a **confidence interval test** to assess discriminant validity between factors. This test involves calculating a confidence interval of plus or minus 2 standard errors around the correlation between factors and determining whether this interval includes 1.0. If it does not include 1.0, support is found for the discriminant validity of factors (Anderson and Gerbing 1988).

For the final measurement model, the information necessary to perform this test is again presented in Output 5.13. This output presents the covariance estimates (correlations, in this case) between all latent factors, along with their associated standard errors.

Output 5.13: Correlations Coefficients between Exogenous Variables, Investment Model Study

Standardized Results for Covariances Among Exogenous Variables					
Var1	Var2	Parameter	Estimate	Standard Error	t Value
F1	F2	CF1F2	0.60873	0.04733	12.86262
F1	F3	CF1F3	0.43995	0.06647	6.61925
F1	F4	CF1F4	-0.01632	0.07311	-0.22319
F1	F5	CF1F5	0.71440	0.04435	16.10878
F1	F6	CF1F6	-0.22331	0.07333	-3.04531
F2	F3	CF2F3	0.53383	0.06245	8.54758
F2	F4	CF2F4	-0.22450	0.07131	-3.14807
F2	F5	CF2F5	0.63479	0.05211	12.18058
F2	F6	CF2F6	-0.37506	0.06911	-5.42735
F3	F4	CF3F4	-0.09224	0.08170	-1.12901
F3	F5	CF3F5	0.51592	0.06898	7.47897
F3	F6	CF3F6	-0.42402	0.07524	-5.63599
F4	F5	CF4F5	0.00805	0.07896	0.10192
F4	F6	CF4F6	0.25464	0.07622	3.34103
F5	F6	CF5F6	-0.30048	0.07710	-3.89728

Once again, you can see that the correlation between F1 and F5 is .71, and the standard error for this estimate is .04. To compute the confidence interval for this correlation, you first multiply this standard error by 2:

$$2 \times .04 = .08$$

The lower boundary for the confidence interval will be two standard errors below the correlation:

$$.71 - .08 = .63$$

The upper boundary for the confidence interval will be two standard errors above the correlation:

$$.71 + .08 = .79$$

So the confidence interval for the relationship between F1 and F5 ranges from .63 to .79. This confidence interval does not include the value of 1.0, meaning that it is unlikely that the actual population correlation between F1 and F5 is 1.0. This finding supports the discriminant validity of the measures.

Finally, discriminant validity may also be assessed with the **variance extracted test**. With this test, you review the variance extracted estimates (as described above) between the two factors of interest and compare these estimates to the square of the correlation between factors. Discriminant validity is demonstrated if both variance extracted estimates are greater than this squared correlation.

In the present study, the correlation between factors F1 and F5 is .71, and this value squared is .50 (i.e., $.71^2$). Variance extracted estimates were calculated earlier and appear in Table 5.3. You can see that the variance extracted estimate is .78 for F1 and .50 for investment size. Because the variance extracted estimate for investment size is not greater than the square of the inter-factor correlation (i.e., .50 = .50), this test does not support the discriminant validity of the two factors.

In summary, your analyses provided mixed support for the discriminant validity of the commitment and investment size measures. The chi-square difference test and the confidence interval test suggested that indicators V1 to V3 and indicators V14 to V16 are measuring two distinct constructs, while the variance extracted test did not. (See Cappeliez and O'Rourke [2006] for an example where each of these steps is performed with actual participant data.)

Characteristics of an Ideal Fit for the Measurement Model

A measurement model provides ideal fit to the data when it displays the following characteristics:

- the Comparative Fit Index (CFI) exceeds .94, both the Standardize Root Mean Square Residual (SRMR) and the Root Mean Square Error of Approximation (RMSEA) are less than .055 (and the upper bound 90% confident limit for the RMSEA is less than .09)
- the model has sufficient statistical power to have confidence in the goodness-of-fit statistics
- the absolute value of the t statistics for each parameter estimate and error term exceed 1.96 (or less than -1.96)
- composite reliabilities for the latent factors should exceed .70, ideally greater than .80
- variance extracted estimates for the latent factors should exceed .50
- discriminant validity for questionable pairs of factors should be demonstrated using the chi-square difference test, the confidence interval test, or the variance extracted test

Remember that the above represent an ideal that very often is often not attained with real-world data even when the measurement model is quite good. Model fit need not meet all of the above criteria in order to be deemed "acceptable." For instance, it is not atypical to accept models which exhibit ideal values for two of three goodness-of-fit statistics (particularly if confidence limits for the RMSEA are within acceptable range of values).

For example, the final measurement model for the investment model study demonstrated all of the preceding characteristics, with the exception of one test that failed to support the discriminant validity of factors F1 and F5. Overall results, however, support the final measurement model.

Having established that you have developed an acceptable measurement model, you may now turn you attention to the analyses of central interest: The test of the theoretical model that specifies directional relationships between latent factors (as illustrated in Figure 5.4). The following chapter shows how to test this theoretical model.

Conclusion: On to Structural Equation Modeling

This chapter has shown how to use the CALIS procedure to perform confirmatory factor analysis. In some cases, your only purpose in performing a confirmatory factor analysis will be to test the factor structure underlying a set of data. This is often the case when a specific theory describes the factor structure that should underlie a derived dataset and you want to empirically test the theory. In this instance, your analysis will essentially begin and end with the confirmatory factor analysis.

In other cases, however, confirmatory factor analysis will merely be the first step in a two-step process of theory testing. This will be the case when you want to test a model that specifies directional relationships among a number of latent variables. To test such models, confirmatory factor analysis is used to develop an acceptable measurement model; this measurement model is then modified to become a structural equation model. The current chapter has shown how to use PROC CALIS to develop measurement models. This material provides a basis for the next chapter describing how PROC CALIS is used to compute structural equation models.

References

Anderson, J. C., and Gerbing, D. W. (1988). Structural equation modeling in practice: A review and recommended two-step approach. *Psychological Bulletin, 103,* 411–423.

Bentler, P. M., and Chou, C. (1987). Practical Issues in structural modeling. *Sociological Methods and Research, 16,* 78–117.

Byrne, B. M. (1998). *Structural equation modeling with LISREL, PRELIS, and SIMPLIS: Basic concepts, applications, and programming.* Mahwah, NJ: Lawrence Erlbaum.

Campbell, D. T. and Fiske, D. W. (1959). Convergent and discriminant validation by the multitrait-multimethod matrix. *Psychological Bulletin, 56,* 81–105.

Cappeliez, P., and O'Rourke, N. (2006). Empirical validation of a comprehensive model of reminiscence and health in later life. *Journals of Gerontology: Psychological Sciences, 61,* P237–P244.

Chou, P. H. B., and O'Rourke, N. (2012). Development and initial validation of the Therapeutic Misunderstanding Scale for use with clinical trial research participants. *Aging and Mental Health, 16,* 45–15.

Cohen, J. (1988). *Statistical power analysis for the behavioral sciences* (2nd Ed.). Hillsdale, NJ: Lawrence Erlbaum.

DeVellis, R. F. (2012). *Scale development theory and applications* (3rd Ed.).Thousand Oaks, CA: Sage.

Finn, J. D. (1974). *A general model for multivariate analysis.* New York: Holt, Rinehart, and Winston.

Floyd, F. J., and Widaman, K. F. (1995). Factor analysis in the development and refinement of clinical assessment instruments. *Psychological Assessment, 7,* 286–299.

Fornell, C., and Larcker, D. F. (1981). Evaluating structural equation models with unobservable variables and measurement error. *Journal of Marketing Research, 18,* 39–50.

Hu, L. T., and Bentler, P. M. (1999). Cutoff criteria for fit indices in covariance structure analysis: Conventional criteria versus new alternatives. *Structural Equation Modeling, 6,* 1–55.

Kline, R. B. (2005). *Principles and practice of structural equation modeling.* (2nd ed.). New York: Guilford Press.

Le, B., and Agnew, C. R. (2003). Commitment and its theorized determinants: A meta-analysis of the investment model. *Personal Relationships, 10,* 37–57.

Lehmann, E.L. (1999). *Elements of large-sample theory.* New York: Springer.

MacCallum, R.C., Browne, M.W., and Cai, L. (2006). Testing differences between nested covariance structure models: Power analysis and null hypotheses. *Psychological Methods, 11,* 19–35.

MacCallum, R. C., Browne, M. W., and Sugawara, H. M. (1996). Power analysis and determination of sample size for covariance structure modeling. *Psychological Methods, 2,* 130–149.

MacCallum, R. C., Widaman, K. F., Zhang, S., and Hong, S. (1999). Sample size in factor analysis. *Psychological Methods, 4,* 84–99.

McDonald, R. P., and Ho, M. R. (2002). Principles and practice in reporting structural equation analyses. *Psychological Methods, 7*, 64–82.

Netemeyer, R. G., Johnston, M.W., and Burton, S. (1990). Analysis of Role conflict and role ambiguity in a structural equations framework. *Journal of Applied Psychology, 75,* 148–157.

Rusbult, C. E. (1980). Commitment and satisfaction in romantic associations: A test of the investment model. *Journal of Experimental Social Psychology, 16,* 172–186.

Sternberg, R. (1994). *In search of the human mind.* New York: Harcourt Brace.

Chapter 6: Structural Equation Modeling

Basic Concepts in Covariance Analyses with Latent Variables

The concept of covariance analyses with latent variables was introduced in Chapter 5, "Developing Measurement Models with Confirmatory Factor Analysis." Chapter 5 indicated that directional models with latent (unobserved) variables are generally referred to as structural equation models. The difference between confirmatory factor analyses (CFA) and structural equation modeling (SEM) is that with CFA, all latent variables are correlated (i.e., covariance is assumed between parings of latent constructs). With SEM, in contrast, directional relationships are assumed between latent variables. SEM models used to be called LISREL-type models or causal models (Jöreskog and Sörbom 2001). The latter term is no longer used because directional associations between latent variables are not sufficient to allow us to conclude that their relationship is, in fact, causal.

Analysis with Manifest Variables versus Latent Variables

In some ways, performing covariance analysis with latent variables is similar to performing path analysis. For example, in Chapter 4 you learned how to use path analysis to test a directional model derived from Rusbult's (1980) investment model (see Le and Agnew 2003 for a review). That model predicted that (a) relationship commitment was determined by satisfaction, investment size, and alternative value, while (b) relationship satisfaction was determined by the rewards and costs associated with the relationship. In this chapter, you will learn how to test the same directional model with latent variables.

One of the differences between the two procedures involves the number of indicator variables that are used to represent the underlying constructs in the SEM model. In path analysis, each construct of interest is measured by just *one* indicator variable. For example, in Chapter 4 the variable "commitment" was measured by just one observed variable (i.e., participants' scores on a single "commitment" scale). With CFA and SEM, on the other hand, each construct of interest is measured by *multiple* indicator variables. For example, in this chapter you will learn how to compute a model in which the latent construct "commitment" is measured by three observed variables, the latent construct "satisfaction" is measured by three different observed variables, and so forth.

Chapter 5 pointed out that analysis of covariance structures has a number of important advantages over path analysis. For example, when you perform CFA, you can estimate measurement error for each latent constructs as well as have the the opportunity to assess the convergent and discriminant validity of your latent constructs.

A Two-Step Approach to Structural Equation Modeling

Chapter 5 indicated that this text will follow a two-step approach for performing analysis of covariance structures as first described by Anderson and Gerbing (1988). The first step of this process involves using confirmatory factor analysis to develop an acceptable measurement model. A **measurement model** is a CFA model in which you identify latent constructs of interest and indicate which observed variables measure each latent construct. In a measurement model, you do not specify any directional relationships between latent constructs; instead, you allow each latent construct to covary (correlate) with every other latent construct. Chapter 5 discussed a number of procedures that you can use to verify that your measurement model displays an acceptable fit to data, and also showed how to modify the model to achieve a better fit.

Once you have developed a measurement model with acceptable fit, you can then to move on to the second step of the two-step procedure. In this phase, you modify the measurement model so that it now specifies directional relationships between latent variables. You make these modifications so that the model comes to represent the theoretical model that you want to test. The resulting theoretical model is a "combined model" that actually consists of two components:

- a **measurement model** that specifies relationships between latent constructs and their indicator variables
- a **structural model** that specifies directional relationships between latent constructs

When you perform SEM, you perform a simultaneous test that determines whether this combined measurement and structural model provides an acceptable fit to data. If it does, then your theoretical model has survived an attempt at disconfirmation; you obtain support for its predictions.

This chapter focuses on the second step of the two-step procedure (Anderson and Gerbing 1988). It shows how to modify a measurement model so that it specifies directional relationships between latent constructs. It reviews a number of procedures and indices that can be used to determine whether the resulting theoretical model provides an acceptable and parsimonious fit to data. It also shows how to use modification indices to achieve a better model fit when necessary.

The Importance of Reading Chapters 4 and 5 First

If you are like most readers, it will be necessary to read Chapters 4 and 5 of this text before reading this chapter. Chapter 4 discusses a number of basic issues in path analysis, and provides an introduction to PROC CALIS. Unless you are already familiar with path analysis and the CALIS procedure, you should read Chapter 4 before proceeding.

Chapter 5 discusses not only CFA but also introduces basic issues in SEM. In fact, Chapter 5 and the current chapter were designed to be used together as a two-part introduction to analysis of covariance structures. The current chapter assumes that you are already familiar with the two-step approach along with all of the other topics introduced in Chapter 5. Even if you are only interested in SEM, you will still need to understand the concepts described in Chapter 5 before beginning this chapter.

Testing the Fit of the Theoretical Model from the Investment Model Study

This chapter shows how to perform SEM in order to test a theoretical model based on Rusbult's (1980) investment model. The model to be tested here is similar to the one first presented in Chapter 4, which specifies the relationship among the following six constructs:

- **Commitment:** the intention to maintain a current romantic relationship
- **Satisfaction:** the emotional response to the current relationship
- **Investment size:** the amount of time and effort that one has put into a current relationship
- **Alternative value:** the perceived attractiveness of alternatives to a current relationship
- **Rewards:** the perceptions of the number of good things association with a current relationship
- **Costs:** the perceptions of the number of bad things associated with the current relationship

The theoretical model to be tested here predicts that (a) commitment is determined by satisfaction, investment size, and alternative value whereas (b) satisfaction is determined by rewards and costs.

This analysis actually began in Chapter 5 with an initial measurement model (illustrated in Figure 5.6) in which the latent construct commitment was measured by four indicator variables, while the remaining five latent constructs were measured by three indicator variables. Eventually, one of the indicator variables was dropped so that the measurement model would achieve a better fit. The resulting revised measurement model displayed a generally acceptable fit to data and is reproduced here as Figure 6.1.

Figure 6.1: The Revised Measurement Model, Investment Model Study (from Chapter 5)

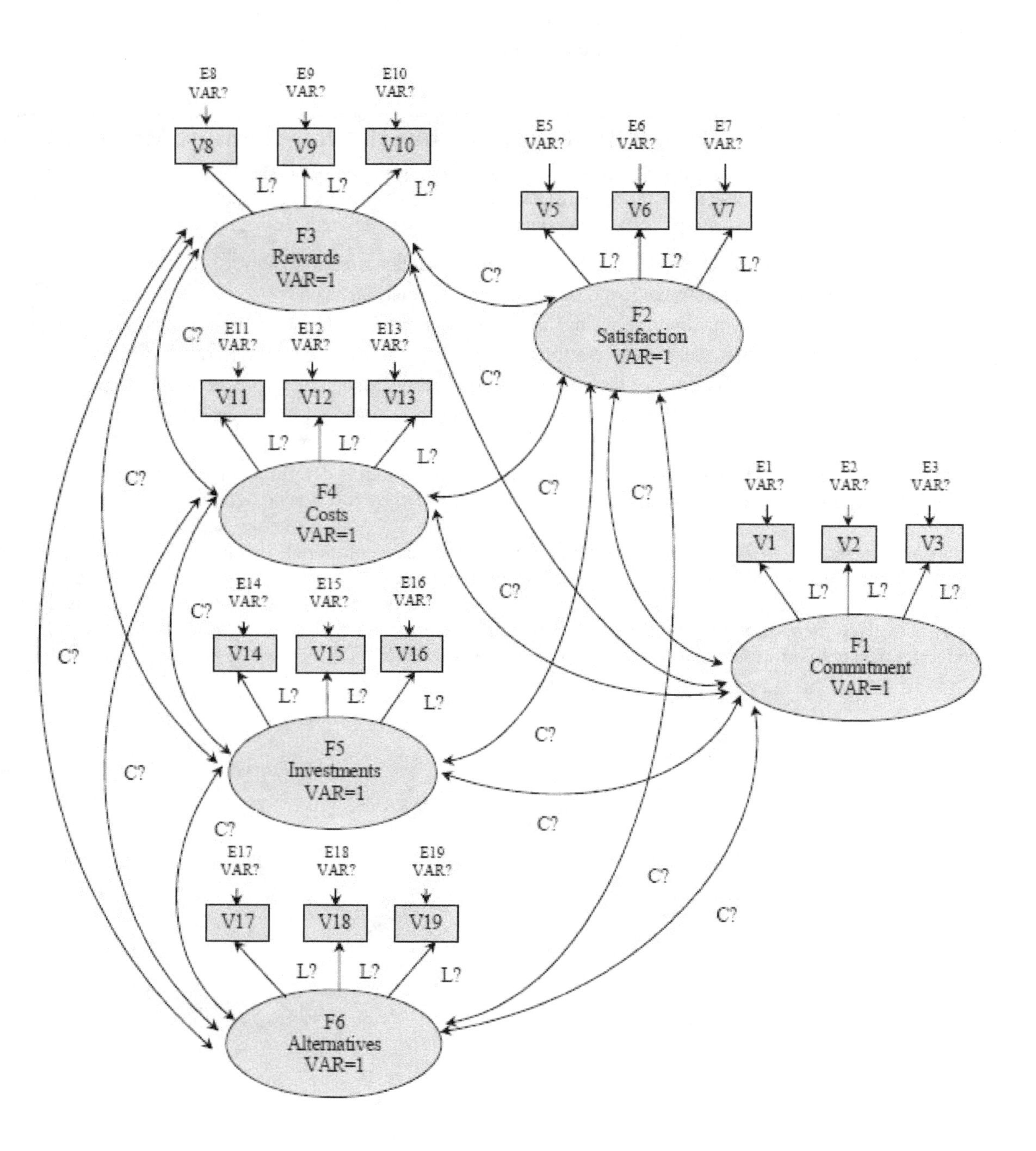

Notice that Figure 6.1 is a standard confirmatory factor analytic model, in that no directional relationships are assumed between any of the latent constructs (i.e., the F variables); instead, each latent construct is allow to freely covary (correlate) with every other latent construct. Covariance is represented by curved, double-headed arrows that connect the various F variables.

In this chapter, you will learn how to convert this measurement model into a theoretical model that predicts directional relationships between some of the F variables. A measurement model is converted into a directional model by replacing double-headed arrows that predict covariance or correlation with straight, single-headed arrows that predict directional effects.

The theoretical model to be tested is reproduced here as Figure 6.2. Notice how commitment (F1) and satisfaction (F2) are no longer connected to the other F variables by curved, double-headed arrows; instead, straight, single-headed arrows now point from satisfaction (F2), investments (F5), and alternatives (F6) to commitment (F1). This represents the prediction that these three latent constructs will have directional effects on commitment. Similarly, straight, single-headed arrows now point from rewards (F3) and costs (F4) to satisfaction (F2), which is consistent with the prediction that these two constructs will have direct effects on satisfaction.

Figure 6.2: The Initial Theoretical Model

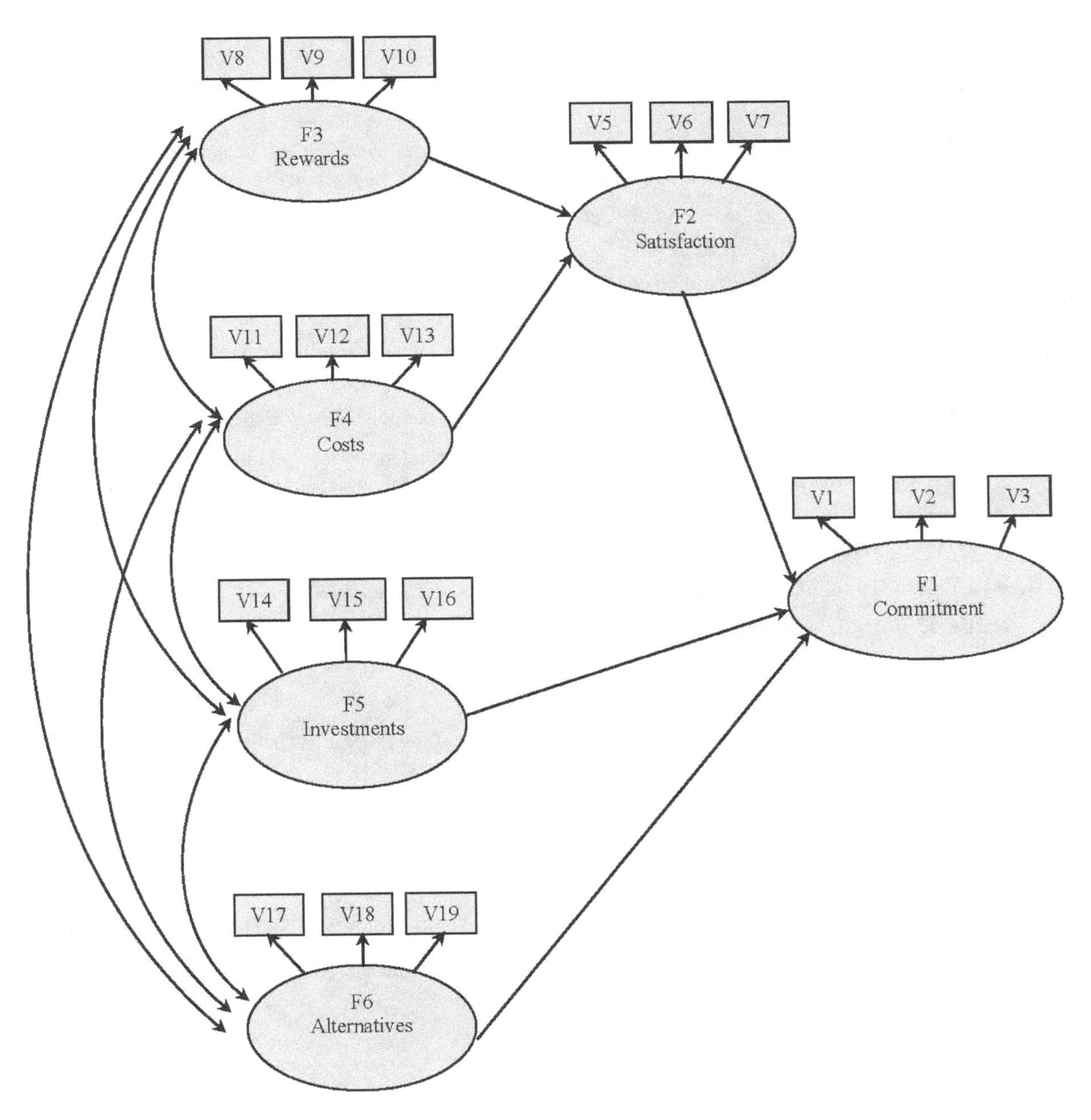

The steps followed in testing the theoretical model of Figure 6.2 are essentially the same as those followed in step 1 (from Chapter 5): You will first prepare a program figure, and then use that figure to modify your PROC CALIS program. Because the steps in doing this should now be familiar to you, this will be done in brief. In reviewing the results of the analysis, you will use the same fit indices discussed earlier to assess overall model fit, and you will learn some new indices to assess fit in just the structural portion of the model.

The Rules for Structural Equation Modeling

Chapters 4 and 5 provided a number of rules to follow when performing either path analysis with manifest variables or confirmatory factor analysis. These rules are again listed below:

RULE 1: Only exogenous variables are supposed to have covariance parameters in the model

RULE 2: A residual term is identified for each endogenous variable in the model.

RULE 3: Exogenous variables do not have residual terms.

RULE 4: Variances should be estimated for every exogenous variable in the model, including residual terms.

RULE 5: In most cases, there should be covariance estimates for every pair of manifest or exogenous variables; covariance estimates are not required for endogenous variables.

RULE 6: For simple recursive models, covariance is not estimated for residual terms. (With more elaborate models; this rule is relaxed; e.g., O'Rourke et al. 2013.)

RULE 7: One equation should be created for each endogenous variable with that variable's name to the left of the equals sign.

RULE 8: Variables that have a direct effect on that endogenous variable are listed to the right of the equals sign.

RULE 9: Exogenous variables, including residual terms, are never listed to the left of the equals sign.

RULE 10: To estimate a path coefficient for a given independent variable, a unique path coefficient name can be created for the path coefficient associated with that independent variable.

RULE 11: The last term in each equation should be the residual (disturbance) term for that endogenous variable; this E (or D) term will have no name for its path coefficient.

RULE 12: To estimate path coefficients, create a name for that parameter.

RULE 13: To fix a parameter at a given numerical value, insert that value in the place of the parameter's name.

RULE 14: To constrain two or more parameters to be equal, use the same name for those parameters.

RULE 15: In confirmatory factor analysis, the variances of the latent F variables are usually fixed at 1 (or other number to provide a metric for each latent variable).

Many of these rules will also apply when performing SEM; specific rules will be mentioned when they become relevant in the sections to follow. This chapter will also introduce three additional rules that are pertinent to the types of analyses to be discussed here. The new rules are presented here for future reference, so all of the rules relevant to structural equation modeling will be grouped together in one location:

RULE 16: In SEM, the variances of the exogenous F variables are free parameters to be estimated.

RULE 17: In SEM, one factor loading for each F variable should be fixed at 1 (or other value).

Preparing the Program Figure

Remember that this chapter deals only with directional models in which the structural portion of model is recursive (versus bidirectional). This means that the model will contain no reciprocal relationships or feedback loops.

When preparing the program figure for a theoretical model, you must first verify that the structural portion of the model is not **saturated** (i.e., is not just-identified). The structural portion of a model is saturated if every structural variable is related to every other structural variable by either a curved arrow or directional path.

For example, consider the model in Figure 6.2. The structural variables in this model are the variables that constitute the structural portion of the system, and in this case the structural variables are the latent F variables displayed in ovals: Commitment, satisfaction, rewards, costs, investments, and alternatives. The structural portion of this model would be saturated if every oval were directly connected to every other oval by either a curved or straight arrow.

Fortunately, you can see that this is not the case. You know that this model is not saturated because:

- the latent variable rewards is not directly connected to commitment in any way
- costs is similarly not directly connected to commitment
- investments is not directly connected to satisfaction
- alternatives is not connected to satisfaction

If the structural portion of the model had been saturated, it would have been possible to estimate the model, but it would not have been possible to test just the structural portion of the model for goodness of fit. Because the four paths described above are not estimated, the structural portion of the model is not saturated and may be tested. You may therefore proceed with the development of the program figure.

Step 1: Identifying Disturbance Terms for Endogenous Variables

The program figure for a theoretical model is prepared by following the same steps used with the measurement model. First, the disturbance terms for all endogenous variables are identified. Earlier, it was noted that a disturbance term represents factors as random shocks, misspecifications, measurement error, and omitted independent variables.

There are generally two types of disturbance terms in SEM: E terms and D terms. These are illustrated in Figure 6.3.

Figure 6.3: The Initial Theoretical Model, Including Disturbance (Residual) Terms for Endogenous Variables

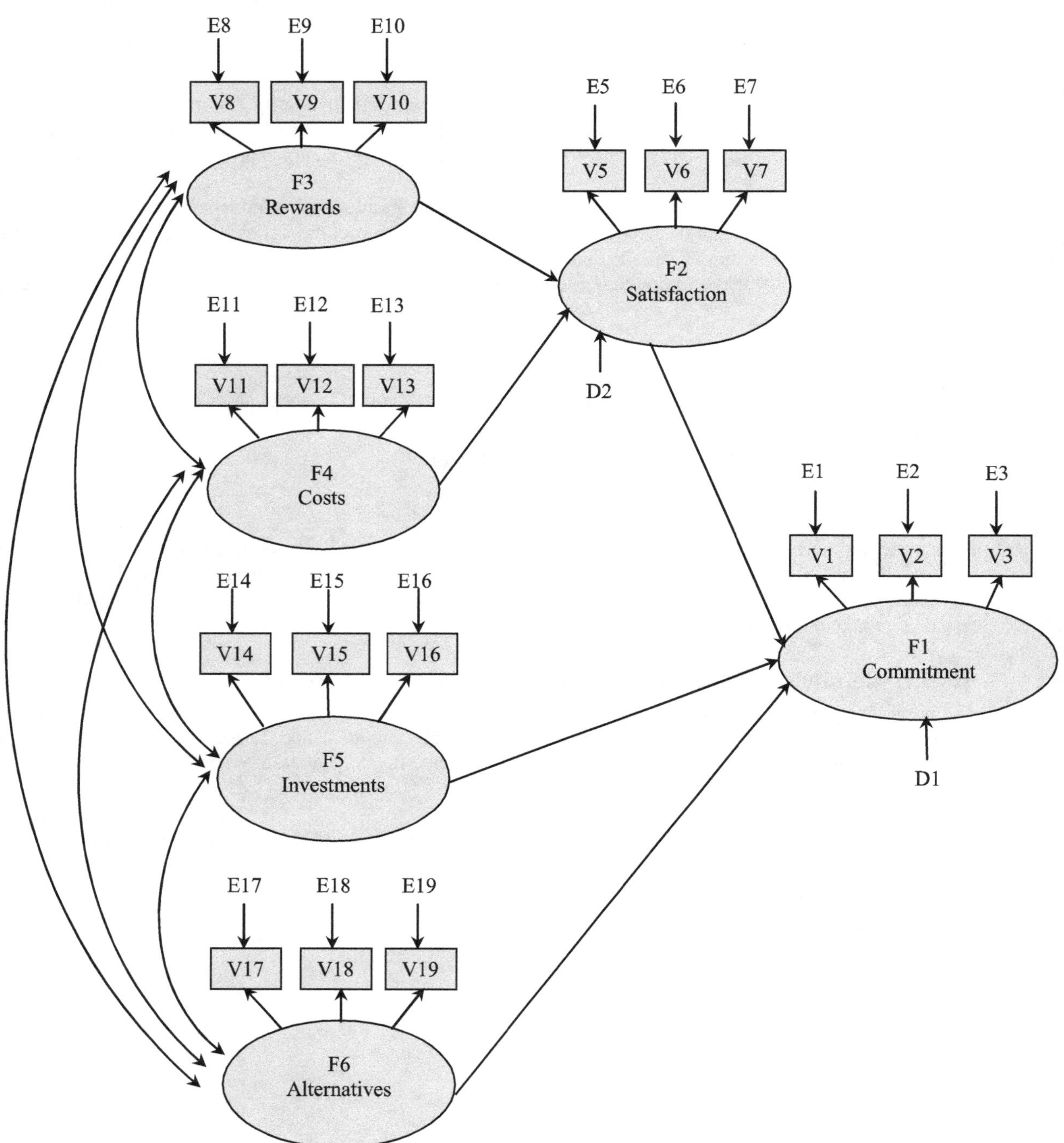

First, E terms are disturbance terms for manifest endogenous variables. These were earlier referred to as residual terms or error terms; the three names are sometimes used interchangeably. In Figure 6.3, the E terms for all manifest endogenous variables have been identified. This was done the same way as when the program figure for the measurement model was prepared.

Second, D terms are disturbance terms for latent endogenous variables (the F variables). Figure 6.3 shows that two of the latent factors are endogenous variables: Commitment (which is affected by three independent variables) and satisfaction (which is affected by two independent variables). Therefore, there is a directional arrow drawn from the disturbance term D1 to F1 (commitment) as well as a directional arrow drawn from D2 to F2 (satisfaction). Notice that there are no disturbance terms for F3, F4, F5, or F6 consistent with Rule 3, which stated that exogenous variables do not have residual (disturbance) terms.

Step 2: Identifying All Parameters to Be Estimated

Figure 6.4 identifies all the parameters to be estimated (or fixed) for the current analysis. Most are the same parameters estimated for the measurement model.

Figure 6.4: The Initial Theoretical Model, after Identifying All Parameters to Be Estimated (Completed Program Figure for the CFA Model)

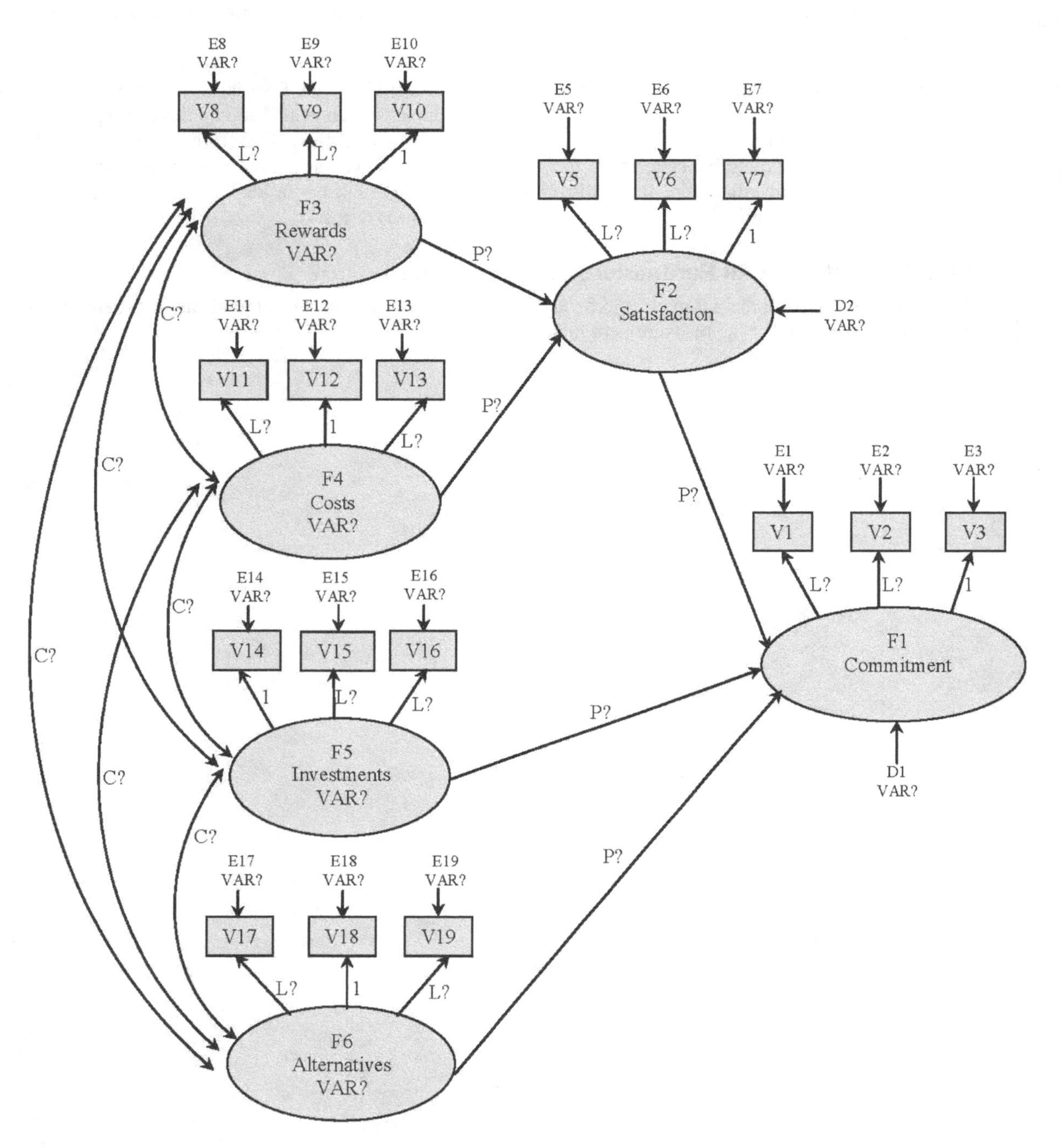

First, you must identify covariances to be estimated with the C? symbol. Rule 5 tells you to estimate covariances for every possible pair of exogenous variables, but Rule 6 says that you generally will not estimate covariances involving residual (disturbance) terms. In practice, this will generally mean estimating covariances for every possible pair of variables that (a) are part of the structural model, and (b) are exogenous variables within that structural model. Figure 6.4 shows that you will therefore estimate all possible pairings of covariance estimates between variables F3, F4, F5, and F6. All of the remaining exogenous variables in this model are disturbance terms (E or D terms), so no covariance involving these terms will be estimated.

Next, place the symbol VAR? just beneath the name of each variable to be estimated. As usual, variance will be estimated for all disturbance terms (notice the VAR? below all E and D terms).

It is at this point, however, that you will note one of the differences between SEM and CFA. When performing CFA, it was noted that the variance of the latent exogenous F factors are generally fixed at 1 in order to solve the problem of scale indeterminacy. When performing SEM, however, variances for these F variables should be free to be estimated. This is reflected in Figure 6.4: Notice that the VAR? symbol appears below the name of each of the exogenous F variables. This point is important enough to be summarized in a rule:

RULE 16: In SEM, the variance for the exogenous F variables are free parameters to be estimated.

But if you estimate the variance for the exogenous F variables, what about the problem of scale indeterminacy? In this type of analysis, the indeterminacy problem is typically solved by adhering to Rule 17:

RULE 17: In SEM, one factor loading for each F variable should be fixed at 1.

Remember that the scale indeterminacy problem (as explained in Chapter 5) involves the fact that an F variable is an unobserved variable that has no established unit of measurement. However, by fixing a path at 1 from the F variable to one of its manifest indicators, the unit of measurement for the F variable is set to the metric of measurement for that indicator variable (minus its error term).

But which indicator best represents the F variable? One way to make this decision is to review the results of the confirmatory factor analysis (CFA) of the measurement model, and identify the indicator that had the largest standardized loading for that factor. In the subsequent SEM, the factor loading for this indicator is fixed at one (e.g., O'Rourke, Cappeliez, and Guindon 2003).

For example, consider the latent variable F1 in Figure 6.4. According to the program figure, the paths from F1 to V1 and V2 are free parameters to be estimated in this analysis (this is signified by the L? next to their directional arrows). However, a "1" appears next to the path from F1 to V3. This means that this path will be fixed at 1 in order to solve the scale determinacy problem. This particular path has been fixed at 1 because, in the CFA of the measurement model (reported in Chapter 5), the indicator V3 displayed the largest standardized coefficient out of all of the variables that were predicted to load on F1.

To verify this, review Output 5.10 from Chapter 5. The loadings of interest are the standardized factor loadings that appear under the heading "Equations with Standardized Coefficients." You can see that three variables (V1, V2, and V3) load on Factor 1 (F1). The standardized factor loadings for these three variables were .89 (for V1), .82 (for V2), and .94 (for V3). Because V3 displayed the largest factor loading for F1 in the confirmatory factor analysis, you fix at 1 the path that goes from F1 to V3 when you compute the SEM.

This process was then repeated for each of the remaining F variables. That is, the results of the CFA (from Output 5.10 in Chapter 5) were reviewed to identify the indicator that displayed the largest standardized loading for each latent factor. These are the factor loadings that have been fixed at 1 (or other numeric value) in Figure 6.4.

With this done, all that remains is to identify the path coefficients to be estimated. This is done by placing the symbol "P?" on each of the paths that constitute the structural model in the figure. In Figure 6.4, this meant placing the symbol on the path from F2 to F1, from F3 to F2, and so forth.

As a final step before preparing the SAS program, you are well-advised to verify that the model is identified. In fact, this should be done each time the model is modified, as any modification has the potential of resulting in an unidentified model (Kline 2005). The preceding chapters provided a number of references on procedures for assessing model identification; in particular, see "Step 8: Verifying That the Model Is Overidentified" from Chapter 4, and "Step 4: Verifying That the Model Is Overidentified" from Chapter 5.

Preparing the SAS Program

In presenting programs, this chapter uses a system of notation based in part on the system developed by Bentler (1989) for the EQS structural equations program (e.g., latent factors are represented by the letter F, and so forth). The modified system used by this text was introduced in Chapter 4 and further developed in Chapter 5.

Below is the entire SAS program (minus the DATA step) that was used to analyze the model portrayed in Figure 6.4:

```
      proc calis modification ;
         lineqs
❶           V1  = LV1F1   F1 + E1,
            V2  = LV2F1   F1 + E2,
            V3  =         F1 + E3,
            V5  = LV5F2   F2 + E5,
            V6  = LV6F2   F2 + E6,
            V7  =         F2 + E7,
            V8  = LV8F3   F3 + E8,
            V9  = LV9F3   F3 + E9,
            V10 =         F3 + E10,
            V11 = LV11F4  F4 + E11,
            V12 =         F4 + E12,
            V13 = LV13F4  F4 + E13,
            V14 =         F5 + E14,
            V15 = LV15F5  F5 + E15,
            V16 = LV16F5  F5 + E16,
            V17 = LV17F6  F6 + E17,
            V18 =         F6 + E18,
❷           V19 = LV19F6  F6 + E19,
❸           F1  = PF1F2   F2 + PF1F5 F5 + PF1F6 F6 + D1,
❹           F2  = PF2F3   F3 + PF2F4 F4            + D2;
         variance
❺           E1-E3   = VARE1-VARE3,
❻           E5-E19  = VARE5-VARE19,
❼           F3-F6   = VARF3-VARF6,
❽           D1-D2   = VARD1-VARD2;
         cov
            F3 F4 = CF3F4,
            F3 F5 = CF3F5,
            F3 F6 = CF3F6,
            F4 F5 = CF4F5,
            F4 F6 = CF4F6,
            F5 F6 = CF5F6;
         var  V1 V2 V3 V5-V19 ;
         run;
```

The easiest way to create the SAS program that will perform SEM is to simply modify the program that had performed the confirmatory factor analysis of the corresponding measurement model (or, better still, to modify a copy of that program). Many aspects of the program that tests the theoretical model are identical to the program used with the measurement model (e.g., the PROC CALIS statement). Therefore, to save time, those aspects will not be reviewed again here. Instead, this section will discuss how the program for the measurement model must be changed to compute the SEM.

For purposes of reference, "Preparing the SAS Program" in Chapter 5 discussed the DATA step and the various statements that constitute a PROC CALIS program. The initial measurement model and the revised measurement model for the investment model study appear in Chapter 5 in "Overview of the PROC CALIS Program" and "Estimating the Revised Measurement Model," respectively. It is the revised measurement model that will be modified here in order to perform SEM.

The LINEQS Statement

The LINEQS statement serves two functions in SEM: (a) It indicates which factor loadings are to be estimated or fixed and (b) it specifies the directional relations between variables in the structural model. Each will be discussed in turn.

Lines between ❶ and ❷ of the preceding program indicate which manifest variables load on which latent factors. Notice that this portion of the program is identical to that used to estimate the revised measurement model, with one important difference: The coefficient names for some of the factor loadings have been blanked out. It is in this way that the factor loadings are fixed at 1.

For example, consider the following equation from the program:

❶ V1 = LV1F1 F1 + E1,

This line estimates the factor loading for the path from F1 to V1. You know that this parameter is estimated because the name for the parameter (LV1F1) appears just before the name of the variable where the path originates (F1).

In contrast, notice how the following line is different:

V3 = F1 + E3

In the preceding equation, the name for the factor loading (LV3F1) has been omitted from the equation; it does not appear just before the F1. When writing the SAS program for path analysis (Chapter 4), we indicated that this has the effect of fixing that parameter at 1. Therefore, you know from reviewing this equation that the factor loading LV3F1 is fixed 1.

The SAS program shows that the factor loadings for all of the following indicators have been fixed at 1: V3, V7, V10, V12, V14, and V18. Note that this is consistent with program figure appearing in Figure 6.4.

The VARIANCE Statement

Rule 4 (presented earlier) states that variance is to be estimated for every exogenous variable in the model, including residual terms; Figure 6.4 adheres to that rule. In that figure, the VAR? symbol is used to identify variables whose variance is to be estimated including all of the residual, or disturbance, terms (E and D variables). In addition, the figure shows that you are also to estimate variances for the exogenous F variables: F3, F4, F5, and F6. The VARIANCE statement that reflects this aspect of the figure is presented below:

```
        variance
❺         E1-E3   = VARE1-VARE3,
❻         E5-E19  = VARE5-VARE19,
❼         F3-F6   = VARF3-VARF6,
❽         D1-D2   = VARD1-VARD2;
```

In some ways, the VARIANCE statement is identical to that used with the CFA of the measurement model. Specifically, it still contains equations to estimate variances of the E terms (lines ❺ and ❻).

However, it also differs from the previous VARIANCE statement in three important ways. First, it no longer contains equations for F1 or F2. This is because F1 and F2 are now endogenous variables, and you do not estimate variance for endogenous variables.

Second, variance for F3, F4, F5, and F6 is now estimated rather than fixed at one (line ❼). As was discussed before, it is now possible to estimate variance for these latent variables because you establish their scale by fixing one factor loading at 1 for each F variable.

Finally, you now estimate variance for the disturbance terms D1 and D2 (line ❽). With this done, your VARIANCE statement will now compute all variance estimates indicated by the program figure.

The COV Statement

One of the ways that the theoretical model 4 differs from the measurement model 1 involves the covariances: With the theoretical model, there are no longer any covariance between F1 or F2 and any of the other F variables. This is consistent with Rule 1, which stated that only exogenous variables are allowed to covary. Because F1 and F2 are now endogenous variables, the SAS program must be modified so that it no longer estimates covariance between F1 or F2 and any other variable.

This is done in the following COV statement:

```
COV
   F3 F4 = CF3F4,
   F3 F5 = CF3F5,
   F3 F6 = CF3F6,
   F4 F5 = CF4F5,
   F4 F6 = CF4F6,
   F5 F6 = CF5F6;
```

Notice that none of the equations above include either F1 or F2. However, the statement does estimate all possible covariance estimates between F3, F4, F5, and F6. This is because these latent variables are still exogenous variables.

Interpreting the Results of the Analysis

The output generated by this SAS program would follow the same general format as the output for analysis of the measurement model (see Output 5.3 in Chapter 5). The following indicates the pages on which various results appear.

- Page 1 lists the endogenous and exogenous variables specified in the LINEQS statement, which is the general form of the structural equations specified in the LINEQS statement.
- Page 2 provides some univariate statistics for the manifest variables.
- Page 3 presents initial parameter estimates.
- Page 4 includes the iteration history.
- Page 5 reports a variety of goodness-of-fit indices (to be discussed below).
- Page 6 includes R^2 values for each endogenous variable.
- Page 7 reports standardized parameter estimates and associated t values.
- Page 8 reports Wald test results.
- Page 9 reports Lagrange Multiplier test results.

Once the SAS program for estimating the theoretical model has been executed, the SAS log and SAS output files should be reviewed to verify that the program ran correctly. This should be done in the usual way, as described in "Making Sure That the SAS Log and Output Files "Look Right" from Chapter 5. The information on the first four pages of output are particularly important for this purpose; these pages are presented here as Output 6.1. (Remember that the analyses reported in this chapter are fictitious and should not be viewed as legitimate tests of the investment model.)

Output 6.1: PROC CALIS Output Pages 1 to 4 for Analysis of the Investment Study Model

The CALIS Procedure
Covariance Structure Analysis: Model and Initial Values

Modeling Information	
Data Set	WORK.D1
N Obs	240
Model Type	LINEQS
Analysis	Covariances

Variables in the Model		
Endogenous	Manifest	V1 V10 V11 V12 V13 V14 V15 V16 V17 V18 V19 V2 V3 V5 V6 V7 V8 V9
	Latent	F1 F2
Exogenous	Manifest	
	Latent	F3 F4 F5 F6
	Error	E1 E10 E11 E12 E13 E14 E15 E16 E17 E18 E19 E2 E3 E5 E6 E7 E8 E9 D1 D2
Number of Endogenous Variables = 20 Number of Exogenous Variables = 24		

Initial Estimates for Linear Equations							
V1	=	.	*	F1	+	1.0000	E1
				LV1F1			
V2	=	.	*	F1	+	1.0000	E2
				LV2F1			
V3	=	1.0000		F1	+	1.0000	E3
V5	=	.	*	F2	+	1.0000	E5
				LV5F2			
V6	=	.	*	F2	+	1.0000	E6
				LV6F2			
V7	=	1.0000		F2	+	1.0000	E7
V8	=	.	*	F3	+	1.0000	E8
				LV8F3			
V9	=	.	*	F3	+	1.0000	E9
				LV9F3			
V10	=	1.0000		F3	+	1.0000	E10
V11	=	.	*	F4	+	1.0000	E11
				LV11F4			
V12	=	1.0000		F4	+	1.0000	E12
V13	=	.	*	F4	+	1.0000	E13
				LV13F4			
V14	=	1.0000		F5	+	1.0000	E14
V15	=	.	*	F5	+	1.0000	E15
				LV15F5			
V16	=	.	*	F5	+	1.0000	E16

Initial Estimates for Linear Equations																	
					LV16F5												
V17	=		.	*	F6	+	1.0000		E17								
					LV17F6												
V18	=		1.0000		F6	+	1.0000		E18								
V19	=		.	*	F6	+	1.0000		E19								
					LV19F6												
F1	=		.	*	F2	+	.	*	F5	+	.	*	F6	+	1.0000		D1
					PF1F2				PF1F5				PF1F6				
F2	=		.	*	F3	+	.	*	F4	+	1.0000		D2				
					PF2F3				PF2F4								

Initial Estimates for Variances of Exogenous Variables			
Variable Type	Variable	Parameter	Estimate
Error	E1	VARE1	.
	E2	VARE2	.
	E3	VARE3	.
	E5	VARE5	.
	E6	VARE6	.
	E7	VARE7	.
	E8	VARE8	.
	E9	VARE9	.
	E10	VARE10	.
	E11	VARE11	.
	E12	VARE12	.
	E13	VARE13	.
	E14	VARE14	.
	E15	VARE15	.
	E16	VARE16	.
	E17	VARE17	.
	E18	VARE18	.
	E19	VARE19	.
Latent	F3	VARF3	.
	F4	VARF4	.
	F5	VARF5	.
	F6	VARF6	.
Disturbance	D1	VARD1	.
	D2	VARD2	.

Initial Estimates for Covariances Among Exogenous Variables			
Var1	Var2	Parameter	Estimate
F3	F4	CF3F4	.
F3	F5	CF3F5	.
F3	F6	CF3F6	.
F4	F5	CF4F5	.
F4	F6	CF4F6	.
F5	F6	CF5F6	.

Covariance Structure Analysis: Descriptive Statistics

Simple Statistics		
Variable	Mean	Std Dev
V1	0	2.48600
V2	0	2.90900
V3	0	2.72400
V5	0	1.92900
V6	0	2.11300
V7	0	2.05600
V8	0	1.41700
V9	0	1.40800
V10	0	1.72400
V11	0	2.59500
V12	0	2.69100
V13	0	2.36000
V14	0	2.10200
V15	0	2.21900
V16	0	1.87400
V17	0	2.00100
V18	0	1.96600
V19	0	2.18500

Covariance Structure Analysis: Optimization

Initial Estimation Methods	
1	Instrumental Variables Method
2	McDonald Method
3	Two-Stage Least Squares

Optimization Start Parameter Estimates			
N	Parameter	Estimate	Gradient
1	LV1F1	0.89069	0.17082
2	LV2F1	0.89461	-0.12280
3	LV5F2	0.92545	0.07090
4	LV6F2	1.04052	0.01153
5	LV8F3	0.84001	0.12907
6	LV9F3	0.63254	-0.07209
7	LV11F4	0.86916	0.00228
8	LV13F4	0.44411	-0.01323
9	LV15F5	0.89697	0.04741
10	LV16F5	0.57209	-0.01743
11	LV17F6	1.04325	0.03868
12	LV19F6	1.16728	0.01552
13	PF1F2	0.52397	-0.02719
14	PF1F5	0.69577	-0.05914
15	PF1F6	0.12538	0.01143
16	PF2F3	0.72184	-0.07532
17	PF2F4	-0.13164	-0.01968
18	VARE1	0.93251	-0.22615
19	VARE2	3.16834	0.02917
20	VARE3	0.80545	-0.15004
21	VARE5	1.06336	-0.04423
22	VARE6	1.10513	0.00960
23	VARE7	1.12406	0.03839
24	VARE8	0.88586	-0.10828
25	VARE9	1.34622	0.03734
26	VARE10	1.38202	-0.01570
27	VARE11	2.36858	-0.00212
28	VARE12	1.46284	-0.00826
29	VARE13	4.42987	0.00123
30	VARE14	1.26888	0.01471
31	VARE15	2.38998	-0.01027

Optimization Start Parameter Estimates			
N	Parameter	Estimate	Gradient
32	VARE16	2.48109	0.00686
33	VARE17	1.93220	-0.01947
34	VARE18	1.96158	0.02807
35	VARE19	2.18051	-0.00692
36	VARF3	1.59016	0.04688
37	VARF4	5.77864	0.0004595
38	VARF5	3.14953	0.01461
39	VARF6	1.90358	-0.0001732
40	VARD1	3.23324	0.02032
41	VARD2	2.12335	0.02179
42	CF3F4	-0.26849	0.00497
43	CF3F5	1.10070	-0.08688
44	CF3F6	-0.75485	0.01407
45	CF4F5	0.04126	0.00154
46	CF4F6	0.86782	0.0002789
47	CF5F6	-0.77760	-0.02361
Value of Objective Function = 1.0477184241			

Covariance Structure Analysis: Optimization

Levenberg-Marquardt Optimization
Scaling Update of More (1978)

Parameter Estimates	47
Functions (Observations)	171

Optimization Start			
Active Constraints	0	Objective Function	1.0477184241
Max Abs Gradient Element	0.2261527842	Radius	1

Iteration	Restarts	Function Calls	Active Constraints	Objective Function	Objective Function Change	Max Abs Gradient Element	Lambda	Ratio Between Actual and Predicted Change
1	0	4	0	0.91784	0.1299	0.0515	0	0.891
2	0	6	0	0.90868	0.00916	0.0161	0	1.053
3	0	8	0	0.90728	0.00140	0.00818	0	1.213
4	0	10	0	0.90700	0.000284	0.00308	0	1.373
5	0	12	0	0.90693	0.000069	0.00209	0	1.458
6	0	14	0	0.90691	0.000018	0.000938	0	1.500
7	0	16	0	0.90691	4.914E-6	0.000559	0	1.516

Iteration	Restarts	Function Calls	Active Constraints	Objective Function	Objective Function Change	Max Abs Gradient Element	Lambda	Ratio Between Actual and Predicted Change
8	0	18	0	0.90690	1.36E-6	0.000276	0	1.524
9	0	20	0	0.90690	3.791E-7	0.000154	0	1.527
10	0	22	0	0.90690	1.061E-7	0.000079	0	1.529
11	0	24	0	0.90690	2.972E-8	0.000043	0	1.529
12	0	26	0	0.90690	8.333E-9	0.000022	0	1.529
13	0	28	0	0.90690	2.338E-9	0.000012	0	1.530

Optimization Results			
Iterations	13	Function Calls	31
Jacobian Calls	15	Active Constraints	0
Objective Function	0.9069032227	Max Abs Gradient Element	0.0000119446
Lambda	0	Actual Over Pred Change	1.5296306541
Radius	0.0002322371		

Convergence criterion (GCONV=1E-8) satisfied.

You may begin your assessment of the fit of the theoretical model by following the same procedures used with the measurement model. Once this is done, however, this chapter will introduce some additional indices that are particularly useful for evaluating the fit of theoretical models.

Step 1: Reviewing Goodness-of-Fit Indices

Output 6.2 shows that the Comparative Fit Index (CFI) for the theoretical model was .95, which is a bit lower than the CFI value of .97 observed for the measurement model but still within ideal range. The Standardized Root Mean Square Residual is also somewhat larger (SRMR = .059 > SRMR = .042) as is the Root Mean Square Error of Approximation (RMSEA = .056 RMSEA = .046). Note also that the 90% confidence limits for this SEM model are somewhat higher than the measurement model (.043 ≤ RMSEA CL_{90} ≤ .068 vs. .032 ≤ RMSEA CL_{90} ≤ .059).

Output 6.2: Goodness-of-Fit Indices for Initial Theoretical Model, Investment Model Study

Fit Summary		
Modeling Info	N Observations	240
	N Variables	18
	N Moments	171
	N Parameters	47
	N Active Constraints	0
	Baseline Model Function Value	9.0702
	Baseline Model Chi-Square	2167.7711
	Baseline Model Chi-Square DF	153
	Pr > Baseline Model Chi-Square	<.0001
Absolute Index	Fit Function	0.9069
	Chi-Square	216.7499
	Chi-Square DF	124
	Pr > Chi-Square	<.0001
	Z-Test of Wilson & Hilferty	4.8756
	Hoelter Critical N	167
	Root Mean Square Residual (RMSR)	0.2684
	Standardized RMSR (SRMSR)	0.0590
	Goodness of Fit Index (GFI)	0.9094
Parsimony Index	Adjusted GFI (AGFI)	0.8750
	Parsimonious GFI	0.7370
	RMSEA Estimate	0.0559
	RMSEA Lower 90% Confidence Limit	0.0434
	RMSEA Upper 90% Confidence Limit	0.0682
	Probability of Close Fit	0.2080
	ECVI Estimate	1.3342
	ECVI Lower 90% Confidence Limit	1.1766
	ECVI Upper 90% Confidence Limit	1.5278
	Akaike Information Criterion	310.7499
	Bozdogan CAIC	521.3399
	Schwarz Bayesian Criterion	474.3399
	McDonald Centrality	0.8243
Incremental Index	Bentler Comparative Fit Index	0.9540
	Bentler-Bonett NFI	0.9000
	Bentler-Bonett Non-normed Index	0.9432
	Bollen Normed Index Rho1	0.8766
	Bollen Non-normed Index Delta2	0.9546
	James et al. Parsimonious NFI	0.7294

Step 2: Reviewing the Significance Tests for Factor Loadings and Path Coefficients

As before, it is good practice to review the standard errors of factor loadings and path coefficients. None of the standard errors appear to be inordinately small.

The factor loadings of Output 6.3 are represented with coefficient names that begin with the "L" prefix (such as "LV1F1"). Results show that all factor loadings have t values greater than 1.96 (or less than –1.96) and therefore differ significantly from zero.

Of interest here are the coefficients for the directional paths that constitute the structural portion of the model. These path coefficients are represented with coefficient names that begin with the "P" prefix (such as "PF1F2"), and appear under the heading "Standardized Results for Linear Equations" in Output 6.3.

These results show that all path coefficients were significant except for the path from F6 (alternative value) to F1 (commitment), which displayed a nonsignificant t value of 0.92. Consistent with this, the standardized path coefficients for the latent variable equations shows that the standardized path coefficient for the path from F6 to F1 was quite small (.06). This is an important finding because if you later decide to modify the model, deleting the path from F6 to F1 may be the place to start.

Output 6.3: PROC CALIS Output Pages 6 and 7 from Analysis of Initial Theoretical Model, Investment Model Study

The CALIS Procedure
Covariance Structure Analysis: Maximum Likelihood Estimation

Linear Equations							
V1	=	0.8609	*	F1	+	1.0000	E1
Std Err		0.0425		LV1F1			
t Value		20.2619					
V2	=	0.9388	*	F1	+	1.0000	E2
Std Err		0.0535		LV2F1			
t Value		17.5628					
V3	=	1.0000		F1	+	1.0000	E3
V5	=	0.8820	*	F2	+	1.0000	E5
Std Err		0.0565		LV5F2			
t Value		15.6052					
V6	=	1.0216	*	F2	+	1.0000	E6
Std Err		0.0608		LV6F2			
t Value		16.8132					
V7	=	1.0000		F2	+	1.0000	E7
V8	=	0.7659	*	F3	+	1.0000	E8
Std Err		0.0975		LV8F3			
t Value		7.8534					
V9	=	0.7348	*	F3	+	1.0000	E9
Std Err		0.0961		LV9F3			
t Value		7.6497					
V10	=	1.0000		F3	+	1.0000	E10
V11	=	0.9082	*	F4	+	1.0000	E11
Std Err		0.1015		LV11F4			
t Value		8.9472					
V12	=	1.0000		F4	+	1.0000	E12
V13	=	0.4668	*	F4	+	1.0000	E13
Std Err		0.0715		LV13F4			
t Value		6.5322					
V14	=	1.0000		F5	+	1.0000	E14
V15	=	0.8467	*	F5	+	1.0000	E15
Std Err		0.0846		LV15F5			
t Value		10.0113					
V16	=	0.5708	*	F5	+	1.0000	E16
Std Err		0.0712		LV16F5			
t Value		8.0190					
V17	=	0.9149	*	F6	+	1.0000	E17
Std Err		0.1061		LV17F6			
t Value		8.6257					

Linear Equations																	
V18	=		1.0000		F6	+	1.0000		E18								
V19	=		1.0722	*	F6	+	1.0000		E19								
Std Err			0.1206		LV19F6												
t Value			8.8935														
F1	=		0.4608	*	F2	+	0.7581	*	F5	+	0.1000	*	F6	+	1.0000		D1
Std Err			0.0910		PF1F2		0.1037		PF1F5		0.1094		PF1F6				
t Value			5.0618				7.3126				0.9138						
F2	=		0.9736	*	F3	+	-0.1213	*	F4	+	1.0000		D2				
Std Err			0.1321		PF2F3		0.0510		PF2F4								
t Value			7.3690				-2.3777										

Estimates for Variances of Exogenous Variables					
Variable Type	Variable	Parameter	Estimate	Standard Error	t Value
Error	E1	VARE1	1.34660	0.17990	7.48522
	E2	VARE2	2.71411	0.29893	9.07950
	E3	VARE3	0.89811	0.19047	4.71524
	E5	VARE5	1.20075	0.14494	8.28467
	E6	VARE6	1.08304	0.15697	6.89947
	E7	VARE7	0.98719	0.14721	6.70597
	E8	VARE8	1.18796	0.13583	8.74565
	E9	VARE9	1.22790	0.13658	8.99061
	E10	VARE10	1.57452	0.19418	8.10874
	E11	VARE11	2.20594	0.49120	4.49092
	E12	VARE12	1.75119	0.56533	3.09762
	E13	VARE13	4.37329	0.42193	10.36494
	E14	VARE14	1.15373	0.24949	4.62428
	E15	VARE15	2.58330	0.29637	8.71638
	E16	VARE16	2.44818	0.24600	9.95215
	E17	VARE17	2.14495	0.26338	8.14404
	E18	VARE18	1.64409	0.25169	6.53218
	E19	VARE19	2.22087	0.31028	7.15765
Latent	F3	VARF3	1.39766	0.26483	5.27753
	F4	VARF4	5.49029	0.84089	6.52914
	F5	VARF5	3.26467	0.45093	7.23990
	F6	VARF6	2.22107	0.37832	5.87089
Disturbance	D1	VARD1	2.92500	0.39030	7.49424
	D2	VARD2	1.77088	0.26889	6.58580

Covariances Among Exogenous Variables					
Var1	Var2	Parameter	Estimate	Standard Error	t Value
F3	F4	CF3F4	-0.26870	0.23677	-1.13488
F3	F5	CF3F5	1.37561	0.22524	6.10720
F3	F6	CF3F6	-0.82443	0.17730	-4.64984
F4	F5	CF4F5	0.02701	0.32839	0.08225
F4	F6	CF4F6	0.89476	0.29326	3.05103
F5	F6	CF5F6	-0.79934	0.23276	-3.43419

Squared Multiple Correlations			
Variable	Error Variance	Total Variance	R-Square
V1	1.34660	5.97089	0.7745
V2	2.71411	8.21337	0.6695
V3	0.89811	7.13775	0.8742
V5	1.20075	3.72104	0.6773
V6	1.08304	4.46477	0.7574
V7	0.98719	4.22714	0.7665
V8	1.18796	2.00789	0.4084
V9	1.22790	1.98246	0.3806
V10	1.57452	2.97218	0.4702
V11	2.20594	6.73402	0.6724
V12	1.75119	7.24148	0.7582
V13	4.37329	5.56960	0.2148
V14	1.15373	4.41840	0.7389
V15	2.58330	4.92396	0.4754
V16	2.44818	3.51188	0.3029
V17	2.14495	4.00400	0.4643
V18	1.64409	3.86516	0.5746
V19	2.22087	4.77422	0.5348
F1	2.92500	6.23964	0.5312
F2	1.77088	3.23994	0.4534

Covariance Structure Analysis: Maximum Likelihood Estimation

Standardized Results for Linear Equations																	
V1	=		0.8800	*	F1	+	1.0000		E1								
Std Err			0.0192		LV1F1												
t Value			45.8871														
V2	=		0.8183	*	F1	+	1.0000		E2								
Std Err			0.0247		LV2F1												
t Value			33.1551														
V3	=		0.9350		F1	+	1.0000		E3								
Std Err			0.0152														
t Value			61.4144														
V5	=		0.8230	*	F2	+	1.0000		E5								
Std Err			0.0260		LV5F2												
t Value			31.6993														
V6	=		0.8703	*	F2	+	1.0000		E6								
Std Err			0.0222		LV6F2												
t Value			39.2494														
V7	=		0.8755		F2	+	1.0000		E7								
Std Err			0.0218														
t Value			40.1732														
V8	=		0.6390	*	F3	+	1.0000		E8								
Std Err			0.0497		LV8F3												
t Value			12.8498														
V9	=		0.6169	*	F3	+	1.0000		E9								
Std Err			0.0511		LV9F3												
t Value			12.0670														
V10	=		0.6857		F3	+	1.0000		E10								
Std Err			0.0468														
t Value			14.6375														
V11	=		0.8200	*	F4	+	1.0000		E11								
Std Err			0.0458		LV11F4												
t Value			17.9227														
V12	=		0.8707		F4	+	1.0000		E12								
Std Err			0.0457														
t Value			19.0327														
V13	=		0.4635	*	F4	+	1.0000		E13								
Std Err			0.0570		LV13F4												
t Value			8.1242														
V14	=		0.8596		F5	+	1.0000		E14								
Std Err			0.0342														
t Value			25.1186														

Standardized Results for Linear Equations																	
V15	=		0.6895	*	F5	+	1.0000		E15								
Std Err			0.0430		LV15F5												
t Value			16.0488														
V16	=		0.5503	*	F5	+	1.0000		E16								
Std Err			0.0522		LV16F5												
t Value			10.5398														
V17	=		0.6814	*	F6	+	1.0000		E17								
Std Err			0.0473		LV17F6												
t Value			14.4054														
V18	=		0.7580		F6	+	1.0000		E18								
Std Err			0.0441														
t Value			17.1969														
V19	=		0.7313	*	F6	+	1.0000		E19								
Std Err			0.0451		LV19F6												
t Value			16.2181														
F1	=		0.3321	*	F2	+	0.5483	*	F5	+	0.0597	*	F6	+	1.0000		D1
Std Err			0.0625		PF1F2		0.0600		PF1F5		0.0651		PF1F6				
t Value			5.3108				9.1363				0.9166						
F2	=		0.6395	*	F3	+	-0.1578	*	F4	+	1.0000		D2				
Std Err			0.0537		PF2F3		0.0651		PF2F4								
t Value			11.9045				-2.4262										

Standardized Results for Variances of Exogenous Variables					
Variable Type	Variable	Parameter	Estimate	Standard Error	t Value
Error	E1	VARE1	0.22553	0.03376	6.68117
	E2	VARE2	0.33045	0.04039	8.18171
	E3	VARE3	0.12583	0.02847	4.41987
	E5	VARE5	0.32269	0.04273	7.55127
	E6	VARE6	0.24257	0.03860	6.28501
	E7	VARE7	0.23354	0.03816	6.12029
	E8	VARE8	0.59164	0.06356	9.30870
	E9	VARE9	0.61938	0.06308	9.81836
	E10	VARE10	0.52975	0.06425	8.24484
	E11	VARE11	0.32758	0.07504	4.36568
	E12	VARE12	0.24183	0.07967	3.03534
	E13	VARE13	0.78521	0.05288	14.84970
	E14	VARE14	0.26112	0.05883	4.43845
	E15	VARE15	0.52464	0.05924	8.85621

Standardized Results for Variances of Exogenous Variables					
Variable Type	Variable	Parameter	Estimate	Standard Error	t Value
	E16	VARE16	0.69712	0.05747	12.12912
	E17	VARE17	0.53570	0.06446	8.31037
	E18	VARE18	0.42536	0.06683	6.36475
	E19	VARE19	0.46518	0.06595	7.05313
Latent	F3	VARF3	1.00000		
	F4	VARF4	1.00000		
	F5	VARF5	1.00000		
	F6	VARF6	1.00000		
Disturbance	D1	VARD1	0.46878	0.05648	8.30022
	D2	VARD2	0.54658	0.06813	8.02242

Standardized Results for Covariances Among Exogenous Variables					
Var1	Var2	Parameter	Estimate	Standard Error	t Value
F3	F4	CF3F4	-0.09700	0.08433	-1.15020
F3	F5	CF3F5	0.64398	0.05914	10.88875
F3	F6	CF3F6	-0.46792	0.07276	-6.43082
F4	F5	CF4F5	0.00638	0.07756	0.08226
F4	F6	CF4F6	0.25623	0.07605	3.36912
F5	F6	CF5F6	-0.29685	0.07672	-3.86934

Step 3: Reviewing R^2 Values for Latent Endogenous Variables

The R^2 values for the study's endogenous variables appear on page 6 of the output. Of particular interest are the R^2 values for the structural model's latent endogenous variables F1 (commitment) and F2 (satisfaction). The results on page 6 show that the independent F variables accounted for 53% of the variance in commitment and 45% of the variance in satisfaction.

The theoretical model presented in Figure 6.4 attempts to explain variability in satisfaction and commitment in romantic associations. It identifies four exogenous variables (rewards, costs, investments, and alternatives) that are assumed to predict levels of satisfaction and commitment. It posits that relationships between these constructs can be accounted for with just five directional paths: The two paths from rewards and costs to satisfaction, and the three paths from satisfaction, investments, and alternatives to commitment. And this model did a fairly good job of accounting for the observed covariances in the data (as is indicated by the relatively low SRMR and RMSEA values).

Step 4: Performing a Chi-Square Difference Test Comparing the Theoretical Model to the Measurement Model

Before moving to the next stage of the analysis, you should perform a chi-square difference test to determine whether there is a significant difference between the fit provided by the theoretical model versus the fit provided by the measurement model. A finding of no significant differences provides support for the nomological validity of the theoretical model (Anderson and Gerbing 1988). If the theoretical model is successful in accounting for the observed associations between the F variables, there will not be a significant difference between the chi-square for the theoretical model and the chi-square for the measurement model. Reviewing how the measurement model differs conceptually from the theoretical model will help make this clear.

Earlier, it was noted that the measurement model is basically a confirmatory factor analysis model in which the relations between all F variables are saturated, or just-identified. In other words, each F variable is connected to every other F variable by a curved or double-headed arrow. Because each F variable is connected to every other F variable, this measurement model does a thorough job of accounting for the covariances between F variables.

In most cases, you hope to develop a theoretical model that (a) is more parsimonious than the measurement model, but at the same time (b) does nearly as good a job of accounting for the covariances between latent F variables. A theoretical model is more parsimonious than a measurement model because the theoretical model is really a *constrained* version of the measurement model. This means that a theoretical model is basically a measurement model in which some of the covariances between F variables have been either (a) replaced with unidirectional paths or (b) fixed at zero. You can see this by comparing Figure 6.1 (the final measurement model) with Figure 6.4 (the theoretical model). In the measurement model, covariance is estimated for F1 (commitment) and F3 (rewards). In the theoretical model, this covariance estimate has been eliminated (fixed at zero). In the new model, there is no direct relationship between F1 and F3. The same is true for several other covariances in the measurement model.

The adequacy of the theoretical model can be determined by performing a chi-square difference test that compares the theoretical model (symbolized as M_t) to the measurement model (symbolized M_m). This is done by simply subtracting the chi-square values for the two models, as they appear in Table 6.1. This is done below:

$$M_t - M_m = 216.75 - 180.87 = 35.88$$

The resulting chi-square difference value (35.88 in this case) also follows a chi-square distribution, and the degrees of freedom for the test may be determined by subtracting the corresponding degrees of freedom for the two models. These degrees of freedom may also be obtainedfrom Table 6.1:

$$df_t - df_m = 124 - 120 = 4$$

With 4 degrees of freedom, the critical value of chi-square is 13.3 at $p < .01$. Your obtained chi-square difference value of 35.88 is clearly greater than this critical value, meaning that there is a significant difference between the fit provided by the theoretical model versus the measurement model. In other words, the theoretical model provides a fit to data that is significantly worse than the fit provided by the measurement model. This finding fails to support the theoretical model's predictions concerning the relationships between the F variables in the structural portion of the model. Apparently, your theoretical model contains misspecifications and will have to be modified if it is to fit the data.

Characteristics of an "Ideal Fit" for the Theoretical Model

Before moving on to the section on model modification, it will be helpful to first briefly summarize the results that you should expect to see if your model provides an ideal fit to data. A theoretical model provides an ideal fit when it displays the following characteristics:

- The sample upon which the model is based should have 200+ data points and statistical power at .80 or above; see SAS syntax to estimate statistical power in Chapter 5 (MacCallum, Browne, and Sugawara 1996).
- The absolute value of t statistics for each factor loading and path coefficient should exceed 1.96 (or less than −1.96).
- Standardized factor loadings should be nontrivial.

- R^2 values for the latent endogenous variables should be relatively large compared to what typically is obtained in research with these variables.
- Three or more goodness-of-fit indices are generally examined and reported including an absolute index (e.g., SRMR < .055), an incremental index (e.g., CFI > .94), and a parsimony index such as the RMSEA. Values for the RMSEA less than .09 suggest adequate fit, whereas values less than.055 suggest good fit. It is ideal if the full 90% range of confidence limits for the RMSEA is within good (i.e., $.09 \geq$ RMSEA $CL_{90} \geq 0$) to ideal parameters (i.e., $.054 \geq$ RMSEA $CL_{90} \geq 0$).
- A chi-square difference test should reveal no significant difference between the theoretical model and the measurement model.

Remember that the above characteristics represent an ideal that is often not attained with real-world data even with a theoretical model that is quite good. A model's fit need not meet all of the above criteria in order to be deemed "acceptable." In particular, requiring the full 90% range of confidence limits for the RMSEA to be within ideal limits would be deemed quite strict in most applied situations (particularly with smaller sample sizes).

However, it *is* important that there be no significant difference between the fit of the (final) theoretical model and the measurement model. This is because a significant difference between the chi-square for the theoretical and measurement models shows that the theoretical model fails to successfully account for the observed covariances between the F variables in the structural portion of the model. For this reason, it will be necessary to modify the model to attain a better fit.

Using Modification Indices to Modify the Present Model

In Chapter 4, "Path Analysis," much was said concerning the dangers of data-driven model modifications. All of those warnings apply to the modification of latent-variable models as well. Whenever you modify models based on the results of an analysis, you run the risk of capitalizing on chance characteristics of sample data, and creating a new model that will not generalize to the population.

Chapter 4 made five recommendations concerning things that you can do to minimize the dangers of over-fitting models:

- obtain large samples
- make few modifications
- make only changes that can be meaningfully interpreted
- compare alternative *a priori* models
- fully describe the limitations of your study

The following sections will describe the steps to be followed in modifying the present model for purposes of illustration. Here, you will see how information from the Wald test and the Lagrange Multiplier may be used to develop a better-fitting model.

The Wald Test

Although models may be modified in any of a number of ways (e.g., by placing equality constraints on parameters), they are most frequently modified by either (a) fixing directional paths at zero (e.g., eliminating a nonsignificant path from the model), or (b) freeing directional paths to be estimated (i.e., adding new paths to the model). Of these alternatives, eliminating a nonsignificant path is less likely to capitalize on chance characteristics of the data, and is therefore less risky. For this reason, you will first review the results of the analysis to identify any nonsignificant paths which can be deleted.

This review should normally begin with the Wald test, as it identifies parameters that may be dropped without causing a significant decrease in model chi-square. The Wald test for your analysis of the theoretical model appears on page 8 of the Output 6.3, and is reproduced here as Output 6.4.

Output 6.4: Wald Test Results for Initial Theoretical Model, Investment Model Study

Covariance Structure Analysis: Maximum Likelihood Estimation

Stepwise Multivariate Wald Test					
	Cumulative Statistics			Univariate Increment	
Parm	Chi-Square	DF	Pr > ChiSq	Chi-Square	Pr > ChiSq
CF4F5	0.00677	1	0.9344	0.00677	0.9344
PF1F6	0.84794	2	0.6544	0.84117	0.3591
CF3F4	2.68895	3	0.4421	1.84102	0.1748

Earlier, you learned that the Wald test estimates the change in model chi-square that would result from fixing a given parameter at zero. The first parameter listed in the preceding Wald test results is the covariance estimate between F4 (costs) and F5 (investments), and the third entry in the table is the covariance estimate between F3 (rewards) and F4 (costs). Covariance is generally estimated for all possible pairs of exogenous F variables in an analysis of this sort (unless there is theoretical reason that they be fixed at zero), so you will disregard these Wald test results for the moment.

In this case, you are more interested in finding directional paths that may be eliminated without significantly affecting the model's fit, and the Wald test has identified just such a path. The parameter PF1F6 represents the path from F1 (commitment) to F6 (alternative value), and the univariate Wald test suggests that model chi-square will change only 0.84 (a nonsignificant amount) if this path were deleted. Remember that this finding is very consistent with what you learned when you reviewed the latent variable equations on page 7 of Output 6.3. There, you learned that the path coefficient for the path from F6 to F1 was nonsignificant (i.e., $t = .92$).

The safest approach to modifying models is to change just one parameter at a time. You will therefore re-estimate your model with PF1F6 fixed at zero, and then review the results to see if any additional modifications are necessary.

Creating Revised Model 1

Figure 6.5 presents the program figure for revised model 1. You can see that this model is identical to the initial theoretical model (Figure 6.4) except that the path from alternatives to commitment has been deleted.

Figure 6.5: Revised Model 1, in Which the Path from F6 (alternatives) to F1 (commitment) Has Been Deleted

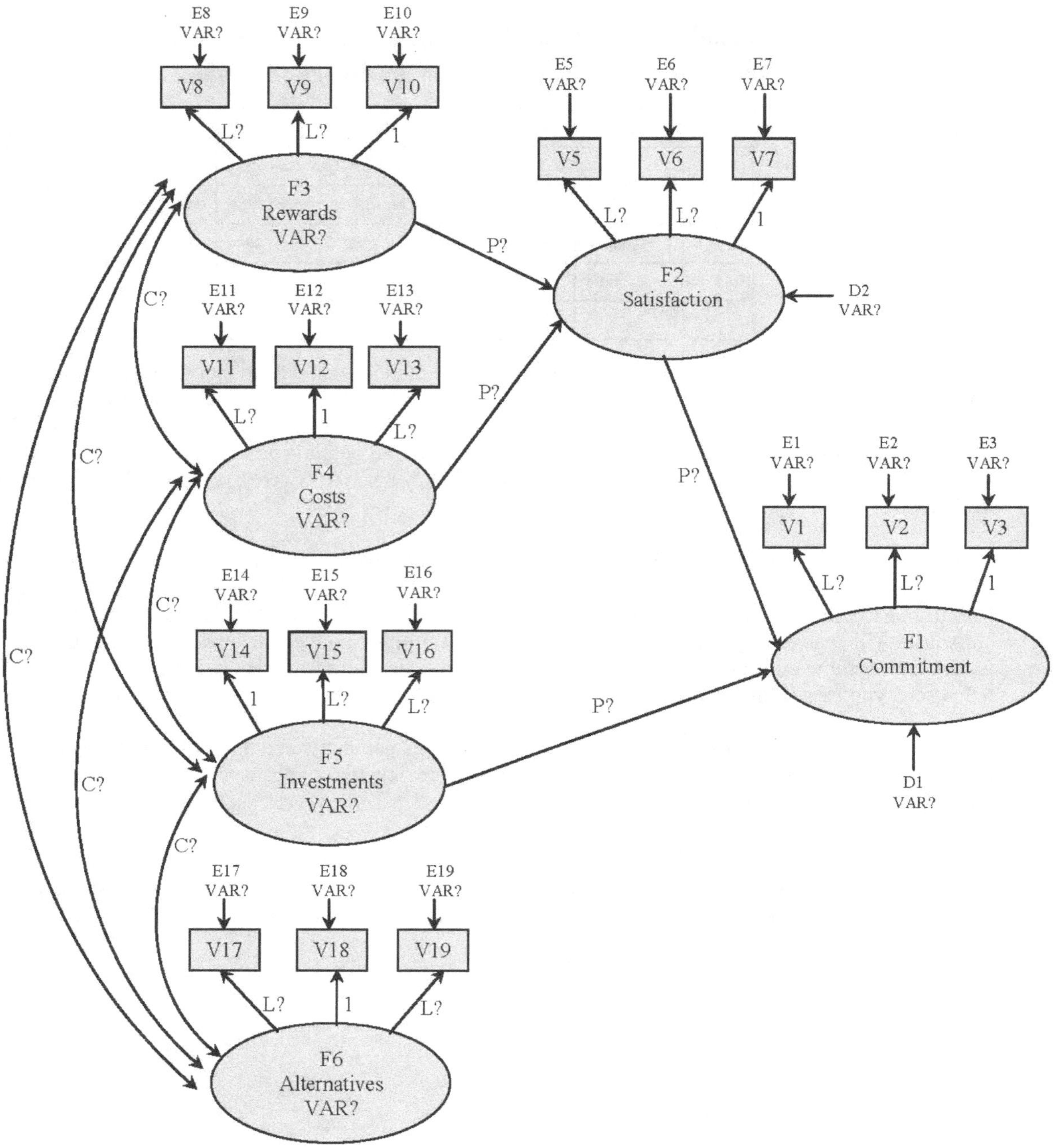

To create the PROC CALIS program that will estimate revised model 1, it is necessary to make just one change in the program that had estimated the initial theoretical model. Specifically, the latent-variable equation for F1 in the LINEQS statement has to be modified so that F6 is no longer specified as an independent variable for F1. This can be easily done by making a copy of the original program, and then blanking out the path coefficient name (PF1F6) and the short name (F6) and plus sign (+) for the alternatives construct.

The PROC CALIS program that estimated the initial theoretical model appeared earlier in the "Preparing the SAS Program" section. In that program, the latent variable equation for F1 appeared on line ❸:

```
❸    F1   = PF1F2 F2 + PF1F5 F5 + PF1F6 F6 + D1,
```

In the PROC CALIS program that estimates revised model 1, the equation takes on the following form; notice that the path coefficient name and the short name for the alternatives construct has been blanked out:

```
❸    F1   = PF1F2 F2 + PF1F5 F5 +             D1,
```

No other changes to the program were necessary. The complete program (minus the DATA step) for estimating revised model 1 appears below:

```
        proc calis modification ;
           lineqs
❶            V1  = LV1F1   F1 + E1,
             V2  = LV2F1   F1 + E2,
             V3  =         F1 + E3,
             V5  = LV5F2   F2 + E5,
             V6  = LV6F2   F2 + E6,
             V7  =         F2 + E7,
             V8  = LV8F3   F3 + E8,
             V9  = LV9F3   F3 + E9,
             V10 =         F3 + E10,
             V11 = LV11F4  F4 + E11,
             V12 =         F4 + E12,
             V13 = LV13F4  F4 + E13,
             V14 =         F5 + E14,
             V15 = LV15F5  F5 + E15,
             V16 = LV16F5  F5 + E16,
             V17 = LV17F6  F6 + E17,
             V18 =         F6 + E18,
❷            V19 = LV19F6  F6 + E19,
❸            F1  = PF1F2 F2 + PF1F5 F5             + D1,
❹            F2  = PF2F3 F3 + PF2F4 F4             + D2;
           variance
❺            E1-E3   = VARE1-VARE3,
❻            E5-E19  = VARE5-VARE19,
❼            F3-F6   = VARF3-VARF6,
❽            D1-D2   = VARD1-VARD2;
           cov
             F3 F4 = CF3F4,
             F3 F5 = CF3F5,
             F3 F6 = CF3F6,
             F4 F5 = CF4F5,
             F4 F6 = CF4F6,
             F5 F6 = CF5F6;
           var  V1 V2 V3 V5-V19 ;
        run;
```

Before reviewing goodness-of-fit indices, you first perform a chi-square difference test comparing the theoretical model (M_t) to revised model 1 (M_{r1}). Finding a significant difference between the two models would indicate that the path from F6 to F1 had been an important path and should not have been deleted. Model chi-square values are obtained from Table 6.1 to perform this test:

$$M_{r1} - M_t = 217.57 - 216.75 = 0.82$$

So the chi-square difference value is 0.82, which is quite close to the value of 0.84 that had been estimated by the Wald test. The degrees of freedom for the test (Δdf) are equal to the difference between the df for the two models:

$$df_{r1} - df_t = 125 - 124 = 1$$

A table of chi-square (in Appendix C) shows the critical chi-square value with 1 degree of freedom is 3.84. Because your observed chi-square difference value of 0.83 is less than this, you conclude that there is not a significant difference between the fit provided by the theoretical model and that provided by revised model 1. Apparently, deleting the path from F6 to F1 did not hurt the model's fit.

However, the critical test of the validity of this revised model is the chi-square difference test comparing the revised model to the measurement model. A significant difference between these two models suggests that revised model 1 is not successfully accounting for the relationships between the latent F variables that constitute the structural portion of the model.

Values of chi-square may be found in Table 6.1 and substituted in the following equation:

$$M_{r1} - M_m = 217.57 - 180.87 = 36.70$$

The degrees of freedom for the test are calculated in the usual way:

$$df_{r1} - df_m = 125 - 120 = 5$$

The critical chi-square value with 5 degrees of freedom is 11.1 at $p < .05$ and 15.1 at $p < .01$. The obtained value of chi-square is greater than these, which indicates a significant difference between chi-squares for the two models. In other words, revised model 1 exhibits a fit to data that is significantly worse than the fit displayed by the measurement model. Apparently, there is still misspecification involving the relationships among F variables in M_{r1}.

As before, your first stop is the Wald test, as it is generally safer to drop parameters than to add them. But the Wald test reveals no path coefficient that may be dropped without affecting model fit, only covariances. This is consistent with the finding that all factor loadings and path coefficients are significant for revised model 1. Given that it apparently is not possible to drop nonsignificant paths from the model, you now turn attention to identifying new paths that might be added. The Lagrange Multiplier is used for this purpose.

The Lagrange Multiplier

The modification option requested in the PROC CALIS statement resulted in four tables of Lagrange Multiplier tests:

- Rank order of the 10 largest LM statistics for paths from endogenous variables
- Rank order of the 10 largest LM statistics for paths from exogenous variables
- Rank order of the 10 largest LM statistics for paths with new endogenous variables
- Rank order of the 10 largest LM statistics for error variances and covariances

Of the Lagrange Multiplier values, the largest indicates that the model chi-square would decrease by 34.35 (a significant amount) if a path were added that went from F5 (investments) to F2 (satisfaction). The chapter on path analysis indicated that such a path could be defended on theoretical grounds as it is consistent with aspects of cognitive dissonance theory. You will therefore add this path and re-estimate the model to see if it results in improved fit.

Creating Revised Model 2

Figure 6.6 displays the resulting model: Revised model 2. This system is identical to revised model 1, except that a directional path has been added that leads from investments to satisfaction.

Figure 6.6: Revised Model 2, in Which the Path from F5 (Investments) to F2 (Satisfaction) Has Been Added

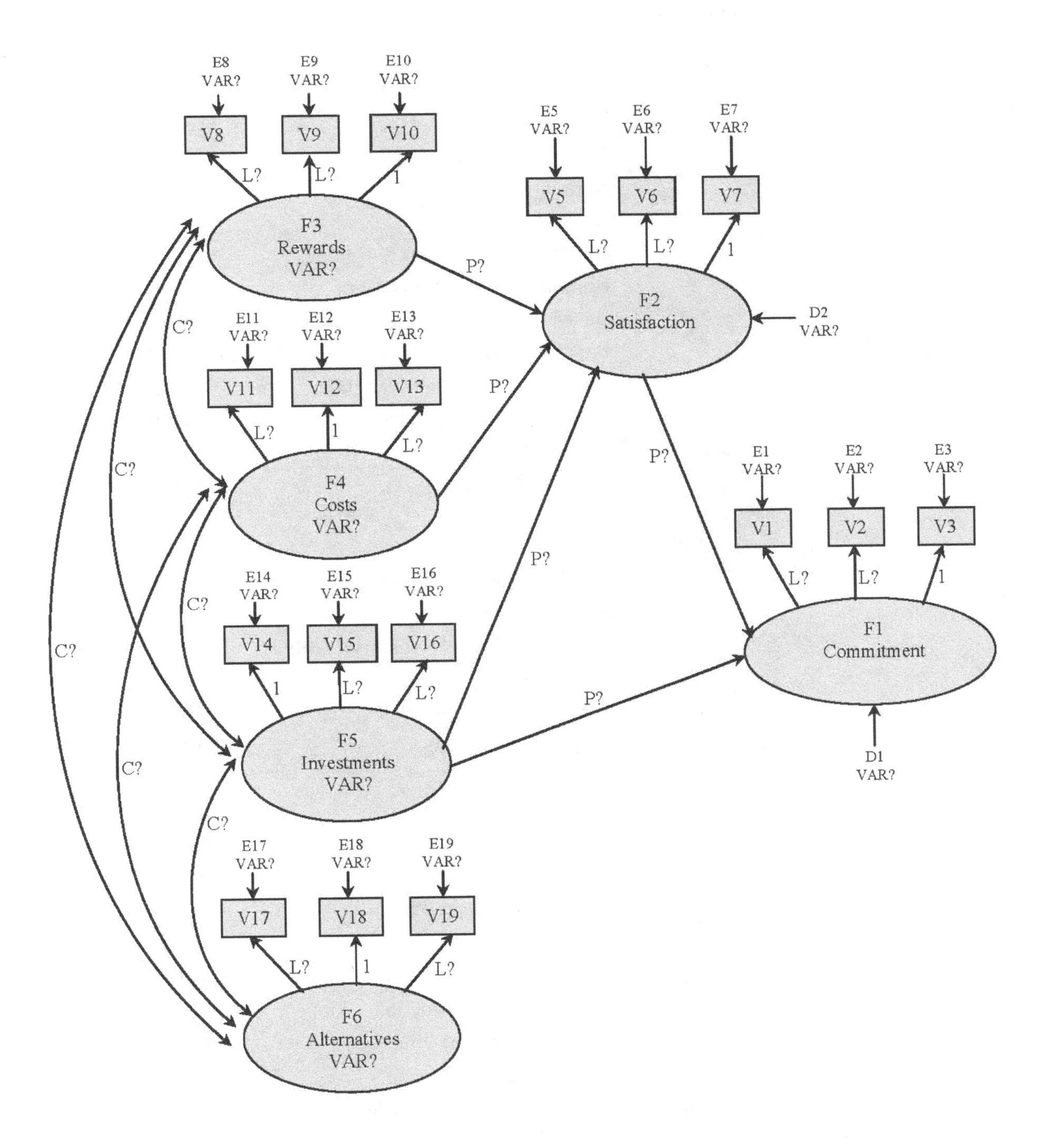

The PROC CALIS program that estimates revised model 2 is identical to the one for revised model 1, except that the latent variable equation for F2 had been changed to reflect the new path from F5 to F2. The original version of this equation had been as follows:

```
❹      F2  = PF2F3 F3 + PF2F4 F4          + D2;
```

while the revised equation took this form:

```
❹        F2  = PF2F3 F3 + PF2F4 F4 + PF2F5 F5 + D2;
```

Revised model 2 was estimated and the goodness-of-fit indices for this model are presented in Output 6.5.

Output 6.5: Goodness-of-Fit Indices for Revised Model 2, Investment Model Study

Fit Summary		
Modeling Info	Number of Observations	240
	Number of Variables	18
	Number of Moments	171
	Number of Parameters	47
	Number of Active Constraints	0
	Baseline Model Function Value	9.0702
	Baseline Model Chi-Square	2167.7711
	Baseline Model Chi-Square DF	153
	Pr > Baseline Model Chi-Square	<.0001
Absolute Index	Fit Function	0.7665
	Chi-Square	183.1915
	Chi-Square DF	124
	Pr > Chi-Square	0.0004
	Z-Test of Wilson & Hilferty	3.3240
	Hoelter Critical N	197
	Root Mean Square Residual (RMR)	0.2077
	Standardized RMR (SRMR)	0.0440
	Goodness of Fit Index (GFI)	0.9243
Parsimony Index	Adjusted GFI (AGFI)	0.8957
	Parsimonious GFI	0.7491
	RMSEA Estimate	0.0447
	RMSEA Lower 90% Confidence Limit	0.0301
	RMSEA Upper 90% Confidence Limit	0.0579
	Probability of Close Fit	0.7314
	ECVI Estimate	1.1938
	ECVI Lower 90% Confidence Limit	1.0574
	ECVI Upper 90% Confidence Limit	1.3667
	Akaike Information Criterion	277.1915
	Bozdogan CAIC	487.7815
	Schwarz Bayesian Criterion	440.7815
	McDonald Centrality	0.8840

Incremental Index	Bentler Comparative Fit Index	0.9706
	Bentler-Bonett NFI	0.9155
	Bentler-Bonett Non-normed Index	0.9638
	Bollen Normed Index Rho1	0.8957
	Bollen Non-normed Index Delta2	0.9710
	James et al. Parsimonious NFI	0.7420

Before proceeding with a detailed assessment of fit, it is first necessary to perform two chi-square difference tests. First, you will compare revised model 2 (M_{r2}) with revised model 1 (M_{r1}). Here, you hope to observe a significant chi-square difference value, as this will indicate that the model with the new path (M_{r2}) provides a fit to data that is significantly better than the fit provided by the more constrained model (M_{r1}). As before, chi-square values for the test may be obtained from Table 6.1.

$$M_{r1} - M_{r2} = 217.57 - 183.19 = 34.38$$

The chi-square difference value for this comparison is 34.38 which is quite close to the value of 34.35 that was predicted by the Lagrange Multiplier. The degrees of freedom for the test are equal to the difference between the degrees of freedom for the two models, or 125 - 124 = 1. The critical value for the chi-square statistic with 1 degree of freedom ($p < .01$) is 6.63, so this chi-square difference test is significant at $p < .01$. In other words, the test shows that revised model 2 (with the new path) provides a fit that is significantly superior to that of revised model 1.

So far, so good. The second chi-square difference test involves comparing revised model 2 to the measurement model (M_m). In this case you hope for a *nonsignificant* chi-square difference value, as this will suggest that M_{r2} does a good job of accounting for the relationships between the F variables that constitute the structural portion of the model. The test is performed as follows:

$$M_{r2} - M_m = 183.19 - 180.87 = 2.32$$

The degrees of freedom for the test are 124 - 120 = 4, and the critical value of the chi-square statistic ($p < .05$) with 4 degrees of freedom is 9.49. The observed chi-square difference value of 2.33 is less than this critical value, meaning that there is no significant difference in the fit provided by the two models. This finding supports the validity of revised model 2. Apparently, revised model 2 provides a fit to data that is essentially as good as the fit provided by the measurement model.

Some findings regarding revised model 2 are summarized below:

- With a sample of 240 participants and 124 degrees of freedom, statistical power for revised model 2 is approximately .99 using the SAS syntax (see Chapter 5).
- The absolute value of t statistics for each factor loading and path coefficient exceed 1.96 (or less than -1.96)
- No standardized factor loadings are trivial; only three are lower than .50.
- R^2 values for the commitment and satisfaction constructs were quite large at .55 and .51, respectively.
- Goodness-of-fit values for the Comparative Fit Index (CFI = .97), the Standardized Root Mean Square Residual (SRMR = .044), and the Root Mean Square Error of Approximation (RMSEA = .045) are each within ideal parameters and the full 90% confidence limits for the RMSEA are within good parameters (just short of ideal). Goodness-of-fit indices for all three models are summarized in Table 6.1

These results indicate that M_{r2} provides very acceptable levels of fit. These results, coupled with the finding that revised model 2 provided a fit to data that was not significantly worse than that of the measurement model, support M_{r2} as the study's "final" model.

Table 6.1: Goodness of Fit Indices for Various Models, Investment Model Study (N = 240)

Model	χ^2	df	$\Delta\chi^2$	Δdf	CFI	SRMR	RMSEA (CL_{90})
Baseline	247.67	137	--	--	.95	.047	.058 (.046-.070)
Re- specified Structural Model	180.87	120	66.80 **	17	.97	.042	.046 (.032-.059)
Initial Structural Model	217.57	125	37.20 **	5	.95	.060	.056 (.043-.068)
Revised Structural Model	216.75	124	-.82	1	.95	.060	.056 (.043-.068)
Finale Structural Model	183.19	124	2.32	4	.97	.044	.045 (.030-.058

Note. χ^2 = chi-square; df = degrees of freedom; CFI = Comparative Fit Index; SRMR = Standardized Root Mean Square Residual; RMSEA = Root Mean Square Error of Approximation; CL_{90} = 90% Confidence Limits.
** $p < .01$

Preparing a Formal Description of Results for a Paper

Figures and Tables

In most cases, it is best use a figure to illustrate the directional model tested in your study. Often a figure is presented to illustrate an initial theoretical model. This figure should follow the standard conventions in which latent variables are portrayed as oval or circles and manifest variables are portrayed as squares or rectangles. Your predictions will be made even more clear if directional paths are labeled with "+" and "-" signs to indicate whether positive or negative relationships are anticipated.

Figure 6.7 illustrates one example of how the theoretical model for the present fictitious study could be presented. Notice that, in order to simplify the reader's task, the model in Figure 6.7 is much simpler than the program figures referred to throughout this chapter (e.g., it does not contain symbols for parameters to be estimated).

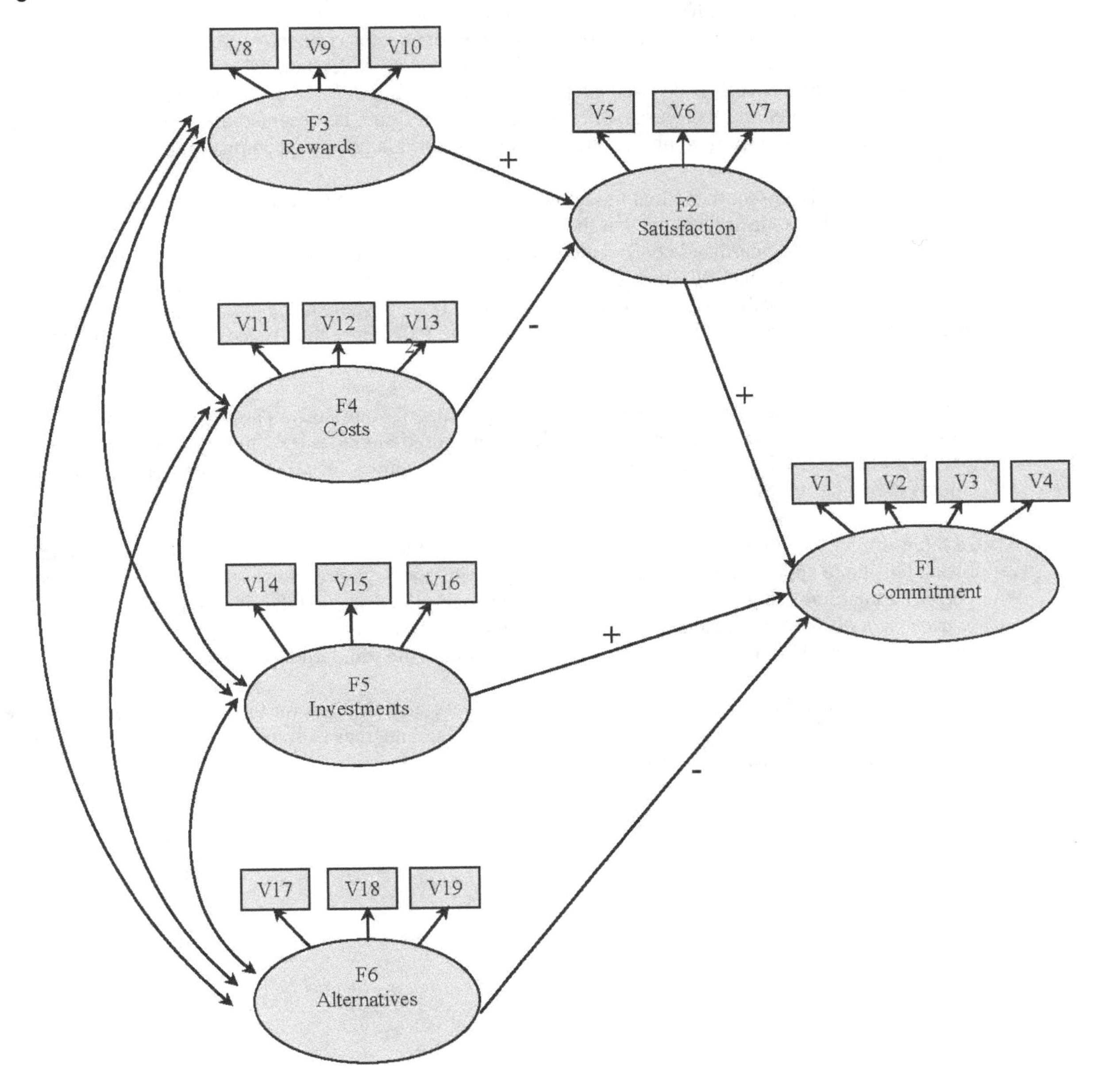

Figure 6.7: The Initial Theoretical Model, Investment Model Study

More importantly, you should present a second figure that illustrates the study's "final" model after modifications have been made. Ideally, it is easiest for the reader if standardized path coefficients (and parenthetically associated t values) appear along with directional paths connecting observed to latent variables and paths connecting the study's F variables. This will help readers understand which independent variables exerted relatively strong effects, and which exerted relatively weak effects. Coefficients for variance estimates are generally not presented in this figure. An example of such a figure is presented as Figure 6.8. If coefficients are not shown along with your final model, these should appear in a table to accompany the figure.

Some authors, instead, present their final model with unstandardized path coefficient along and error terms; we do not recommend this. Our rationale is that the reader is unable to directly examine the relative strength of association between unstandardized coefficients. Because standardized path coefficients share the same metric (i.e., each range from -1.00 to 1.00), they can be directly compared. As shown in Figure 6.8, for instance, we see that the standardized path coefficient connecting F5 to F1 = .56 whereas the path connecting F2 to F1 = .25. This indicates that the strength of association between F5 and F1 is more than twice the strength of association between F2 and F1.

You should also prepare several tables to summarize the results of your study. The first is usually a table of descriptive statistics showing the mean, standard deviation, range of values (i.e., highest and lowest), Cronbach's alpha (internal consistency), skewness, and kurtosis values for all interval and ration scales used. This allows the reader to see whether or not responses to your scales appear normally distributed.

Also, a table similar to Table 5.1 should be included to display the correlation coefficients between manifest variables. (This table appeared in Chapter 5, in the "Preparing the SAS Program" section.) Many researchers who perform structural equation modeling omit this information, but we feel that it is important to include such a table because it allows other researchers to replicate your analyses (and even test competing models!). As always, if your matrix is a correlation matrix, be sure to include the standard deviations.

Third, a table similar to Table 6.1 should summarize goodness-of-fit indices for the various models estimated. This should include chi-square values, change in chi-square (and degrees of freedom, change in degrees of freedom) to support selection of models.

Figure 6.8: Revised Model 2, Investment Model Study

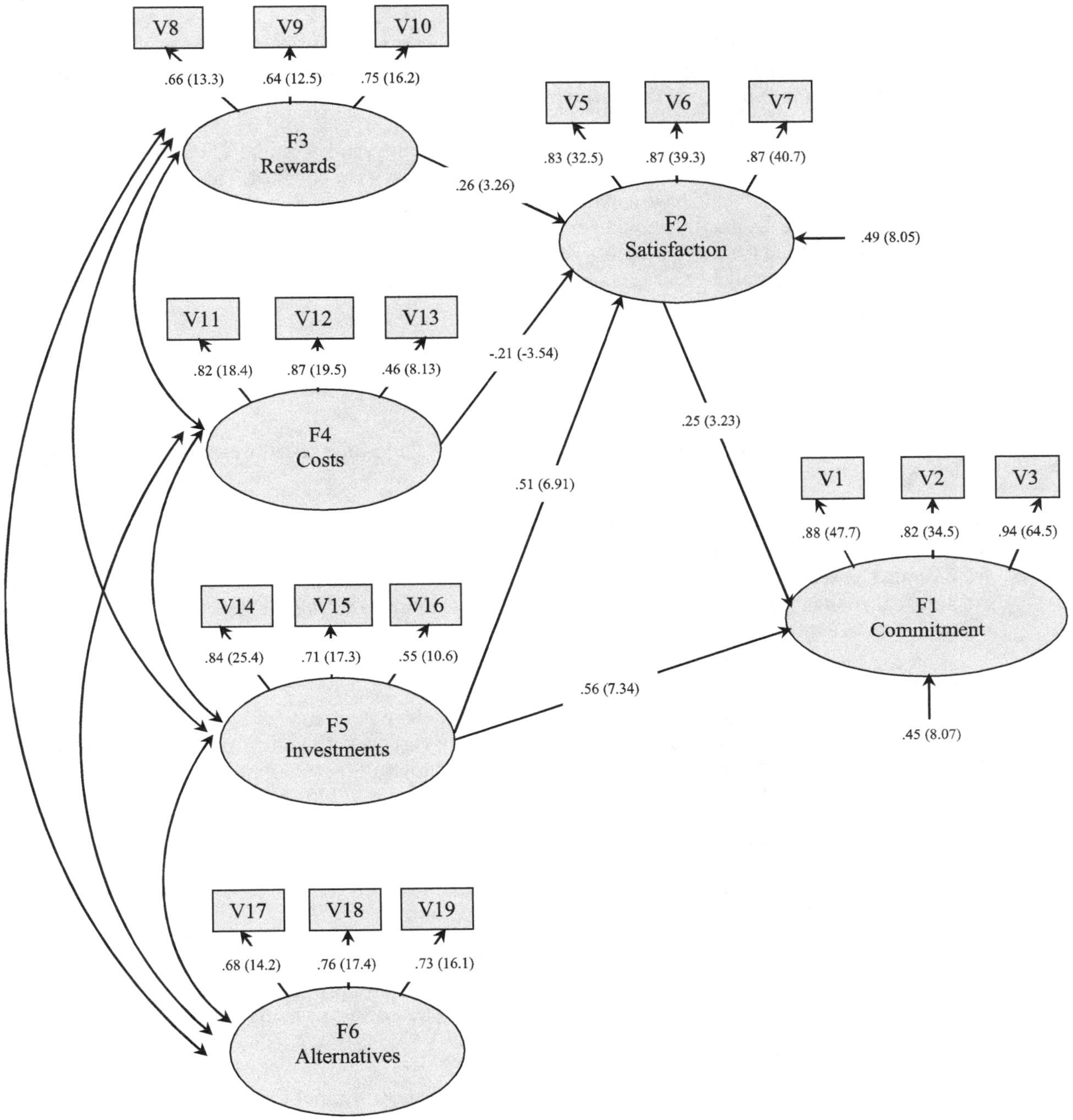

NOTE: Parameters expressed as maximum likelihood estimates (standardized solution). Parenthetical numbers indicate significance levels for parameter estimates (statistically significant t values > +/-1.96).

Preparing Text for the Results Section of the Paper

There is a good deal of variability in the way that research reports describe SEM procedures and findings. The way that you report results will depend, in part, on the purpose of your research. If your research was designed to test hypotheses related to the measurement model, then it is appropriate to discuss in some detail the tests that dealt specifically with the measurement model, such as the procedures that assess the convergent and discriminant validity of measures and constructs (e.g., Cappeliez and O'Rourke 2006). If measurement concerns are not a central focus of the research, less time can be spent on these matters (e.g., O'Rourke, Cappeliez, and Claxton 2011).

Below, we present one approach to describing the analyses reported in this chapter. The following touches on several different aspects of the analysis; you should feel free to deal with any of these topics in greater or lesser detail, depending on the purpose of the research.

One word of warning: The following results section is very lengthy, and goes into a great deal of detail in describing the procedures that were performed and the results that were obtained. This was done for the sake of completeness, so that you would have a model to follow when reporting the results of your analyses (regardless of which aspects of the results you chose to emphasize). In its entirety, the following section is actually much longer and more detailed than would be allowed in many scholarly journals. When you write for an academic journal, you should generally be more concise, and explain in great detail only those aspects of the analysis that are particularly relevant to your research questions.

Results

Overview of the Analysis

Data were analyzed using the SAS 9.3 CALIS procedure, and the models tested were covariance structure models with multiple indicators for all latent constructs.

The present analysis followed a two-step procedure based on the approach described by Anderson and Gerbing (1988). In the first step, confirmatory factor analysis was used to develop a measurement model that demonstrated an acceptable fit to data. In the second step, the measurement model was modified to become a structural equation model representing the theoretical model of interest. This theoretical model was then tested and revised until a theoretically meaningful and statistically acceptable model was achieved.

The Measurement Model

A measurement model describes the nature of the relationship between a number of latent variables, or factors and the manifest indicator variables that measure those latent variables. The model investigated in this study consisted of six latent variables corresponding to the six constructs of the investment model: Commitment, satisfaction, rewards, costs, investment size, and alternative value (N = 240). Each of the six latent variables was measured by at least three manifest or indictor variables as recommended for analyses of covariance structures (i.e., latent variable models).

The Initial Measurement Model

We followed Bentler's (1989) convention of identifying latent variables with the letter "F" (for factor), and labeling manifest variables with the letter "V" (for variable). Figure 6.7 uses these conventions in identifying the six latent constructs investigated in this study, as well as the indicators that measure these constructs. The figure shows that the commitment construct (F1) is measured by manifest variables V1 through V4, the satisfaction construct (F2) is measured by manifest variables V5 through V7, and so forth.

The measurement model computed in the first stages of this analysis was not identical to the model in Figure 6.7, because the model in that figure posits certain directional relationships between the latent constructs. The measurement model, in contrast, posits no directional paths between latent variables. Instead, in a measurement model, covariance is estimated to connect each latent variable with every other latent variable. In a figure, this would be indicated by a curved, two-headed arrow connecting each F variable to every other F variable. In other words, a measurement model is equivalent to a confirmatory factor analysis model in with each latent construct is allowed to covary with every other latent construct.

This measurement model was estimated using the maximum likelihood method, $\chi^2(df=137) = 247.68$, $p < .01$. Statistical power was estimated at .99 for this model using the SAS syntax provided by MacCallum and colleagues (1996). The Lagrange Multiplier showed that one manifest indicator, V4, was apparently affected by both the alternative value construct (F6) as well as the construct that it was hypothesized to measure (commitment, F1). Because V4 appears to be a complex variable, it was eliminated from the measurement model, and the measurement model was re-estimated.

The Revised Measurement Model

Standardized factor loadings for the indicator variables are presented in Table 6.2. Factor loadings range from .46 to .94, only two are less than .60; none of these values are trivial. PROC CALIS provides estimates standard errors for these coefficients which allow large-sample t tests of the null hypothesis that the coefficients are equal to zero in the population. All factor loadings are statistically significant ($p < .01$). These findings provide support for the convergent validity of responses to indicator variables (Anderson and Gerbing 1988).

Goodness-of-fit indices for the respecified measurement model (M_m) are presented in Table 6.1. Here, we report an absolute index (SRMR; Standardized Root Mean Square Residual), an incremental index (CFI; Confirmatory Fit Index), and a parsimony index (RMSEA; Root Mean Square Error of Approximation) as well as the 90% confidence limits for the RMSEA (CL_{90}). In accord with the threshold values recommended by Hu and Bentler (1999), the CFI value for this revised model is ideal ($.97 > .94$) as arethe SRMR ($.042 < .055$) and the RMSEA ($.046 < .055$); moreover, 90% confidence limits for the RMSEA are within acceptable parameters ($.032 \leq$ RMSEA $CL_{90} \leq .068$). Therefore model M_m was tentatively accepted as the study's "final" measurement model.

Table 6.2 also provides reliability estimates for responses to each observed or indicator variable (the square of the factor loadings), along with the composite reliability for latent constructs. (Composite reliability is a measure of internal consistency comparable to coefficient alpha.) Responses to all six indicator variables demonstrate adequate reliability with coefficients in excess of .70.

Table 6.2: Properties of the Revised Measurement Model (Table 5.3 from Chapter 5)

Construct and Indicators	Standardized Loading	t[a]	Reliability	Variance Extracted Estimate
Commitment (F1)			.91[b]	.78
V1	.89	48.11	.78	
V2	.82	34.52	.68	
V3	.94	64.15	.88	
Satisfaction (F2)			.89[b]	.73
V5	.83	32.63	.68	
V6	.87	39.08	.75	
V7	.88	40.89	.77	
Rewards (F3)			.72[b]	.47
V8	.67	13.36	.44	
V9	.63	12.32	.40	
V10	.75	16.22	.56	
Costs (F4)			.77[b]	.55
V11	.83	18.37	.68	
V12	.86	19.27	.75	
V13	.46	8.12	.21	
Investment size (F5)			.75[b]	.50
V14	.84	25.27	.71	
V15	.71	17.28	.50	
V16	.55	10.61	.30	
Alternative value (F6)			.77[b]	.53

V17	.68	14.48	.47
V18	.76	17.27	.58
V19	.73	16.11	.53

[a] All t tests were significant at p < .01.
[b] Denotes composite reliability.

The final column of Table 6.2 provides the variance extracted estimate for each indicator variable. This is a measure of the amount of variance captured by a construct, relative to the variance due to random measurement error. Five of the six constructs demonstrated variance extracted estimates in excess of .50, the level recommended by Fornell and Larcker (1981).

Combined, these findings provide general support for the reliability and validity of constructs and their indicators. The revised measurement model (M_m) was therefore retained as the study's final measurement model against which successive models are compared.

The Structural Model

The Initial Theoretical Model

The theoretical model tested in the present study is identical to the one presented in Figure 6.7, with the exception that V4 has been deleted as a measure of commitment (consistent with our findings when analyzing the measurement model). As we now specify directional paths between latent factors, this is a structural equation model (SEM).

Although goodness-of-fit indices for this SEM are within good to ideal parameters, one of the paths linking two latent constructs are nonsignificant: The standardized path coefficient from alternative value (F6) to commitment (F1) is only .06 (t = .92, ns).

The nomological validity of a theoretical model is tested by performing a chi-square difference test in that the theoretical model is compared to the measurement model. A finding of no significant difference indicates that the theoretical model is successful in accounting for the observed relationships among latent constructs. The chi-square value for the measurement model was subtracted from the chi-square value for the theoretical model with the resulting difference of 216.75 - 180.87 = 35.88. The degrees of freedom for the test are equal to the difference between models (i.e., $\Delta df = 124 - 120$). The critical chi-square value with 4 df is 13.3 ($p < .01$), and so this chi-square difference was significant. In other words, the theoretical model was unsuccessful in accounting for the relationships among latent constructs. A specification search was conducted to arrive at a better-fitting model.

Revised Model 1

When conducting a specification search there is a danger that data-driven model modifications will capitalize on chance characteristics of the sample data and result in a final model that will not generalize to other samples the or population. We therefore began the search by attempting to identify parameters that could be dropped from the model without significantly affecting the model's fit; as it is generally safer to drop parameters than to add new parameters when modifying models (Bentler and Chou 1987).

The Wald test (Bentler 1989) suggested that it was possible to delete the path from alternative value to commitment without a significant increase in model chi-square. This is consistent with the nonsignificant path coefficient between these latent constructs. Therefore, this path was deleted, and the resulting model, revised model 1 (M_{r1}) was then computed.

Dropping the alternatives-commitment path would be acceptable only if it did not result in a significant increase in model chi-square. A significant increase would indicate that M_{r1} provided a fit that was significantly worse than M_t. Therefore, a chi-square difference test was conducted, comparing M_t to M_{r1} (see Table 6.1 for model chi-square values). The chi-square difference for this comparison was equal to 217.57 - 216.75 = 0.82 which, with 1 df, was nonsignificant ($p > .05$).

M_{r1} was next compared to M_m to determine whether it successfully accounted for the relationships between the latent constructs. The chi-square difference was calculated as 217.57 - 180.87 = 36.70, which, with 5 df, was statistically significant ($p < .01$). Once again, the model failed to provide an acceptable fit.

Revised Model 2

Wald tests conducted in the course of analyzing M_{r1} did not reveal any additional causal paths between latent constructs that could be deleted without affecting the model's fit. We therefore reviewed results of Lagrange Multipliers to identify new paths that should be added to the model.

A Lagrange Multiplier estimated that model chi-square for M_{r1} could be reduced by 34.35 if a directional path were added between investment size (F5) and satisfaction (F2). Adding such a path would be consistent with the prediction from cognitive dissonance theory (Festinger 1957) that individuals often adjust their attitudes (i.e., become more satisfied) so that their attitudes will be consistent with their behaviors (i.e., the behavior of investing time and effort in a relationship). Because this model revision can be justified on theoretical grounds, a path from investment size to satisfaction was added to M_{r1}. The resulting model, revised model 2 (M_{r2}), was then estimated.

Statistical power for this revised model 2 remained at .99. All path coefficients, including the new path between investment size and satisfaction, are nontrivial and statistically significant (i.e., $|t| > 1.96$). Figure 6.8 depicts standardized path coefficients for revised model 2.

R^2 values showed that satisfaction, investment size and alternative value accounted for 55% of variance in commitment, while rewards, costs, and investment size accounted for 51% of variance in satisfaction. As shown in Table 6.1, all goodness-of-fit indices for revised model 2 are in ideal parameters.

A chi-square difference test comparing M_{r2} to M_{r1} revealed a significant difference value of 217.57 - 183.19 = 34.38 ($\Delta df = 1$, $p < .01$). This finding shows that revised model 2 provides a fit to data that is significantly better than the fit provided by revised model 1.

As a final test, a chi-square difference test compares the fit of M_{r2} with that of M_m resulting in a difference of 183.19 - 180.87 = 2.32, which, with 4 df is nonsignificant ($p > .05$). This nonsignificant chi-square value indicates that M_{r2} provided a fit that was not significantly worse than that provided by the measurement model in which all F variables were free to covary. In other words, this finding indicates that the relationships described in revised model 2 are successful in accounting for the observed relationships between the latent constructs.

Combined, these findings provide support for revised model 2 over the other models tested. M_{r2} was therefore retained as this study's "final" model as depicted in Figure 6.8.

It should be noted, however, that revised model 2 was obtained on the basis of modification indices versus theory specific to Rusbult's (1980) original investment model. The results of this study should be replicated with other samples derived from this and other populations, ideally randomly derived samples.

Additional Example: A SEM Predicting Victim Reactions to Sexual Harassment

This final section of the chapter will very briefly present another theoretical model to be analyzed. It will also provide the program figures and PROC CALIS programs needed to conduct the analyses. All material needed to compute the programs have already been provided. The purpose of this section is to quickly apply these concepts to another model. As an informal exercise, you are encouraged to:

- Study the models as they appear in the text's figures.
- Prepare the PROC CALIS program that will analyze the model.
- Compare your output to that presented in this text.

Chapter 4 described a fictitious analogue study in which a sample of women read a description of a woman who was being sexually harassed on the job. The scenarios that the women read varied with regard to the seriousness of the offence that the theoretical woman experienced. After reading the scenario randomly assigned to them, participants were asked to imagine how they would feel if they had been the victim of the harassment, and to respond to a questionnaire that assessed the following constructs:

- Their **intention to report** the harassment to senior manager at the organization. With this variable, higher scores indicate greater intention to report.
- The **expected outcomes** of reporting the harassment, where higher scores indicate stronger belief that reporting the harassment will result in positive results for the victim.
- **Feminist ideology**, where higher scores indicate egalitarian attitudes about sex roles.
- **Seriousness of the offense**, where higher scores indicate a stronger belief that the woman in the scenario experienced a serious form of harassment.
- **Normative expectations,** where higher scores reflect a stronger belief that the victim's family, friends, and coworkers would support her if she reported the harassment.

Figure 6.9 presents a theoretical model that describes predicted relationships among these five constructs. According to the model, intention to report is expected to be directly affected by expected outcomes, feminist ideology, and normative expectations. Expected outcomes, in turn, are expected to be affected by all three of the study's exogenous variables. (Remember that this model was constructed for purposes of illustration, and should not necessarily be regarded as a *bona fide* model of victim reactions to sexual harassment.)

Figure 6.9: Initial Theoretical Model, Sexual Harassment Study

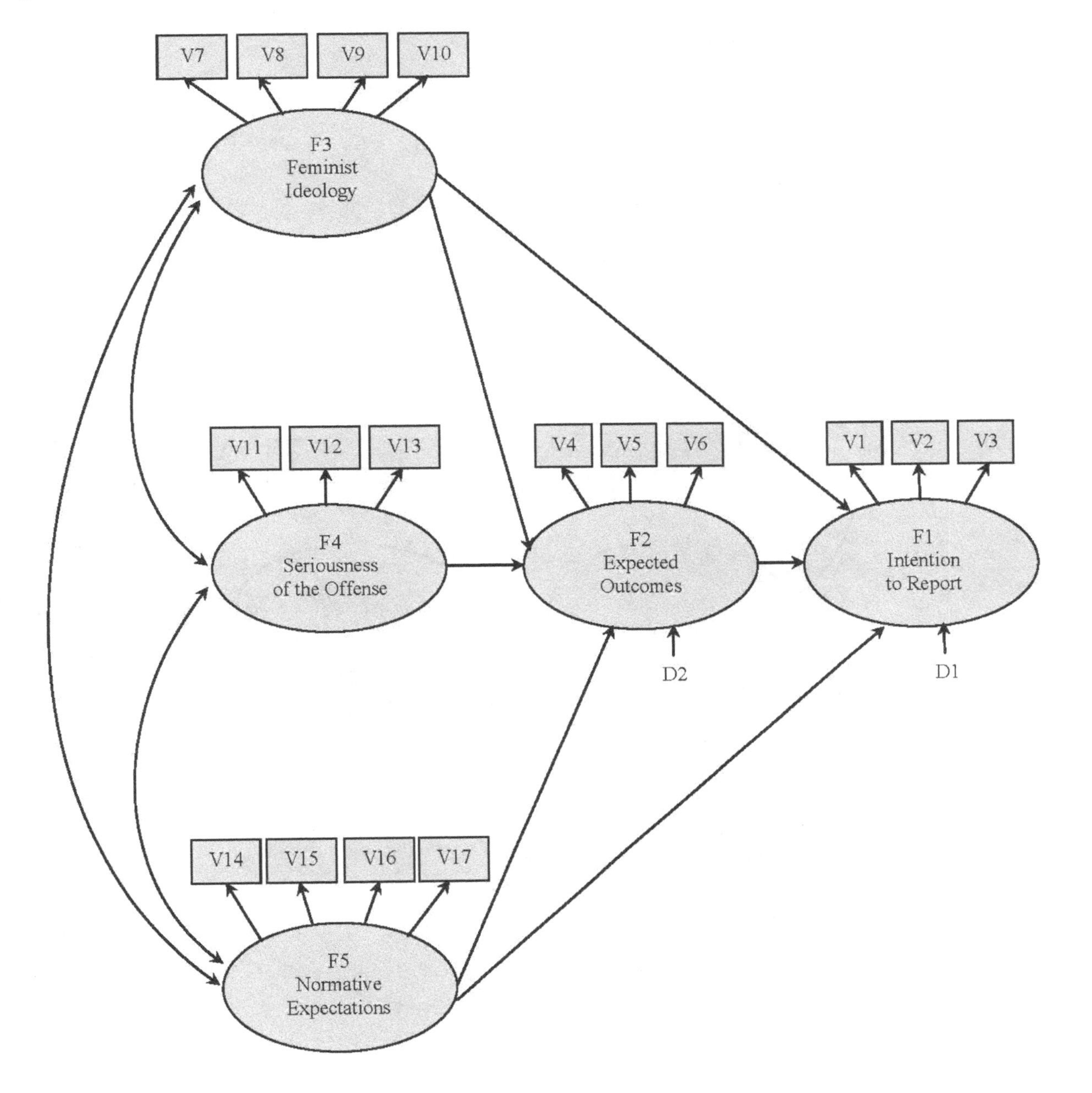

Figure 6.10 is a standard SEM model, in that all five constructs are represented as F variables with multiple indicators. Assume that the V variables are responses to individual items on a hypothetical questionnaire.

In the first phase of testing this model, confirmatory factor analysis would be performed to develop a satisfactory measurement model. The program figure for this CFA is presented in Figure 6.10. Notice that all structural variables in the model (F variables, in this case) are allowed to freely covary with each other.

Figure 6.10: Program Figure for CFA Measurement Model, Sexual Harassment Study

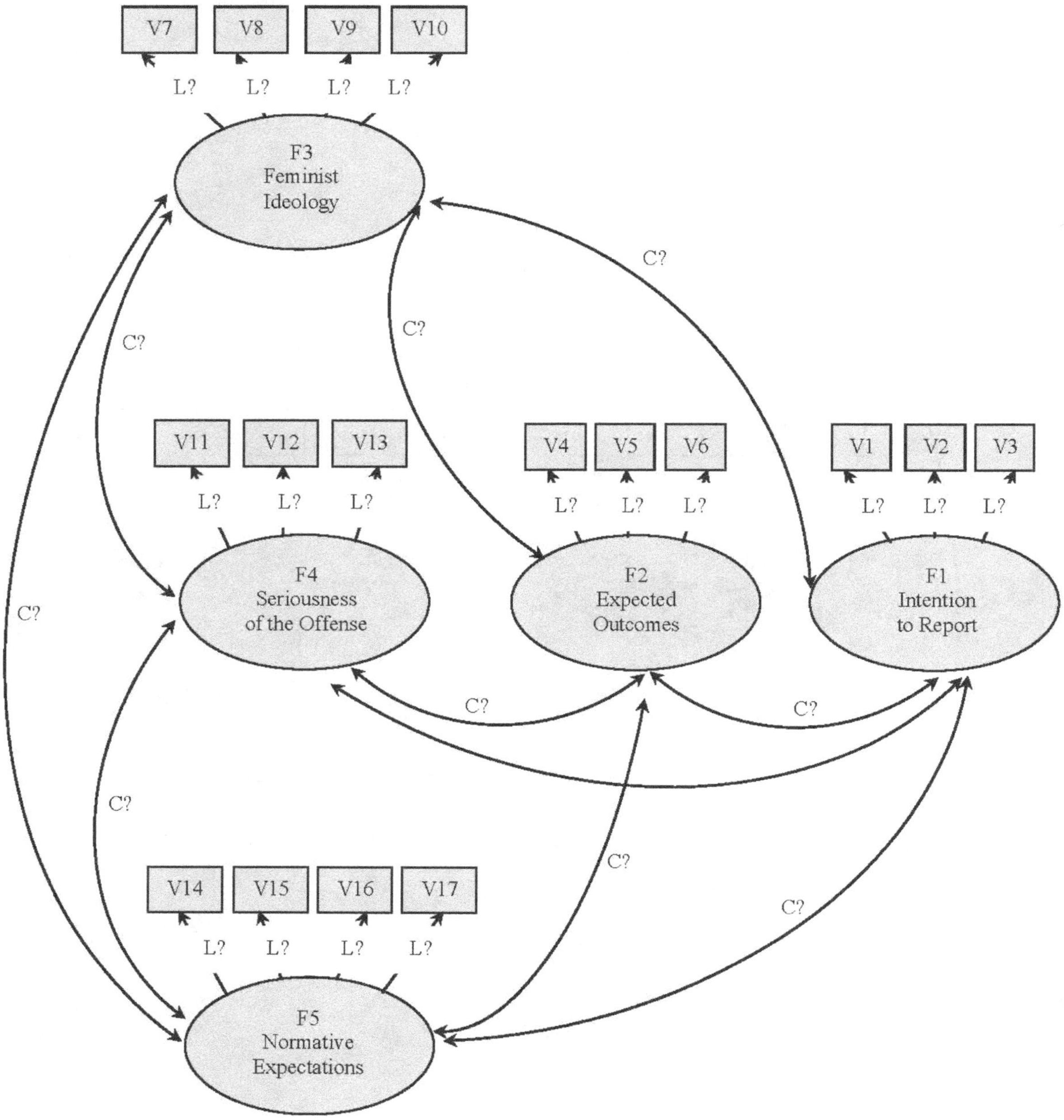

Below is the PROC CALIS program that would estimate the measurement model of Figure 6.10:

```
proc calis  modification ;
   lineqs
      V1   = LV1F1   F1 + E1,
      V2   = LV2F1   F1 + E2,
      V3   = LV3F1   F1 + E3,
      V4   = LV4F2   F2 + E4,
      V5   = LV5F2   F2 + E5,
      V6   = LV6F2   F2 + E6,
      V7   = LV7F3   F3 + E7,
      V8   = LV8F3   F3 + E8,
      V9   = LV9F3   F3 + E9,
      V10  = LV10F3  F3 + E10,
      V11  = LV11F4  F4 + E11,
      V12  = LV12F4  F4 + E12,
      V13  = LV13F4  F4 + E13,
      V14  = LV14F5  F5 + E14,
      V15  = LV15F5  F5 + E15,
      V16  = LV16F5  F5 + E16,
      V17  = LV17F5  F5 + E17;
   variance
      E1-E17 = VARE1-VARE17,
      F1 = 1,
      F2 = 1,
      F3 = 1,
      F4 = 1,
      F5 = 1;
   cov
      F1  F2 = CF1F2,
      F1  F3 = CF1F3,
      F1  F4 = CF1F4,
      F1  F5 = CF1F5,
      F2  F3 = CF2F3,
      F2  F4 = CF2F4,
      F2  F5 = CF2F5,
      F3  F4 = CF3F4,
      F3  F5 = CF3F5,
      F4  F5 = CF4F5;
    var  V1-V17 ;
  run;
```

Assume that the initial measurement model did not provide an acceptable fit, and that the indicator V17 had to be dropped from F5 (normative expectations), in order to achieved the desired fit. Having developed an adequate measurement model, the study now moves to step 2, and the estimation of the theoretical model. Figure 6.11 displays the completed program figure.

Figure 6.11: SEM Program Figure, Sexual Harassment Study

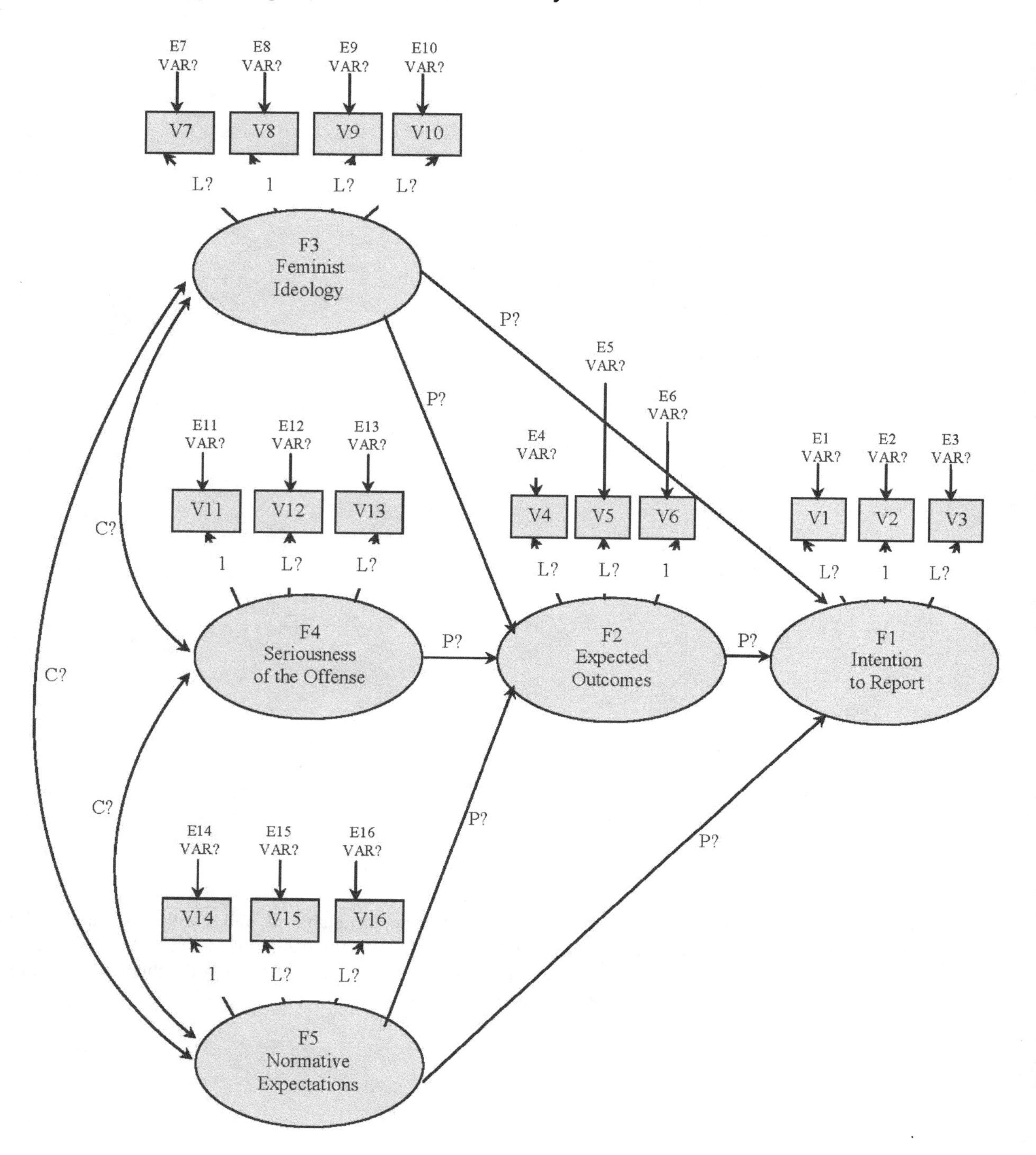

The PROC CALIS program for estimating this model appears below:

```
proc calis  modification ;
   lineqs
      V1  = LV1F1  F1 + E1,
      V2  =        F1 + E2,
      V3  = LV3F1  F1 + E3,
      V4  = LV4F2  F2 + E4,
      V5  = LV5F2  F2 + E5,
      V6  =        F2 + E6,
      V7  = LV7F3  F3 + E7,
      V8  =        F3 + E8,
      V9  = LV9F3  F3 + E9,
      V10 = LV10F3 F3 + E10,
      V11 =        F4 + E11,
      V12 = LV12F4 F4 + E12,
      V13 = LV13F4 F4 + E13,
      V14 =        F5 + E14,
      V15 = LV15F5 F5 + E15,
      V16 = LV16F5 F5 + E16,
      F1  = PF1F2 F2 + PF1F3 F3 + PF1F5 F5 + D1,
      F2  = PF2F3 F3 + PF2F4 F4 + PF2F5 F5 + D2;
   variance
      E1-E16 = VARE1-VARE16,
      F3 = VARF3,
      F4 = VARF4,
      F5 = VARF5,
      D1 = VARD1,
      D2 = VARD2;
   cov
      F3 F4 = CF3F4,
      F3 F5 = CF3F5,
      F4 F5 = CF4F5;
   var  V1-V16 ;
  run;
```

Notice that the preceding program adheres to the usual conventions for testing a standard theoretical model (e.g., the variances of the exogenous F variables are free parameters to be estimated, one factor loading for each F variable has been fixed at 1). If a dataset were actually analyzed using this program, the results would be interpreted according to the same guidelines presented earlier in this chapter.

Conclusion: To Learn More about Latent Variable Models

The chapter on path analysis stated that the best way to learn was to practice performing the procedure; the same is true for SEM. Your next step should be to locate books or articles reporting the results of CFA and SEM analyses that provide the correlation or covariance matrix analyzed along with the results of the analysis. Re-analyze the datasets you find using PROC CALIS. If you obtain the same results reported by the authors, you have performed the analysis correctly (assuming the original researchers conducted *their* analyses correctly!).

This chapter has introduced some basic concepts in the computation and analysis of structural equation models, but has been somewhat narrow in focusing only on recursive models with cross-sectional data. You should now be ready to move on to more complex models, such as time-series designs (e.g., O'Rourke et al. 2003). Byrne (2009, 1998) and Kline (2005) are examples of useful resources when learning about more complex applications.

References

Anderson, J. C., and Gerbing, D. W. (1988). Structural equation modeling in practice: A review and recommended two-step approach. *Psychological Bulletin, 103,* 411–423.

Bentler, P. M. (1989). *EQS structural equations program manual.* Los Angeles: BMDP Statistical Software.

Bentler, P. M., and Chou, C. (1987). Practical Issues in structural modeling. *Sociological Methods and Research, 16,* 78–117.

Byrne, B. M. (2009). *Structural equation modeling with AMOS: Basic concepts, applications, and programming* (2nd Ed.). Mahwah, NJ: Lawrence Erlbaum.

Byrne, B. M. (1998). *Structural equation modeling with LISREL, PRELIS, and SIMPLIS: Basic concepts, applications, and programming.* Mahwah, NJ: Lawrence Erlbaum.

Cappeliez, P., and O'Rourke, N. (2006). Empirical validation of a comprehensive model of reminiscence and health in later life. *Journals of Gerontology: Psychological Sciences, 61,* P237–P244.

Festinger, L. (1957). *A theory of cognitive dissonance.* Stanford, CA: Stanford University Press.

Fornell, C., and Larcker, D. F. (1981). Evaluating structural equation models with unobservable variables and measurement error. *Journal of Marketing Research, 18,* 39–50.

Hu, L. T., and Bentler, P. M. (1999). Cutoff criteria for fit indices in covariance structure analysis: Conventional criteria versus new alternatives. *Structural Equation Modeling, 6,* 1–55.

Jöreskog, K. G., and Sörbom, D. (2001). *LISREL 8: User's reference guide [Computer software manual].* Lincolnwood, IL: Scientific Software.

Kenny, D. A., (1979). *Correlation and causality.* New York: John Wiley and Sons.

Kline, R. B. (2005). *Principles and practice of structural equation modeling.* (2nd Ed.). New York: Guilford Press.

Le, B., and Agnew, C.R. (2003). Commitment and its theorized determinants: A meta-analysis of the investment model. *Personal Relationships, 10,* 37–57.

MacCallum, R. C., Browne, M. W., and Sugawara, H.M. (1996). Power analysis and determination of sample size for covariance structure modeling. *Psychological Methods, 2,* 130–149.

O'Rourke, N., Cappeliez, P., and Claxton, A. (2011). Functions of reminiscence and the psychological well-being of young-old and older adults over time. *Aging and Mental Health, 15,* 272–281.

O'Rourke, N., Cappeliez, P., and Guindon, S. (2003). Depressive symptoms and physical health of caregivers of persons with cognitive impairment: Analysis of reciprocal effects over time. *Journal of Aging and Health, 15,* 688–712.

O'Rourke, N., Carmel, S., Chaudhury, H., Polchenko, N., & Bachner, Y.G. (2013). A cross-national comparison of reminiscence functions between Canadian and Israeli older adults. *Journals of Gerontology: Psychological Sciences, 68,* 184-199. doi:10.1093/geronb/gbs058

Williams, L. J., and Hazer, J. T. (1986). Antecedents and consequences of satisfaction and commitment in turnover models: A reanalysis using latent variable structural equation methods. *Journal of Applied Psychology, 71,* 219–231.

Appendix A.1: Introduction to SAS Programs, SAS Logs, and SAS Output

What Is SAS?

SAS is a modular, integrated, and hardware-independent system of statistical software. It is a particularly powerful tool for social scientists because it allows us to easily perform a myriad of statistical analyses that may be required in the course of conducting research. SAS is sufficiently comprehensive to perform the most sophisticated multivariate analyses (i.e., multiple dependent variables), but is also easy to use so that undergraduates can perform basic analyses after only a short period of instruction.

In a sense, SAS may be viewed as a library of prewritten statistical algorithms. By submitting a short SAS program, you can access a prewritten procedure to analyze a set of data. For example, below are the SAS statements used to call up the algorithm that calculates the Pearson correlation coefficient:

```
proc corr   data=D1;
   run;
```

The preceding statements enable SAS to compute correlation coefficients for all numeric variables in your dataset. The ability to call up complex procedures with such a simple statement makes this system both powerful and easy to use.

Three Types of SAS Files

The purpose of this appendix is to provide a general sense of what it entails to submit a SAS program and interpret the results. This appendix presents a short SAS program and discusses the resulting output. You are encouraged to copy the program that appears in the following example, submit it for analysis, and verify that the resulting output matches the output reproduced here. This exercise will provide you with the SAS big picture and this perspective will facilitate learning the programming details presented in this text.

Briefly, you will work with three types of files when using SAS: One file contains the SAS program; one file contains the SAS log; and one file contains the SAS output. The following sections discuss the differences among these files.

The SAS Program

A SAS program consists of a set of user-written statements. These statements provide the data to be analyzed (or where these data can be found), tell SAS about the nature of these data, and indicate which statistical analyses should be performed.

This section illustrates a simple SAS program by analyzing data from a fictitious study. Assume that six high school students have taken the *SAT Reasoning Test*. This test provides three scores for each student: Critical reading, writing, and mathematics. Scores range from 200 to 800 for each test, with higher scores indicating higher performance.

Assume that you now want to obtain some simple descriptive statistics regarding the six students' scores on each test. For example, what is their *average* score on the critical reading, writing, or SAT math test? What is the *standard deviation* of scores for these three tests?

To perform these analyses, prepare the following SAS program:

```
data D1;
input PARTICIPANT SATREAD SATWRITE SATMATH ;

datalines;
1 520 580 490
2 610 640 590
3 470 430 450
4 410 400 390
5 510 490 460
6 580 510 350
;
run;

proc means   data=D1;
  var  SATREAD SATWRITE SATMATH;
run;
```

The first eleven lines of code make up the **DATA step**, which is used to read data and create a SAS dataset. The last three lines are the **PROC step**, which is used to process or analyze the data. The differences between these steps are described in the next two sections.

The DATA Step

In the DATA step, programming statements create and/or modify a SAS dataset. Among other features, these statements might:

- provide a name for the dataset
- provide a name for the variables to be included in the dataset
- indicate where to find the dataset
- provide the actual data to be analyzed

In the preceding program, the DATA step begins with the DATA statement and ends with a semicolon and RUN statement; these items precede the PROC MEANS statement.

The first statement of the preceding program begins with the word DATA and specifies that SAS should create a dataset named D1. The next line contains the input statement, which indicates that four variables will be contained in this dataset. The first variable is named PARTICIPANT; this variable specifies the participant number of each student. The second variable is named SATREAD (for the SAT critical reading test), the third variable is named SATWRITE (SAT writing test), and the fourth variable is named SATMATH (SAT math test).

The DATALINES statement indicates that lines containing your data are to follow. The first line after the DATALINES statement contains the data (test scores) for participant 1. This first data line contains the numbers 520, 580, and 490, which means that participant 1 received a score of 520 on the critical reading test, 580 on the writing test, and 490 on the math test. The next data line shows that participant 2 received a score of 610 for critical reading, 640 for writing, and 590 for the math test. The semicolon and RUN statement after the last data line signal the end of the data.

The PROC Step

In contrast to the DATA step, the PROC step includes programming statements that request specific statistical analyses. For example, the PROC step might request that correlations be performed between all quantitative variables, or the PROC step might request that a t test be performed. In the preceding example, the PROC step consists of the last three lines of the program.

The first line after the DATA step is the PROC MEANS statement. This requests that SAS use a procedure called MEANS to analyze the data. The MEANS procedure computes means, standard deviations, and some other descriptive statistics for numeric variables in the dataset. Immediately after the words PROC MEANS are the words DATA =D1. This tells the system that the data to be analyzed are in a dataset named D1. (Remember that D1 is the name of the dataset just created.)

Following the PROC MEANS statement is the VAR statement, which includes the names of three variables: SATREAD, SATWRITE, and SATMATH. This requests that descriptive statistics be computed for SATREAD, SATWRITE, and SATMATH.

Finally, the last line of the program is the RUN statement, which signals the end of the PROC step. If a SAS program requests multiple PROCS (or procedures), you have two options for using the RUN statement:

- you may place a separate RUN statement following each PROC statement
- you may place a single RUN statement following the last PROC statement

What is the single most common programming error? For new SAS users, the single most common error involves leaving off a required semicolon (;). Remember that every SAS statement must end with a semicolon. In the preceding program, notice that the DATA statement ends with a semicolon as does the INPUT statement, the DATALINES statement, the PROC MEANS statement, and the RUN statement. When you obtain an error in running a SAS program, one of the first things you should look for is missing semicolons in the program.

Once you submit the preceding program for analysis, SAS creates two types of files reporting the results of the analysis. One file is named the **SAS log** or **log file**. This file contains notes, warnings, error messages, and other information related to the execution of the SAS program. The other file is referred to as the **SAS output file**. The SAS output file contains the results of the requested statistical analyses.

The SAS Log

The SAS log is a listing of notes and messages that should help you verify that your SAS program was executed successfully. Specifically, the log provides:

- a reprinting of the SAS program that was submitted
- a listing of notes indicating how many variables and observations are contained in the dataset
- a listing of any errors made in the execution of the SAS program

Log A.1.1: SAS Log for the Preceding Program

```
NOTE: Copyright (c) 2002-2010 by SAS Institute Inc., Cary, NC, USA.
NOTE: SAS (r) Proprietary Software 9.3 (TS1M0)
      Licensed to NORM OROURKE, Site 70027498.
NOTE: This session is executing on the W32_7PRO  platform.



NOTE: SAS initialization used:
      real time           1.42 seconds
      cpu time            0.70 seconds

1     data D1;
2        input PARTICIPANT SATREAD SATWRITE SATMATH;
3
4     datalines;

NOTE: The data set WORK.D1 has 6 observations and 4 variables.
NOTE: DATA statement used (Total process time):
      real time           0.01 seconds
      cpu time            0.03 seconds


11    ;
12    run;
13
14
15    proc means    data=D1;
16       var   SATREAD SATWRITE SATMATH ;
17    run;

NOTE: Writing HTML Body file: sashtml.htm
NOTE: There were 6 observations read from the data set WORK.D1.
NOTE: PROCEDURE MEANS used (Total process time):
      real time           1.29 seconds
      cpu time            0.25 seconds
```

Notice that the statements constituting the SAS program are assigned line numbers and are reproduced in the SAS log. The datalines are not normally reproduced as part of the SAS log unless they are specifically requested.

About halfway down the log, a note indicates that the dataset contains six observations and three variables. You should check this note to verify that the dataset contains all of the variables that you intended to input (in this case four), and that it contains data from all of your participants (in this case six). So far, everything appears to be correct.

If you made errors when writing the SAS program, there would also be ERROR messages in the SAS log. Often, these error messages enable you to determine what is wrong with the program. For example, a message might indicate that SAS was expecting a program statement that was not included. Whenever you encounter an error message, read it carefully and review all of the program statements that precede it. Often, the error appears in the program statements that immediately precede the error message, but in other cases the error might be hidden much earlier in the program.

If more than one error message is listed, do not panic; there may still be only one error. Sometimes a single error will cause a cascade of subsequent error messages.

Once the error or errors have been identified, revise the SAS program and resubmit it for analysis. Review the new SAS log to see if the errors have been rectified. If the log indicates that the program ran correctly, then review the results of the analyses in the SAS output file.

The SAS Output File

The SAS output file contains the results of the statistical analyses requested in the SAS program. Because the program in the previous example requested the MEANS procedure, the corresponding output file will contain means and other descriptive statistics for the variables analyzed. In this text, the SAS output file is sometimes referred to as the **lst file**. "Lst" is used as an abbreviation for "listing of results."

Following is the SAS output file produced by the preceding SAS program:

Output A.1.1: Results of the MEANS Procedure

The MEANS Procedure

Variable	N	Mean	Std Dev	Minimum	Maximum
SATREAD	6	516.6666667	72.5718035	410.0000000	610.0000000
SATWRITE	6	508.3333333	90.2034737	400.0000000	640.0000000
SATMATH	6	455.0000000	83.3666600	350.0000000	590.0000000

Below the heading "Variable," SAS prints the names of each of the analyzed variables. In this case, the variables are called SATREAD, SATWRITE, and SATMATH. To the right of the heading SATREAD, descriptive statistics for the critical reading test are found. The corresponding figures for the writing test appear to the right of SATWRITE, and descriptive statistics for the math test appear to the right of SATMATH.

Below the heading "N," the number of observations or participants is reported. The average score on each variable is reproduced under "Mean," and standard deviations appear in the column headed "Std Dev." Minimum and maximum scores for the three variables appear in the remaining two columns. The mean score for the critical reading test is 516.67, and the standard deviation of these scores is 72.57. For the writing test, the mean is 508.33 with a standard deviation of 90.20, and for the math test the mean is 455.00 with a standard deviation of 83.37. (Note: SAS output will generally report findings to several decimal places. In this text, however, numbers will be reported to only two decimal places in most instances, and rounded as necessary in accordance with the publication manual of the American Psychological Association [APA, 2009].)

The statistics included in the preceding output are printed by default (i.e., without asking for them specifically). There are many additional statistics that you can request.

SAS Customer Support

Although this text provides examples of many of the analyses performed in the social sciences, specific questions to certain problems may arise. One alternative for registered SAS users (or their institutions) is use of the SAS Customer Support (http://support.sas.com). SAS maintains a comprehensive website with up-to-date information. In the left column, one option that is particularly useful for the novice (and not-so-novice) SAS user appears under the heading "Knowledge Base." Click the *Documentation* link, which will take you to a page where you can search an array of topics including examples and SAS syntax. Here, you can obtain information regarding specific statistical procedures covered, and not covered, in this text.

It is also possible to pose specific questions (*Support → Submit a Problem*) to SAS Customer Support. To use this feature, you will need to provide an e-mail address to which replies can be sent, identify your institution, and provide a customer site number/license information. This latter information can be found in any SAS log file. (See log A.1.1 where the release version and license number are specified in the first section.)

Conclusion

Regardless of the computing environment in which you work, the basics of using SAS remain the same: You prepare the SAS program, submit it for analysis, review the resulting log for any errors, and examine the output files to view the results of your analyses. For further information about the fundamentals of creating SAS datasets, refer to Appendix A.2, "Data Input."

Reference

American Psychological Association (2009). *Publication manual of the American Psychological Association* (6th Ed.). Washington, DC: Author.

Appendix A.2: Data Input

Introduction: Inputting Questionnaire Data versus Other Types of Data

This appendix shows how to create SAS datasets in a number of different ways by illustrating how to input various types of data that are commonly obtained in **questionnaire research**. Questionnaire or survey research generally involves administering standardized instruments to a sample of participants and asking them to select among fixed responses. For example, participants may be asked to indicate the extent to which they agree or disagree with a series of items by selecting a response along a 7-point scale where 1 = "strongly disagree," 4 = "neither agree nor disagree," and 7 = "strongly agree." These are known as Likert-type scales (DeVellis 2012).

Because this appendix (and much of the entire text, for that matter) focuses on questionnaire research, some readers may be concerned that it may not be useful for analyzing data that are obtained using other methods. This concern is understandable, because the social sciences are so diverse and many different types of variables are examined. These variables might be as different as the number of aggressive acts performed by a child, preferences for laundry detergents, or levels of serotonin metabolites in the cerebrospinal fluid of chimpanzees.

Because of the flexibility of the basic principles outlined in this appendix, you can expect to input virtually any type of quantitative data obtained in social science research upon completing this appendix. The same may be said for the remaining appendices and chapters of this text; although this book emphasizes analyses of questionnaire data, the concepts can be readily applied to many types of data. This should become clear as the mechanics of using SAS are presented.

This text emphasizes the analyses of questionnaire data for two reasons. First, for better or for worse, many social scientists rely on questionnaire data almost exclusively when conducting their research. This text provides examples that will be applicable to most readers. Secondly, questionnaire responses often create special data entry and analysis problems that are not generally encountered with other research methods (e.g., large numbers of variables, "check all that apply" variables). This text addresses some of the most common of these difficulties.

Entering Data: An Illustrative Example

Before data can be entered and analyzed by SAS, they must be compiled in some systematic way. There are a number of different approaches to entering data, but to keep things simple, this appendix presents only the fixed format approach. With the **fixed format** method, each variable is assigned to a specific column (or set of columns) in the dataset. The fixed format method has the advantage of being very general; you can use it for almost any type of research application. An additional advantage is that you are less likely to make errors when entering data in this format.

In the following example, you will actually enter some made-up data from a fictitious study. Assume that you have developed a survey to measure attitudes toward *volunteerism*. A copy of the survey appears here:

Volunteerism Survey

Please indicate the extent to which you agree or disagree with each of the following statements by circling the appropriate number to the left of each statement using the response key below:

5 = Agree Strongly

4 = Agree Somewhat

3 = Neither Agree nor Disagree

2 = Disagree Somewhat

1 = Disagree Strongly

For example, if you "Disagree Strongly" with the first question, circle the "1" to the left of that statement. If you "Agree Somewhat," circle the "4," and so on.

Circle Your Response

----

1 2 3 4 5 1. I feel a personal responsibility to help needy people in my community.

1 2 3 4 5 2. I feel I am personally obligated to help homeless families.

1 2 3 4 5 3. I feel no personal responsibility to work with poor people in my community.

1 2 3 4 5 4. Most of the people in my community are willing to help the needy.

1 2 3 4 5 5. A lot of people around here are willing to help homeless families.

1 2 3 4 5 6. The people in my community feel no personal responsibility to work with poor people.

1 2 3 4 5 7. Everyone should feel the responsibility to perform volunteer work in his/her community.

What is your Date of Birth (dd/mm/yyyy)? ______________

Assume that you administer this survey to 10 participants. For each of these individuals, you also obtain their Intelligence Quotient or IQ scores.

All of the survey responses and information about participant 1 appear on the first line of your file. All responses and information about participant 2 appear on the second line of this file, and so forth. You keep the data aligned so that responses to question 1 appear in column 1 for *all* participants, responses to question 2 appear in column 2 for all participants, and so forth. When you enter data in this fashion, your dataset should look similar to this:

```
2234243 22  98  1
3424325 20 105  2
3242424 32  90  3
3242323  9 119  4
3232143  8 101  5
3242242 24 104  6
4343525 16 110  7
3232324 12  95  8
1322424 41  85  9
5433224 19 107 10
```

You can think of the preceding dataset as a matrix consisting of 10 rows and 17 columns. The rows run horizontally (from left to right), and each row represents data for a single participant. The columns run vertically (up and down). For the most part, a given column represents a different variable that you measured or created. (Though, in some cases, a given variable is more than one column wide, but more on this later.)

For example, look at the last column in the matrix: The vertical column on the right side that goes from 1 (at the top) to 10 (at the bottom). This column codes the Participant Number variable. In other words, this variable simply tells us *which* participant's data are included on that line. For the top line, the assigned value of Participant Number is 1, so you know that the top line includes data for participant 1. The second line down has the value 2 in the participant number column, so this second line includes data for participant number 2, and so forth.

The *first* column of data includes participant responses to survey question 1. It can be seen that participant 1 selected "2" in response to this item, while participant 2 selected "3." The *second* column of data includes participants' responses to survey question 2, the *third* column codes question 3, and so forth. After entering responses to question 7, column 8 was left blank. Then, in columns 9 and 10, you enter each participant's age. We can see that participant 1 is 22 years old, while participant 2 is 20 years old. You left column 11 blank, and then entered the participants' IQs in columns 12, 13, and 14. (IQ can be a 3-digit number, so it required three columns to enter it.) Column 15 is left blank; participant identification numbers are entered in columns 16 and 17.

Note, we recommend you always ask participants' their date of birth instead of asking them to report their age. For example, assume you ask two participants their age and both Noah and Yaacov report that they are 22. However, Noah had his birthday yesterday, and Yaacov has his birthday tomorrow; both are 22 but Noah is nearly one year older than Yaacov (363 days to be exact). Without recording their date of birth, your data is less accurate and measurement error is (unnecessarily) greater. Yet keep in mind that you need to be clear how participants should report their date of birth. More precisely, you need to be very clear if they should report as day/month/year, month/day/year, and so forth.

Table A.2.1 presents a brief coding guide to summarize how you entered your data.

Table A.2.1

	Variable	
Column	**Name**	**Explanation**
1	Q1	Responses to survey question 1
2	Q2	Responses to survey question 2
3	Q3	Responses to survey question 3
4	Q4	Responses to survey question 4
5	Q5	Responses to survey question 5
6	Q6	Responses to survey question 6
7	Q7	Responses to survey question 7
8	blank	
9–10	AGE	Participant's age
11	Blank	
12–14	IQ	Participant's IQ score
15	blank	
16–17	NUMBER	Participant's number

Guides similar to Table A.2.1 will be used throughout this text to explain how datasets are arranged, so a few words of explanation are in order. This table simply identifies the specific columns in which variable values are assigned. For example, the first line of the preceding table indicates that in column 1 of the dataset, the values of a variable called Q1 are stored; this variable includes responses to question 1. The next line shows that in column 2, the values of variable Q2 are stored; this variable includes responses to question 2. The remaining lines of the guide are interpreted the same way. You can see, therefore, that it is necessary to read down the lines of this table to learn what is in each column of the dataset.

A few important notes about how you should enter data that you will analyze using SAS:

- **Make sure that you enter variables in the correct column**. For example, make sure that the data are aligned so that responses to question 6 always appear in column 6. If a participant happened to leave question 6 blank, then you should leave column 6 blank when entering your data. (Leave this column blank by simply pressing the space bar.) Then go on to enter that participant's response to question 7 in column 7. Do *not* enter a zero if the participant did not answer a question; simply leave the space blank.

  It is also acceptable to enter a period (.) instead of a blank space to represent missing data. When using this convention, if a participant has a missing value on a variable, enter a single period in place of that missing value. If this variable happens to be more than one column wide, you should still enter just one period. For example, if the variable occupies columns 12 to 14 (as does IQ in the table), enter just one period in column 14; do not enter three periods in columns 12, 13, and 14.

- **Right-justify numeric data**. You should align numeric variables to the right side of columns in which they appear. For example, IQ is a 3-digit variable (it could assume values such as 112 or 150). However, the IQ score for many individuals is a 2-digit number (half, by definition). Therefore, the 2-digit IQ scores should appear to the right side of this 3-digit column of values. A correct example of how to right-justify your data follows:

```
 99
109
100
 87
118
```

The following is *not* right-justified, and so is less preferable:

```
99
109
100
87
118
```

There are a few exceptions to this rule. For example, if numeric data contain decimal points, it is generally preferable to align the decimal points when entering the data so that the decimals appear in the same column. If there are no values to the right of the decimal point for a given participant, you may enter zeros to the right of the decimal point. Here is an example of this approach:

```
  3.45
 12.00
  0.13
144.75
  0.00
```

The preceding dataset includes scores for five participants for just one variable. Assume that possible scores for this variable range from 0.00 to 200.00. Participant 1 had a score of 3.45, participant 2 had a score of 12, and so forth. Notice that the scores have been entered so that the decimal points are aligned in the same vertical column.

Notice also that if a given participant's score does not include any digits to the right of the decimal point, zeros have been added. For example, participant 2 has a score of 12; this participant's score has been entered as 12.00 so that it is aligned with the other scores.

Technically, it is not always necessary to align participant data in this way to include in a SAS dataset. However, arranging data in an orderly fashion generally decreases the likelihood of making errors when entering data.

- **Left-justify character data**. Character variables may include letters of the alphabet. In contrast to numeric variables, you typically should left-justify character variables. This means that you align entries to the left, rather than the right.

  For example, imagine that you are going to enter two character variables for each participant. The first variable will be called FIRST and will include each participant's first name. You will enter this variable in columns 1 to 15. The second variable will be called LAST and will include each participant's last name (surname, or family name). You will enter this variable in columns 16 to 25. Data for five participants are reproduced here:

  ```
  Francis    Smith
  Ishmael    Khmali
  Michel     Hébert
  Jose       Lopez
  ```

  The preceding shows that the first participant is named Francis Smith, the second is named Ishmael Khmali, and so forth. Notice that the value "Francis" has been moved to the left side of the columns that include the FIRST variable (columns 1 to 15). The same is true for "Ishmael" as well as the remaining first names. In the same way, "Smith" has been moved over to the left side of the columns that include the LAST variable (columns 16 to 25). The same is true for the remaining surnames.

- **Use of blank columns can be helpful but is not necessary**. Recall that when you entered your data, you left a blank column between Q7 and the AGE variable, and another blank column between AGE and IQ. Leaving blank columns between variables can be helpful because it makes it easier to look at your data and see if something has been entered out of place. However, leaving blank columns is not necessary for SAS to accurately read your data, so this approach is optional (though recommended).

Inputting Data Using the DATALINES Statement

Now that you know how to enter your data, you are ready to learn about the SAS statements that actually allow the computer to *read* the data and put them into a SAS dataset. There are a variety of ways that you can input data, but this text focuses on only two: Use of the **DATALINES statement** that allows you to include the data within the SAS program itself; and the **INFILE statement** that allows you to include the data lines within an external file.

There are also a number of different ways in which data can be read by SAS with regard to the instructions you provide concerning the location and format of your variables. Although SAS allows for list input, column input, and formatted input, this text presents only formatted input because of its ability to accommodate many different types of data.

Here is the general form for inputting data using the DATALINES statement and the formatted input style:

```
data dataset-name;
   input  #line-number    @column-number   variable-name  column-width.
                          @column-number   variable-name  column-width.
                          @column-number   variable-name  column-width. ;
datalines;
data are placed here
;
run;

proc name-of-desired-statistical-procedure     data=dataset-name ;
run;
```

The following example shows a SAS program to analyze the preceding dataset. In this example, the numbers on the far left side are not actually part of the program; instead, they are provided to make it easier to refer to specific lines of the program when explaining the meaning of the program in subsequent sections.

```
❶  data D1;
❷     input   #1   @1   Q1      1.
                   @2   Q2      1.
                   @3   Q3      1.
                   @4   Q4      1.
                   @5   Q5      1.
                   @6   Q6      1.
                   @7   Q7      1.
                   @9   AGE     2.
                   @12  IQ      3.
❸                  @16  NUMBER  2. ;
❹  datalines;
❺  2234243 22  98  1
    3424325 20 105  2
    3242424 32  90  3
    3242323  9 119  4
    3232143  8 101  5
    3242242 24 104  6
    4343525 16 110  7
    3232324 12  95  8
    1322424 41  85  9
    5433224 19 107 10
❻  ;
❼  run;

❽  proc means   data=D1;
❾  run;
```

A few important notes about these data INPUT statements:

- **The data statement**. Line ❶ from the preceding program included the data statement, where the general form is:

  ```
  data dataset-name;
  ```

  In this case, you gave your dataset the name D1, so the statement reads:

  ```
  data D1;
  ```

- **Dataset names and variable names**. The preceding section indicated that your dataset was assigned the name D1 on line ❶ of the program. The dataset's variables are assigned names such as Q1, Q2, AGE, and IQ.

  You are free to assign a dataset or variable any name so long as it conforms to the following rules:

  - it must begin with a letter (rather than a number);
  - it may contain no special characters such as "*" or "#";
  - it may contain no blank spaces.

  Although the preceding dataset was named D1, it could have been given an almost infinite number of other names. Below are examples of other acceptable names for SAS datasets:

  ```
  SURVEY
  PARTICIPANT
  RESEARCH
  VOLUNTEER
  ```

- **The INPUT statement**. The INPUT statement has the following general form:

  ```
  input  #line-number  @column-number  variable-name  column-width.
                       @column-number  variable-name  column-width.
                       @column-number  variable-name  column-width. ;
  ```

  Compare this general form to the actual INPUT statement of the preceding SAS program and note the values that were filled in to read your data. In the actual program, the word "input" appears on line ❷ and tells SAS that the INPUT statement has begun. SAS assumes that all of the instructions that follow are data input directions *until* it encounters a semicolon (;). At that semicolon, the INPUT statement ends. In this example, the semicolon appears on line ❸.

- **Line number controls**. To the right of the word "input" is the following:

  ```
  #line-number
  ```

  This tells SAS what line it should read in order to find specific variables. In some cases there may be two or more lines of data for each participant; more on this in a later section. For the present example, the situation is fairly simple: There is only one line of data for each participant so your program includes the following line number control (from line ❷ of the program example):

  ```
  input   #1
  ```

  Technically, it is not necessary to include line number controls when there is only one line of data for each participant (as in the present example). In this text, however, line number controls appear throughout for consistency.

- **Column location, variable name, and column width directions**. To the right of the line number directions, you place the column location, variable name, and column width directions. The general form for this is as follows:

```
@column-number   variable-name   column-width.
```

Where "column-number" appears, you enter the number of the column in which a specific variable appears. If the variable occupies more than one column, such as IQ in columns 12, 13 and 14, you should enter the number of the column in which it begins (e.g., column 12). Where "variable-name" appears, you will enter the name that you have given to that variable. And where "column-width" appears, you will enter how many columns are occupied by that variable. In the case of the preceding data, the first variable is Q1, which appears in column 1 and is only one column wide. This program example, therefore, provides the following column location controls (from line ❷):

```
@1   Q1   1.
```

The preceding line tells SAS to go to column 1. In that column you will find a variable called Q1. It is a number and one column wide.

IMPORTANT: Note that you must follow the column width with a period. So for column 1, the width is (1.). It is important that you include this period; later you will learn how the period provides information about decimal places.

Now that variable Q1 has been read, you must give SAS the directions required to read the remaining variables in the dataset. The completed INPUT statement appears as follows. Note that the line number controls are given only once because all of these variables come from the same line (for a given participant). There are different column controls for the different variables, however. Note also how column widths are different for AGE, IQ, and NUMBER:

```
input   #1   @1    Q1       1.
             @2    Q2       1.
             @3    Q3       1.
             @4    Q4       1.
             @5    Q5       1.
             @6    Q6       1.
             @7    Q7       1.
             @9    AGE      2.
             @12   IQ       3.
             @16   NUMBER   2. ;
```

IMPORTANT: Notice the semicolon that appears after the column width entry for the last variable (NUMBER). You must always end your INPUT statement with a semicolon. It is easy to omit, so always check for this semicolon if you get an error message following the INPUT statement.

- **The DATALINES statement**. The DATALINES statement appears after the INPUT statement and tells SAS that raw data are to follow. Do not forget the semicolon after the word "datalines." In the preceding program example, the DATALINES statement appears on line ❹.

- **The data lines**. The data lines, of course, are the lines that contain participants' values for the numeric and/or character variables.

  The data lines should begin on the very next line after the DATALINES statement; there should be no blank lines. These data lines begin on line ❺ in the preceding program example. On the very first line after the last of the data lines (line ❻, in this case), you should add another semicolon to let SAS know that the data have ended. Do *not* place this semicolon at the end of the last line of data (i.e., on the *same line* as the data), as this may cause an error. After this semicolon, a RUN statement should appear at the end of the data lines. In the preceding program example, this appears on line ❼.

The most important thing to remember when it comes to the data lines is that you must enter a given variable in the column specified by the INPUT statement. For example, if your INPUT statement contains the following line:

```
@9   AGE   2.
```

then make sure that the variable AGE really is a 2-digit number found in columns 9 and 10.

- **PROC and RUN statements**. There is little to say about PROC and RUN statements at this point because most of the text will be concerned with using such SAS procedures. Suffice to say that a PROC (or procedure) statement asks SAS to perform some statistical analysis. To keep things simple, this section uses a procedure called PROC MEANS. PROC MEANS asks SAS to calculate means, standard deviations, and other descriptive statistics for numerical variables. The preceding program includes the PROC MEANS statement on line ❽.

  In most cases, your program will end with a RUN statement. In the preceding example, a second RUN statement appears on line ❾. A RUN statement executes any previously entered SAS statements; RUN statements are typically placed after every PROC statement. If your program includes a number of PROC statements in sequence, it is acceptable to place just one RUN statement after the final PROC statement.

  If you submitted the preceding program for analysis, PROC MEANS would produce the results presented here as Output A.2.1:

Output A.2.1: Results of the MEANS Procedure

The MEANS Procedure

Variable	N	Mean	Std Dev	Minimum	Maximum
Q1	10	3.0000000	1.0540926	1.0000000	5.0000000
Q2	10	2.6000000	0.8432740	2.0000000	4.0000000
Q3	10	3.2000000	0.7888106	2.0000000	4.0000000
Q4	10	2.6000000	0.8432740	2.0000000	4.0000000
Q5	10	2.9000000	1.1972190	1.0000000	5.0000000
Q6	10	2.6000000	0.9660918	2.0000000	4.0000000
Q7	10	3.7000000	0.9486833	2.0000000	5.0000000
AGE	10	20.3000000	10.2745641	8.0000000	41.0000000
IQ	10	101.4000000	9.9241568	85.0000000	119.0000000
NUMBER	10	5.5000000	3.0276504	1.0000000	10.0000000

Additional Guidelines

Inputting String Variables with the Same Prefix and Different Numeric Suffixes

In this section, **prefix** refers to the first part of a variable's name, while a **suffix** refers to the last part. For example, think about our variables Q1, Q2, Q3, Q4, Q5, Q6, and Q7. These are multiple variables with the same prefix (Q) and different numeric suffixes (i.e., 1, 2, 3, 4, 5, 6, and 7). Earlier, this appendix provided one way of inputting these variables; the original INPUT statement is repeated here:

```
INPUT    #1    @1    Q1       1.
               @2    Q2       1.
               @3    Q3       1.
               @4    Q4       1.
               @5    Q5       1.
               @6    Q6       1.
               @7    Q7       1.
               @9    AGE      2.
               @12   IQ       3.
               @16   NUMBER   2. ;
```

However, with string variables named in this way, there is an easier way of writing the INPUT statement. You could have written it in this way:

```
input    #1    @1    Q1-Q7    1.
               @9    AGE      2.
               @12   IQ       3.
               @16   NUMBER   2.   ;
```

The first line of this INPUT statement gives SAS the following directions: "Go to line #1. Once there, go to column 1. Beginning in column 1 you will find variables Q1 through Q7. Each of these numeric variables is one column wide." With this second INPUT statement, SAS will read the data in exactly the same way that it would have using the original INPUT statement.

As an additional example, imagine you had a 50-item survey instead of a 7-item survey. You called your variables Q1, Q2, Q3, and so forth. You entered your data in the following way:

	Variable	
Column	Name	Explanation
1–50	Q1–Q50	Responses to survey questions
51	blank	
52–53	AGE	Participants' age
54	blank	
55–57	IQ	Participants' IQ score
58	blank	
59–60	NUMBER	Participants' identification number

You could use the following INPUT to read these data:

```
input    #1    @1    Q1-Q50   1.
               @52   AGE      2.
               @55   IQ       3.
               @59   NUMBER   2.   ;
```

Inputting Character Variables

This text deals with two types of basic variables: Numeric and character variables. A **numeric variable** consists entirely of numbers; it contains no letters. For example, all of your variables from the preceding dataset were numeric variables: Q1 could assume only the values of 1, 2, 3, 4, or 5. Similarly, AGE could take on only numeric values. On the other hand, a **character variable** may consist of either numbers or alphabetic characters (letters), or both.

Remember that responses to the seven questions of the Volunteerism Survey are entered in columns 1 to 7 in this dataset, AGE is entered in columns 9 to 10, IQ is entered in columns 12 to 14, and participants' identification numbers are in columns 16 to 17.

You could include the sex of each participant and create a new variable called SEX that codes the sex of each participant. If a participant is male, SEX would assume the value "M." If a participant is female, SEX would assume the value "F." (If neither apply, SEX could be coded as "O" for other.) In the following, the new SEX variable appears in column 19 (the last column):

```
2234243 22  98  1 M
3424325 20 105  2 M
3242424 32  90  3 F
3242323  9 119  4 F
3232143  8 101  5 F
3242242 24 104  6 M
4343525 16 110  7 F
3232324 12  95  8 M
1322424 41  85  9 M
5433224 19 107 10 F
```

You can see that participants 1 and 2 are males whereas participants 3, 4, and 5 are females, and so forth.

IMPORTANT: You must use a special command within the INPUT statement to input a character variable. Specifically, in the column width region for the character variable, precede the column width with a dollar sign ("$"). For the preceding dataset, you would use the following INPUT statement. Note the dollar sign in the column width region for the SEX variable:

```
INPUT   #1   @1   Q1-Q7    1.
             @9   AGE      2.
             @12  IQ       3.
             @16  NUMBER   2.
             @19  SEX     $1.   ;
```

Using Multiple Lines of Data for Each Participant

Very often a researcher obtains so much data from each participant that it is impractical to enter all data on just one line. For example, imagine that you administer a 100-item questionnaire to a sample, and that you plan to enter responses to question 1 in column 1, responses to question 2 in column 2, and so forth. Following this process, you are likely to run into difficulty because you will need 100 columns to enter all responses from each participant. If you continue this way, your data may *wrap around* or appear in some way that makes it difficult to verify that you are entering a given value in the correct column.

In situations in which you require a very large number of columns for your data, it is often best to divide each participant's data so that they appear on more than one line. In other words, it is often best to have multiple lines of data for each participant. To do this, it is necessary to modify your INPUT statement.

To illustrate, assume that you obtained three additional variables for each participant in your study: Their SAT critical reading, writing, and mathematics test scores. You decide to enter your data so that there are two lines of data for each participant. On line 1 for a given participant, you enter Q1 through Q7, AGE, IQ, NUMBER, and SEX (as above). On line 2 for that participant, you enter SATREAD (critical reading) in columns 1 through 3, SATWRITE (writing) in columns 5 through 7, and SATMATH (mathematics) in columns 9 through 11:

```
2234243 22  98  1 M
520 490 465
3424325 20 105  2 M
440 410 460
3242424 32  90  3 F
390 420 410
3242323  9 119  4 F

3232143  8 101  5 F

3242242 24 104  6 M
330 340 375
4343525 16 110  7 F

3232324 12  95  8 M

1322424 41  85  9 M
380 410 400
5433224 19 107 10 F
640 590 625
```

SATREAD score for participant 1 is 520, SATWRITE is 490, and the SATMATH score is 465.

IMPORTANT: When a given participant has no data for a variable that would normally appear on a given, your dataset must still include a line for that participant, even if it is blank. For example, participant 4 is only 9 years old, so she has not yet taken the SAT Reasoning Test, and obviously has no scores. Nonetheless, you still had to include a second line for participant 4 even though it was blank. Notice that blank lines also appear for participants 5, 7, and 8, who were also too young to take the SAT Reasoning Test.

The following coding guide tells us where each variable appears. Notice that this guide indicates the line on which a variable is located, as well as the column where it is located.

		Variable	
Line	Column	Name	Explanation
1	1–7	Q1–Q7	Survey questions 1–7
	8	blank	
	9–10	AGE	Participant's age
	11	blank	
	12–14	IQ	Participant's IQ score
	15	blank	
	16–17	NUMBER	Participant's number
	18	Blank	
	19	SEX	Participant's sex
2	1–3	SATREAD	Critical Reading test score
	5–7	SATWRITE	Writing test score
	9–11	SATMATH	Mathematics test score

IMPORTANT: When there are multiple lines of data for each participant, the INPUT statement must indicate on which line a given variable is located. This is done with the line number command ("#") that was described earlier. You could use the following INPUT statement to read the preceding dataset:

```
input   #1    @1    Q1-Q7          1.
              @9    AGE            2.
              @12   IQ             3.
              @16   NUMBER         2.
              @19   SEX           $1.
        #2    @1    SATREAD        3.
              @5    SATWRITE       3.
              @9    SATMATH        3.  ;
```

This INPUT statement tells SAS to begin at line #1 for a given participant and to go to column 1 and find variables Q1 through Q7. It also tells SAS where it will find each of the other variables located on line #1. After reading the SEX variable, SAS is told to move on to line #2. There, it is to go to column 1 and find the variable SATREAD that is three columns wide, SATWRITE begins in column 5 and is three columns wide, and SATMATH begins in column 9 and is also three columns wide. In theory, it is possible to have any number of lines of data for each participant so long as you use the line number command correctly.

Creating Decimal Places for Numeric Variables

Assume that you have obtained high school grade point averages (GPAs) for a sample of five participants. You could create a SAS dataset containing these GPAs using the following program:

```
data D1;
   input   #1   @1   GPA   4.  ;
datalines;
3.56
2.20
2.11
3.25
4.00
;
run;

   proc means   data=D1;
run;
```

The INPUT statement tells SAS to go to line 1, column 1, to find a variable called GPA that is four columns wide. Within the dataset itself, values of GPA were entered using a period as a decimal point, with two digits to the right of the decimal point.

This same dataset could have been entered in a slightly different way. For example, what if the data had been entered without a decimal point, as follows?

```
356
220
211
325
400
```

It is still possible to have SAS insert a decimal point where it belongs, in front of the last two digits in each number. You do this in the column width command of the INPUT statement. With this column width command, you indicate how many columns the variable occupies, enter a period, and then indicate how many columns of data should appear to the right of the decimal place. In the present example, the GPA variable was three columns wide, and two columns of data should have appeared to the right of the decimal place. So you would modify the SAS program in the following way. Notice the column width command:

```
data D1;
   input    #1    @1    GPA    3.2   ;
datalines;
356
220
211
325
400
;
run;

   proc means    data=D1;
run;
```

Inputting "Check All That Apply" Questions as Multiple Variables

A "check all that apply" question is a special type of questionnaire item that is often used in social science research. These items generate data that must be input in a special way. The following is an example of a "check all that apply" item that could have appeared on your volunteerism survey:

> Below is a list of activities. Please place a check mark next to each activity in which you have engaged over the past six months.
>
> Check [√] below
>
> -----
>
> _____ 1. Did volunteer work at a shelter for the homeless.
>
> _____ 2. Did volunteer work at a domestic abuse shelter
>
> _____ 3. Did volunteer work at a hospital or hospice.
>
> _____ 4. Did volunteer work for any other community agency or organization.
>
> _____ 5. Donated money to the United Way.
>
> _____ 6. Donated money to a congregation-affiliated charity.
>
> _____ 7. Donated money to any other charitable cause.

The novice researcher might think of the preceding as a single question with seven possible responses and try to enter the data in a single column in the dataset (say, in column 1). But this would lead to big problems. What would you enter in column 1 if a participant checked more than one category?

One way out of this difficulty is to treat the seven possible responses above as seven different questions. Once entered, each of these questions will be treated as a separate variable and will appear in a separate column. For example, whether or not a participant checked activity 1 may be coded in column 1, whether the participant checked activity 2 may be coded in column 2, and so forth.

Researchers may code these variables by placing any values they like in these columns, but you should enter a two ("2") if the participant did not check that activity and a one ("1") if the participant did check it. Why code the variables using ones and twos? The reason is that this makes it easier to perform some types of analyses that you may later wish to perform. A variable that may assume only two values is called a **dichotomous variable**, and the process of coding dichotomous variables with ones and twos is known as **dummy coding**. With dummy coding, we recommend that you not use zeros to avoid the possibility that these might be confused with missing values.

Once a dichotomous variable has been dummy coded, it can be analyzed using a variety of SAS procedures such as PROC REG to perform multiple regression analysis, a procedure that allows one to assess the nature of the relationship between a single criterion variable and multiple predictor variables. If a dichotomous variable has been dummy-coded properly, it can be used as a predictor variable in a multiple regression analysis. For these and other reasons, it is good practice to code dichotomous variables using ones and twos.

The following coding guide summarizes how you could enter responses to the preceding question:

		Variable	
Line	**Column**	**Name**	**Explanation**
1	1–7	ACT1–ACT7	Responses regarding activities 1 through 7. For each activity, a 2 was recorded if the participant did not check the activity, and a 1 was recorded if the participant did check the activity.

When participants have responded to a "check all that apply" item, it is often best to analyze the data using the FREQ (frequency) procedure. PROC FREQ indicates the actual numbers that appear in each category. In this case, PROC FREQ will indicate the number of people who did not check a given activity versus the number who did. It also indicates the percentage of people who appear in each category, along with some additional information.

The following program inputs some fictitious data and requests frequency tables for each activity using PROC FREQ:

```
    data D1;
         input   #1   @1   ACT1-ACT7   1.  ;

    datalines;
❶   2212222
❷   1211111
    2221221
    2212222
    1122222
    ;
    run;

    proc freq    data=D1;
    run;
```

Data for the first participant appears on line ❶ of the program. Notice that a 1 is entered in column 3 for this participant, indicating that she did perform activity 3 ("did volunteer work at a hospital or hospice") and that 2s are recorded for the remaining 6 activities, meaning that the participant did not perform those activities. The data entered for participant 2 on line ❷ shows that this participant performed all of the activities except for activity 2.

Inputting a Correlation or Covariance Matrix

There are times when, for reasons of either necessity or convenience, you may choose to analyze a correlation matrix or covariance matrix rather than raw data (e.g., very large datasets). SAS allows you to enter such a matrix as data, and some (but not all) SAS procedures may then be used to analyze the dataset. For example, a correlation or covariance matrix can be analyzed using PROC REG, PROC FACTOR, or PROC CALIS, as well as some other procedures.

Inputting a Correlation Matrix

This type of data input is sometimes necessary when a researcher obtains a correlation matrix from an earlier study (e.g., from an article published in a research journal) and wishes to perform further analyses on the data. You could input the published correlation matrix as a dataset and analyze it in the same way you would analyze raw data.

For example, imagine that you have read an article that tested a social psychology theory called the *investment model* (Rusbult, 1980). The investment model identifies a number of variables that are believed to influence a person's satisfaction with, and commitment to, a romantic relationship (Le and Agnew, 2003). The following are short definitions for the variables that constitute the investment model:

Commitment
the person's intention to remain in the relationship;

Satisfaction
the person's affective (emotional) response to the relationship;

Rewards
the number of good things or benefits associated with the relationship;

Costs
the number of bad things or hardships associated with the relationship;

Investment size
the amount of time, energy, and personal resources put into the relationship;

Alternative value
the attractiveness of alternatives to the relationship (e.g., attractiveness of alternate romantic partners).

One interpretation of the investment model predicts that commitment to a relationship is determined by satisfaction, investment size, and alternative value, while satisfaction with the relationship is determined by rewards and costs. The predicted relationships among these variables are presented in Figure A.2.1.

Figure A.2.1: Predicted Relationships among Investment Model Variables

Assume that you have read an article that reports an investigation of the investment model and that the article included the (fictitious) table represented as Table A.2.1.

Table A.2.1 Standard Deviations and Intercorrelations for All Variables

		Intercorrelations					
Variable	**SD**	**1**	**2**	**3**	**4**	**5**	**6**
1. Commitment	2.3192	1.0000					
2. Satisfaction	1.7744	.6742	1.0000				
3. Rewards	1.2525	.5501	.6721	1.0000			
4. Costs	1.4086	-.3499	-.5717	-.4405	1.0000		
5. Investments	1.5575	.6444	.5234	.5346	-.1854	1.0000	
6. Alternatives	1.8701	-.6929	-.4952	-.4061	.3525	-.3934	1.0000

Note: N = 240.

Supplied with this information, you may now create a SAS dataset that includes just these correlation coefficients and standard deviations. Here are the necessary data input statements:

```
data D1(TYPE=CORR) ;
  input _type_ $ _name_ $ V1-V6 ;
label
     V1 ='COMMITMENT'
     V2 ='SATISFACTION'
     V3 ='REWARDS'
     V4 ='COSTS'
     V5 ='INVESTMENTS'
     V6 ='ALTERNATIVES' ;
datalines;
n      .    240      240      240      240      240      240
std    .  2.3192   1.7744   1.2525   1.4086   1.5575   1.8701
corr  V1  1.0000     .        .        .        .        .
corr  V2    .6742  1.0000     .        .        .        .
corr  V3    .5501    .6721  1.0000     .        .        .
corr  V4  -.3499   -.5717   -.4405   1.0000     .        .
corr  V5    .6444    .5234    .5346  -.1854   1.0000     .
corr  V6  -.6929   -.4952   -.4061    .3525  -.3934   1.0000
;
run;
```

The following shows the general form for this DATA step in which six variables are to be analyzed. The program would, of course, be modified if the analysis involved a different number of variables.

```
data dataset-name(type=corr) ;
 input _type_ $ _name_ $ variable-list ;
label
     V1 ='long-name'
     V2 ='long-name'
     V3 ='long-name'
     V4 ='long-name'
     V5 ='long-name'
     V6 ='long-name' ;
datalines;
n      .     n        n        n        n        n        n
std    .    std      std      std      std      std      std
corr  V1  1.0000     .        .        .        .        .
corr  V2     r     1.0000     .        .        .        .
corr  V3     r        r     1.0000     .        .        .
corr  V4     r        r        r     1.0000     .        .
corr  V5     r        r        r        r     1.0000     .
corr  V6     r        r        r        r        r     1.0000
;
run;
```

where:

variable-list =	List of variables (e.g., V1, V2,);
long-name =	Full name for the given variable. This will be used to label the variable when it appears in the SAS output. If this is not desired, you can omit the entire label statement;
n =	Number of observations contributing to the correlation matrix. Each correlation in this matrix should be based on the same observations and hence the same number of observations;
std =	Standard deviation obtained for each variable. These standard deviations are needed if you are performing an analysis on the correlation matrix so that SAS can convert the correlation matrix into a variance-covariance matrix. Instead, if you wish to perform an analysis on a variance-covariance matrix, then standard deviations are not required;
r =	Correlation coefficients between pairs of variables.

The observations that appear in the preceding program are easiest to understand if you think of the observations as a matrix with eight rows and eight columns. The first column in this matrix (running vertically) contains the _type_ variable (notice that the INPUT statement tells SAS that the first variable it will read is a character variable named "_type_"). If an "N" appears as a value in this _type_ column, then SAS knows that sample sizes will appear on that line. If "std" appears as a value in the _type_ column, then the system knows that standard deviations will appear on that line. Finally, if "CORR" appears as a value in the _type_ column, then SAS knows that correlation coefficients will appear on that line.

The second column in this matrix contains short names for the observed variables. These names should appear only on the CORR lines. Periods (for missing data) should appear where the n and std lines intersect with this column (i.e., above the diagonal).

Looking at the matrix from the other direction, you see eight rows running horizontally. The first row is the n row (or "line") that should contain:

- the n symbol;
- a period for the missing variable name;
- the sample sizes for the variables, each separated by at least one blank space.

The preceding program shows that the sample size was 240 for each variable.

The std row (or line) should contain:

- the std symbol;
- the period for the missing variable name;
- the standard deviations for the variables, each separated by at least one blank space. If the std line is omitted, the analysis can only be performed on covariances, not correlation coefficients.

Finally, where rows 3 to 8 intersect with columns 3 to 8, the correlation coefficients should appear. These coefficients appear below the diagonal, ones should appear on the diagonal (i.e., the correlation coefficient of a number with itself is always equal to 1.0), and periods appear above the diagonal (where redundant correlation coefficients would again appear if this were a full matrix). Be very careful in entering these correlations; one missing period can cause an error in reading the data.

You can see that the columns of data in this matrix are aligned in an organized fashion. Technically, neatness was not really required as this INPUT statement is in free format. You should try to be equally organized when preparing your matrix as this will minimize the chance of leaving out an entry and causing an error.

Inputting a Covariance Matrix

The procedure for inputting a covariance matrix is very similar to that used with a correlation matrix. An example is presented here:

```
data D1(type=cov) ;
  input _type_ $ _name_ $ V1-V6 ;
label
     V1 ='COMMITMENT'
     V2 ='SATISFACTION'
     V3 ='REWARDS'
     V4 ='COSTS'
     V5 ='INVESTMENTS'
     V6 ='ALTERNATIVES' ;
datalines;
n     .    240      240      240      240      240      240
cov   V1 11.1284     .        .        .        .        .
cov   V2  5.6742   9.0054     .        .        .        .
cov   V3  4.5501   3.6721   6.8773     .        .        .
cov   V4 -3.3499  -5.5717  -2.4405  10.9936     .        .
cov   V5  7.6444   2.5234   3.5346  -4.1854   7.1185     .
cov   V6 -8.6329  -3.4952  -6.4061   4.3525  -5.3934   9.2144
;
run;
```

Notice that the data statement now specifies type=COV rather than type=CORR. The line providing standard deviations is no longer needed and has been removed. The matrix itself now provides variances on the diagonal and covariances below the diagonal; the beginning of each line now specifies COV to indicate that this is a covariance matrix. The remaining sections are identical to those used to input a correlation matrix.

Inputting Data Using the INFILE Statement Rather Than the DATALINES Statement

When working with very large datasets, it is often more convenient to input data using the INFILE statement rather than the DATALINES statement. This involves:

- adding an INFILE statement to your program;
- placing your data lines in a second computer file, rather than in the same file that contains your SAS program;
- deleting the DATALINES statement from your SAS program.

Your INFILE statement should appear *after* the data statement but *before* the INPUT statement. The general form for a SAS program using the INFILE statement is as follows:

```
data dataset-name;
   infile  'name-of-data-file' ;
   input  #line-number    @column-number   variable-name  column-width.
                          @column-number   variable-name  column-width.
                          @column-number   variable-name  column-width. ;

proc name-of-desired-statistical-procedure     data=dataset-name;
run;
```

Notice that the above is identical to the general form for a SAS program presented earlier except that an INFILE statement has been added and the DATALINES statement and data lines have been deleted.

To illustrate the use of the INFILE statement, consider Dr. Lafleur's volunteerism study. The dataset itself is reproduced here:

```
2234243 22  98  1 M
3424325 20 105  2 M
3242424 32  90  3 F
3242323  9 119  4 F
3232143  8 101  5 F
3242242 24 104  6 M
4343525 16 110  7 F
3232324 12  95  8 M
1322424 41  85  9 M
5433224 19 107 10 F
```

If you were to input these data using the INFILE statement, you would enter the data in a separate computer file, giving it any name you like. Assume, in this case, that the preceding data file is named "volunteer.dat."

Note: You must enter these data lines beginning on line 1 of the computer file; do *not* leave any blank lines at the top of the file. Similarly, there should be no blank lines at the end of the file (unless a blank line is appropriate because of missing data for the last participant).

Once the data are entered and saved in the file called volunteer.dat, you could enter the SAS program itself in a separate file. Perhaps you would give this file a name such as survey.sas. A SAS program which would input the preceding data and calculate means for the variables appears here:

```
data D1;
  infile 'A:/volunteer.dat';
  input   #1   @1   Q1-Q7    1.
               @9   AGE      2.
               @12  IQ       3.
               @16  NUMBER   2.
               @19  SEX     $1.   ;

proc means    data=D1;
run;
```

Conclusion

The material presented in this appendix has prepared you to input most types of data commonly encountered in social science research. Even when the data have been entered successfully, however, they are not necessarily ready to be analyzed. Perhaps you have entered raw data, and need to transform the data in some way before they can be analyzed. This is often the case with questionnaire data, as responses to multiple questions are often summed or averaged to create new variables to be analyzed. Or perhaps you have data from a large, heterogeneous sample and you wish to perform analyses on only a subgroup of that sample (such as the female, but not the male, respondents). In these situations, some form of *data manipulation* or data subsetting is requited and the following appendix shows how to do this.

References

DeVellis, R. F. (2012). *Scale development theory and applications* (3rd Ed.). Thousand Oaks, CA: Sage.

Rusbult, C. E. (1980). Commitment and satisfaction in romantic associations: A test of the investment model. *Journal of Experimental Social Psychology, 16,* 172–186.

Le, B., & Agnew, C. R. (2003). Commitment and its theorized determinants: A meta-analysis of the investment model. *Personal Relationships, 10,* 37–57.

Appendix A.3: Working with Variables and Observations in SAS Datasets

Introduction: Manipulating, Subsetting, Concatenating, and Merging Data

Often, researchers obtain a dataset in which the data are not yet in a form appropriate for analyses. For example, imagine that you are conducting research on job satisfaction. Perhaps you wish to compute the correlation between participant age and an index of job satisfaction. You administer a 10-item questionnaire to 200 employees to assess job satisfaction, and you enter their responses to the 10 individual questionnaire items. You now need to add together each participant's responses to those 10 items to arrive at a single composite score that reflects that participant's overall level of satisfaction. This computation is easy to perform by including a number of data-manipulation statements in the SAS program. **Data-manipulation statements** are SAS statements that transform the data in some way. They may be used to recode negatively keyed variables, create new variables from existing variables, and perform a wide range of possible tasks.

At the same time, your original dataset may contain observations that you do not wish to include in your analyses. Perhaps you administered the questionnaire to hourly as well as salaried employees, and you wish to only analyze data from the former. In addition, you may wish to analyze data only from participants who have usable data on all of the study's variables. In these situations, you may include data-subsetting statements to set aside unwanted responses from the sample. **Data-subsetting statements** are SAS statements that eliminate unwanted observations from a sample so that only a specified subgroup is included in the resulting dataset.

In other situations, it may be necessary to concatenate or merge datasets before you can perform the analyses you desire. When you **concatenate** datasets, you combine two previously existing datasets that contain data on the same variables but from different participants. The resulting concatenated dataset contains aggregate data from all participants. In contrast, when you **merge** datasets, you combine two datasets that involve the same participants but contain different variables. For example, assume that dataset D1 contains variables V1 and V2, while dataset D2 contains variables V3 and V4. Assume further that both datasets have a variable called ID (identification number) that will be used to merge data from the same participants. Once D1 and D2 have been merged, the resulting dataset (D3) contains V1, V2, V3, and V4 as well as ID.

The SAS programming language is so comprehensive and flexible that it can perform virtually any type of manipulation, subsetting, concatenating, or merging task. A complete treatment of these capabilities would easily fill a book. This appendix reviews some basic statements that can be used to solve a wide variety of problems that are commonly encountered in social science research, particularly in research that involves the analysis of questionnaire data.

Placement of Data-Manipulation and Data-Subsetting Statements

The use of data-manipulation and data-subsetting statements is illustrated here with reference to the fictitious study described in the preceding appendix. In Appendix A.2, you were asked to imagine that you had developed a 7-item questionnaire dealing with volunteerism, as shown in the following example.

Volunteerism Survey

Please indicate the extent to which you agree or disagree with each of the following statements by circling the appropriate number to the left of each. The following format shows what each response alternative represents:

5 = Agree Strongly

4 = Agree Somewhat

3 = Neither Agree nor Disagree

2 = Disagree Somewhat

1 = Disagree Strongly

For example, if you "Disagree Strongly" with the first question, circle the "1" to the left of that statement. If you "Agree Somewhat," circle the "4," and so on.

Circle Your Response

----

1 2 3 4 5 1. I feel a personal responsibility to help needy people in my community.

1 2 3 4 5 2. I feel I am personally obligated to help homeless families.

1 2 3 4 5 3. I feel no personal responsibility to work with poor people in my community.

1 2 3 4 5 4. Most of the people in my community are willing to help the needy.

1 2 3 4 5 5. A lot of people around here are willing to help homeless families.

1 2 3 4 5 6. The people in my community feel no personal responsibility to work with poor people.

1 2 3 4 5 7. Everyone should feel the responsibility to perform volunteer work in his/her community.

What is your date of birth (dd/mm/yyyy)? _______________

Assume that you administer this survey to a number of participants and you also obtain information concerning sex, IQ scores, and SAT critical reading, writing, and mathematics test scores for each participant. Once the data are entered, you may wish to write a SAS program that includes some data-manipulation or data-subsetting statements to transform the raw data. But where within the SAS program should these statements appear?

In general, these statements should only appear within the *DATA step*. Remember that the DATA step begins with the data statement and ends as soon as SAS encounters a procedure or PROC statement. This means that if you prepare the DATA step, end the DATA step with a procedure, and then place some manipulation or subsetting statements immediately after the procedure, you will receive an error message. (And remember to ask participants their date of birth, not age.)

To avoid this error (and keep things simple), place your data-manipulation and data-subsetting statements in one of two locations within a SAS program:

- immediately following the INPUT statement;
- immediately following the creation of a new dataset.

Immediately Following the INPUT Statement

The first of the two preceding guidelines indicates that the statements may be placed immediately following the INPUT statement. This guideline is illustrated again by referring to the volunteerism study. Assume that you prepare the following SAS program to analyze data obtained in your study. In the following program, lines ❶ and ❷ indicate where you can place data-manipulation or data-subsetting statements. (To conserve space, only some of the data lines are presented in this example.)

```
data D1;
   input   #1   @1    Q1-Q7        1.
                @9    AGE          2.
                @12   IQ           3.
                @16   NUMBER       2.
                @19   SEX         $1.
           #2   @1    SATREAD      3.
                @5    SATWRITE     3.
                @9    SATMATH      3.  ;

❶ place data-manipulation statements and
❷ data-subsetting statements here

datalines;
2234243 22  98  1 M
520 490 465
3424325 20 105  2 M
440 410 460
.
.
5433224 19 107 10 F
640 590 625
;
run;

proc means  data=D1;
run;
```

Immediately after Creating a New Dataset

The second option for placement of data-manipulation or data-subsetting statements is immediately following program statements that create a new dataset. A new dataset may be created at virtually any point in a SAS program (even after procedures have been requested).

At times, you may want to create a new dataset so that, initially, it is identical to an existing dataset (perhaps the one created with a preceding INPUT statement). If data-manipulation or data-subsetting statements follow the creation of this new dataset, the new set displays the modifications requested by those statements.

To create a new dataset that is identical to an existing dataset, the general form is:

```
data  new-dataset-name;
   set  existing-dataset-name;
```

To create such a dataset, use the following statements:

```
data D2;
   set D1;
```

These lines tell SAS to create a new dataset called D2 and to make this new dataset identical to D1. Now that a new dataset has been created, you are free to write as many manipulation and subsetting statements as you like. Once you write a procedure, however, that effectively ends the DATA step and you cannot write any more manipulation or subsetting statements beyond that point unless you create another dataset later in the program.

The following is an example of how you might write your program so that the manipulation and subsetting statements follow the creation of the new dataset:

```
data D1;
   input   #1   @1    Q1-Q7          1.
                @9    AGE            2.
                @12   IQ             3.
                @16   NUMBER         2.
                @19   SEX           $1.
           #2   @1    SATREAD        3.
                @5    SATWRITE       3.
                @9    SATMATH        3.   ;
datalines;
2234243 22  98  1 M
520 490 465
3424325 20 105  2 M
440 410 460
.
.

5433224 19 107 10 F
640 590 625
;
run;

data D2;
   set D1;

place data manipulation statements and
data subsetting statements here

❸   proc means   DATA=D2;
run;
```

SAS creates two datasets according to the preceding program: D1 contains the original data; and D2 is identical to D1 except for modifications requested by the data-manipulation and data-subsetting statements.

Notice that the MEANS procedure on line ❸ requests the computation of simple descriptive statistics. It is clear that these statistics are performed on the data from dataset D2 because data=D2 appears in the PROC MEANS statement. If the statement, instead, specified data=D1, the analyses would have been performed on the original dataset.

The INFILE Statement versus the DATALINES Statement

The preceding program illustrates the use of the DATALINES statement rather than the INFILE statement. The guidelines regarding the placement of data-modifying statements are the same regardless of which approach is followed. The data-manipulation or data-subsetting statement should either immediately follow the INPUT statement or the creation of a new dataset. When a program is written using the INFILE statement rather than the DATALINES statement, data-manipulation and data-subsetting statements should appear *after* the INPUT statement but *before* the first procedure. For example, if your data are entered into an external file called volunteer.dat, you can write the following program. (Notice where the manipulation and subsetting statements are placed.)

```
data D1;
   infile 'A:/volunteer.data';
   input    #1    @1     Q1-Q7          1.
                  @9     AGE            2.
                  @12    IQ             3.
                  @16    NUMBER         2.
                  @19    SEX           $1.
            #2    @1     SATREAD        3.
                  @5     SATWRITE       3.
                  @9     SATMATH        3.   ;

   place data manipulation statements and
   data subsetting statements here

   proc means    data=D1;
   run;
```

In the preceding program, the data-modifying statements again come immediately after the INPUT statement but before the first procedure, consistent with earlier recommendations.

Data Manipulation

Data manipulation involves performing some type of transformation on one or more variables in the DATA step. This section discusses several types of transformations that are frequently required in social science research. These include creation of duplicate variables with new variable names, creation of new variables from existing variables, recoding reversed or negatively-keyed items, and using IF-THEN statements as well as with other related procedures.

Creating Duplicate Variables with New Variable Names

Suppose that you give a variable a certain name when it is inputted, but then you want the variable to have a different, perhaps more meaningful, name, when it appears later in the SAS program or in the SAS output. This can easily be accomplished with a statement written according to the following general form:

```
new-variable-name   =   existing-variable-name;
```

For example, in the preceding dataset, the first 7 questions are given variables names of Q1 through Q7. Item 1 in the questionnaire reads, "I feel a personal responsibility to help needy people in my community." In the INPUT statement, this item was given a SAS variable name of Q1, which is not very meaningful. RESNEEDY, which stands for "responsible for the needy," is a more meaningful name. Similarly, RESHOME is more meaningful than Q2, and NORES is more meaningful than Q3.

One way to rename an existing variable is to create a new variable that is identical to the existing variable and assign a new, more meaningful name to this new variable. The following program renames Q1, Q2, and Q3 in this way.

Note: This and later examples show only a portion of the entire program. However, enough of the program appears to illustrate where the remaining statements should be placed.

```
.
.
5433224 19 107 10 F
640 590 625
;
run;

data D2;
  set D1;

❶ RESNEEDY = Q1;
   RESHOME  = Q2;
   NORES    = Q3;

proc means   data=D2;
run;
```

Line ❶ tells SAS to create a new variable called RESNEEDY and for it to be identical to the existing variable, Q1. Variables RESNEEDY and Q1 now have identical data, but RESNEEDY has a more meaningful name to facilitate the reading of printouts when statistical analyses are later performed.

Note: When creating a new variable name, conform to the rules for naming SAS variables discussed in Appendix A.2 (e.g., begins with a letter). Also, note that each statement that creates a duplicate of an existing variable must end with a semicolon.

Duplicating Variables versus Renaming Variables

Technically, the previous program did not really rename variables Q1, Q2, and Q3. Rather, the program created duplicates of these variables and assigned new names to these duplicate variables. Therefore, the resulting dataset contains both the original variables under their old names (Q1, Q2, and Q3) as well as the duplicate variables under their new names (RESNEEDY, RESHOME, and NORES). If, for some reason, you want to rename the existing variables so that the old variable names no longer exist in the dataset, you can use the rename statement.

Creating New Variables from Existing Variables

It is often necessary to perform mathematical operations on existing variables and use the results to create a new variable. With SAS, the following symbols are used in arithmetic operations:

+ (addition)

- (subtraction)

* (multiplication)

/ (division)

= (equals)

When writing formulae, you should make extensive use of parentheses. Remember that operations enclosed within parentheses are performed first; operations outside of the parentheses are performed later. To create a new variable by performing a mathematical operation on an existing variable, use the following general form:

```
new-variable-name  =  formula-including-existing-variables;
```

For example, three existing variables in your dataset are SATREAD (critical reading test scores), SATWRITE (writing test scores), and SATMATH (math test scores). Suppose you wanted to create a new variable called

SATCOMB; this variable includes each participant's combined SAT score. For each participant, you need to add together SATREAD, SATWRITE, and SATMATH scores. The program repeats this operation for each participant in the sample, using just one statement:

```
SATCOMB = (SATREAD + SATWRITE + SATMATH);
```

The preceding statement tells SAS to create a new variable called SATCOMB and set it equal to the sum of SATREAD, SATWRITE, and SATMATH.

Suppose you wanted to calculate the *average* of SATREAD, SATWRITE, and SATMATH scores. The new variable might be called SATAVG. The program repeats this operation for each participant in the sample using the following statement:

```
SATAVG = (SATREAD + SATWRITE + SATMATH) / 3;
```

The preceding statement tells SAS to create a new variable called SATAVG by adding together the values of SATREAD, SATWRITE, and SATMATH, then dividing this sum by 3. The resulting quotient is labeled SATAVG. You can also arrive at the same result by using two statements instead of one, as shown here:

```
SATCOMB = (SATREAD + SATWRITE + SATMATH);
SATAVG  = SATCOMB/3;
```

Often, researchers need to calculate the average of several items on a questionnaire. For example, look at items 1 and 2 in the questionnaire shown previously. Both items seem to be measuring participants' sense of personal responsibility to help the needy.

Rather than analyze responses to the items separately, it may be more useful to calculate the average of responses to those items. This average could then serve as participants' scores on some "personal responsibility" variable. For example, consider the following:

```
RESPONSE = (Q1 + Q2) / 2;
```

The preceding statement tells SAS to create a new variable called RESPONSE by adding together participants' scores for Q1 and Q2, then dividing the resulting sum by 2. The resulting quotient creates the new RESPONSE variable.

Note: When creating new variables in this manner, be sure that all variables on the right side of the equals sign are *existing* variables. This means that they already exist in the dataset, either because they are listed in the INPUT statement or because they were created with earlier data-manipulation statements.

Priority of Operators in Compound Expressions

A SAS expression (e.g., a formula) that contains just one operator is known as a **simple expression**. The following statement contains a simple expression. Notice that there is only one operator (+ sign) to the right of the = sign:

```
RESPONSE = Q1 + Q2;
```

In contrast, a **compound expression** contains more than one operator. A compound expression is illustrated in the following example. Notice that several different operators appear to the right of the = sign:

```
RESPONS = Q1 + Q2 - Q3 / Q4 * Q5;
```

When an expression contains more than one operator, SAS follows a set of rules that determine which operations are performed first, which are performed second, and so forth. The rules that pertain to mathematical operators (+, -, /, and *) are summarized here:

- multiplication and division operators (* and /) have equal priority, and they are performed first;
- addition and subtraction operators (+ and -) have equal priority, and they are performed second.

One point made in the preceding rules is that multiplication and division are performed prior to addition or subtraction. For example, consider the following statement:

```
RESPONS = Q1 + Q2 / Q3;
```

Since division has priority over addition, the operations in the preceding statement would be executed in this sequence:

- Q2 would first be divided by Q3;
- the resulting quotient would then be added to Q1.

Notice that division is performed first, even though the addition appears earlier in the formula (reading from left to right).

But what if multiple operators having equal priority appear in the same statement? In this situation, SAS reads the formula from left to right, and performs the operations in that sequence. For example, consider the following:

```
RESPONS = Q1 + Q2 - Q3;
```

The preceding expression contains only addition and subtraction: Operations that have equal priority. SAS therefore reads the statement from left to right: First Q1 is added to Q2; then Q3 is subtracted from the resulting sum.

Because different priority is given to different operators, it is all too easy to write a statement that results in operations being performed in some sequence other than that intended. For example, imagine that you want to create a new variable called RESPONSE. Each participant's score for RESPONSE is created by adding responses to Q1, Q2, and Q3 and by dividing this sum by 3. Imagine further that you attempt to achieve this with the following statement:

```
RESPONSE = Q1 + Q2 + Q3 / 3;
```

The preceding statement will not create the RESPONSE variable as you had intended. Because division has priority over addition, SAS performs the operations in the following order:

1. Q3 is divided by 3;
2. the resulting quotient is then added to Q1 and Q2.

Obviously, this is not what you intended.

To avoid such mistakes, it is important to use parentheses when writing formulae. Because operations that are included inside parentheses are performed first, the use of parentheses gives you control over the sequence in which operations are executed. For example, the following statement creates the RESPONSE variable in the way originally intended because the lower priority operations (adding together Q1 plus Q2 plus Q3) are now included within parentheses:

```
RESPONSE = (Q1 + Q2 + Q3) / 3;
```

This statement tells SAS to add together Q1 plus Q2 plus Q3; the sum of these operations is then divided by 3.

Recoding Reversed Variables

Very often, a questionnaire contains a number of reversed items. A reversed- or negatively-keyed item is a question or statement written so that its meaning is the opposite of the meaning of other items in that group. For example, consider the meaning of the following items from the volunteerism survey:

1 2 3 4 5 1. I feel a personal responsibility to help needy people in my community.

1 2 3 4 5 2. I feel I am personally obligated to help homeless families.

1 2 3 4 5 3. I feel no personal responsibility to work with poor people in my community.

In essence, all of these questions are measuring the same thing (i.e., whether the participant feels some sense of personal responsibility to help the needy). Items 1 and 2 are stated so that the more strongly you agree with these statements, the greater your sense of personal responsibility. This means that scores of 5 indicate a strong sense of responsibility and scores of 1 indicate a weak sense of responsibility. Item 3, however, is a *reversed* or negatively keyed item. It is stated so that the more strongly you agree, the weaker your sense of personal responsibility. Here, a response of 1 indicates a strong sense of responsibility whereas a response of 5 indicates a weak sense of responsibility (which is just the reverse of items 1 and 2).

For later analyses, all three items must be consistent so that scores of 5 always indicate a strong sense of responsibility whereas scores of 1 always indicate a weak sense of responsibility. This requires that you recode item 3 so that those who select 5 are instead given a score of 1; those who circle 4 are given a score of 2; those who select 2 are given a score of 4; and those who select 1 are given a score of 5. This can be done easily with the following statement:

```
Q3 = 6 - Q3;
```

The preceding statement tells SAS to create a new version of variable Q3, then take the number 6 and subtract from it the participants' existing (old) scores for Q3. The result is a new score for Q3. Notice that with this statement, if an initial score for Q3 was 5, the new score becomes 1; and if the initial score was 1, the new score is 5.

The general form for this recoding statement is as follows:

```
existing-variable  =  constant  -  existing-variable;
```

The constant is always equal to the number of response points on your survey plus 1. For example, the volunteerism survey included 5 response points: Participants could circle "1" for "Disagree Strongly" all the way through "5" for "Agree Strongly." It was a 5-point scale, so the constant was 5 + 1 = 6. What would the constant be if the following 7-point scale had been used instead?

```
7 = Agree Very Strongly
6 = Agree Strongly
5 = Agree Somewhat
4 = Neither Agree nor Disagree
3 = Disagree Somewhat
2 = Disagree Strongly
1 = Disagree Very Strongly
```

It would be 8 because 7 + 1 = 8, and the recoding statement would read:

```
Q3 = 8 - Q3;
```

Where **should the recoding statements go**? In most cases, reversed- or negatively-keyed items should be recoded before other data manipulations are performed. For example, assume that you want to create a new variable called RESPONSE, which stands for "personal responsibility." With this scale, higher scores indicate higher levels of perceived personal responsibility. Scores on this scale are the average of participant responses to items 1, 2, and 3 from the survey. Because item 3 is a reversed- or negatively-keyed item, it is important that it be recoded before it is added to items 1 and 2 when calculating the overall scale score. Therefore, the correct sequence of statements is as follows:

```
Q3 = 6 - Q3;

RESPONSE = (Q1 + Q2 + Q3) / 3;
```

The following sequence is *not* correct:

```
RESPONSE = (Q1 + Q2 + Q3) / 3;

Q3 = 6 - Q3;
```

Using IF-THEN Control Statements

An IF-THEN control statement allows you to make sure that operations are performed on data only if certain conditions are true. The following comparison operators may be used with IF-THEN statements:

```
        =    is equal to
       ne    is not equal to
  gt or >    is greater than
       ge    is greater than or equal to
  lt or <    is less than
       le    is less than or equal to
```

The general form for an IF-THEN statement is as follows:

```
if  expression  then  statement ;
```

The expression usually consists of some comparison involving existing variables. The statement usually involves some operation performed on existing variables or new variables. For example, assume that you want to create a new variable called SATCRGRP for "critical reading group." This variable will be created so that:

- if you do not know participants' critical reading test scores, they will be assigned a score of "." for missing data;
- if participants' scores are less than 500 on the critical reading test, they will be assigned a score of 1 for SATCRGRP;
- if the participant's score is 500 or greater on the critical reading test, the participant will have a score of 2 for SATCRGRP.

Assume that the variable SATREAD already exists in your dataset and that it contains each participant's score for the SAT critical reading test. You can use it to create the new variable SATCRGRP by writing the following statements:

```
SATCRGRP = .;
if SATREAD lt 500 then SATCRGRP = 1;
if SATREAD ge 500 then SATCRGRP = 2;
```

The preceding statements tell SAS to create a new variable called SATCRGRP and begin by setting everyone's score equal to "." (i.e., missing). If participants' scores for SATREAD are less than 500, then their score for SATCRGRP will be equal to 1. If participants' scores for SATREAD are greater than or equal to 500, then their score for SATCRGRP will be equal to 2.

Using ELSE Statements

In reality, you can perform the preceding operations more efficiently by using the ELSE statement. The general form for using the ELSE statement, in conjunction with the IF-THEN statement, is presented as follows:

```
if  expression  then  statement  ;
   else if expression  then  statement;
```

The ELSE statement provides alternate actions that SAS may take when the original IF expression is not true. For example, consider the following:

```
❶     SATCRGRP = .;
❷     if SATREAD lt 500 then SATCRGRP = 1;
❸     else if SATREAD ge 500 then SATCRGRP = 2;
```

The preceding tells SAS to create a new variable called SATCRGRP and initially assign all participants a value of "missing." If a given participant has a SATREAD score less than 500, the system assigns that participant a score of 1 for SATCRGRP. Otherwise, if the participant has a SATREAD score greater than or equal to 500, then the system assigns that participant a score of 2 for SATCRGRP.

Obviously, the preceding statements are identical to the earlier statements that created SATCRGRP, except that the word *else* has been added to the beginning of line ❸. In fact, these two approaches actually result in assigning exactly the same values for SATCRGRP to each participant. So what is the advantage of including the ELSE statement? The answer has to do with efficiency. When an ELSE statement is included, the actions specified by that statement are executed only if the expression in the preceding IF statement is not true.

For example, consider the situation in which participant 1 has a SATREAD score less than 500. Line ❷ in the preceding statements assigns that participant a score of 1 for SATCRGRP. SAS then ignores line ❸ (because it contains the ELSE statement). If line ❸ did not contain the word *else*, SAS would have executed the command, checking to see whether the SATREAD score for participant 1 is greater than or equal to 500 (which is actually unnecessary, given what was learned in line ❷).

A word of caution regarding missing data is relevant at this point. Notice that line ❷ of the preceding program assigns participants to group 1 (under SATCRGRP) if their values for SATREAD are less than 0. Unfortunately, a value of "missing" (i.e., a value of ".") for SATREAD is viewed as being less than 500 (actually, it is viewed as being less than 0) by SAS. This means that participants with missing data for SATREAD are assigned to group 1 under SATCRGRP by line ❷ of the preceding program. This is not desirable.

To prevent this from happening, you may rewrite the program in the following way:

```
❶     SATCRGRP = .;
❷     if SATREAD GT 0 and SATREAD lt 500 then SATCRGRP = 1;
❸     else if SATREAD ge 500 then SATCRGRP = 2;
```

Line ❷ of the program now tells SAS to assign participants to group 1 only if their values for SATREAD are both greater than 0 and less than 500. This modification involves the use of the conditional AND statement, which is discussed in greater detail in the following section.

Finally, remember that the ELSE statement should only be used in conjunction with a preceding IF statement. In addition, always remember to place the ELSE statement *immediately* following the relevant IF statement.

Using the Conditional Statements AND and OR

As the preceding section indicates, you can also use the conditional statement AND within an IF-THEN statement or an ELSE statement. For example, consider the following:

```
SATCRGRP = .;
if SATREAD gt 0 and SATREAD lt 500 then SATCRGRP = 1;
else if SATREAD ge 500 then SATCRGRP = 2;
```

The second statement in the preceding program tells SAS that if SATREAD is greater than 0 and less than 500, then a score of 1 is given to participants for the SATCRGRP variable. This means that all are given a value of 1 *only* if they are both over 0 and under 500. What happens to those who have a score of 0 or less for SATREAD? They are given a value of "." for SATCRGRP. That is, they are classified as having a missing value for SATCRGRP. This is because they (along with everyone else) were initially given a value of "." in the first statement, and neither of the later statements replaces that "." with 1 or 2. However, those with SATREAD scores greater than 0, one of the subsequent statements replaces "." with either 1 or 2.

You can also use the conditional statement OR within an IF-THEN statement or an ELSE statement. For example, assume that you have a variable in your dataset called ETHNIC. With this variable, participants were assigned the value 5 if they are Caucasian, 6 if they are African-American, or 7 if they are Asian-American. Assume that you now wish to create a new variable called MAJORITY. Participants will be assigned a value of 1 for this variable if they are in the majority group (i.e., if they are Caucasians in North America), and they will be assigned a value of 2 for this variable if they are in a minority group (i.e., if they are either African-Americans or Asian-Americans in North America). This variable is created with the following statements:

```
MAJORITY=.;
if ETHNIC = 5 then MAJORITY = 1;
else if ETHNIC = 6 or ETHNIC = 7 then MAJORITY = 2;
```

In the preceding statements, all participants are first assigned a value of "missing" for MAJORITY. If their value for ETHNIC is 5, their value for MAJORITY changes to 1 and SAS ignores the following ELSE statement. If their value for ETHNIC is not 5, then SAS proceeds to the ELSE statement. There, if participants' value for ETHNIC is either 6 or 7, then they are assigned a value of 2 for MAJORITY.

Working with Character Variables

When working with character variables (i.e., variables in which the values consist of letters rather than numbers), you must enclose values within single quotation marks (or apostrophes) in the IF-THEN and ELSE statements. For example, suppose you want to create a new variable called SEXGRP. With this variable, males are given a score of 1 and females are given a score of 2. The variable SEX already exists in your dataset, and it is a character variable in which males are coded with the letter M, females are coded with the letter F, and others are coded as "O." You can create the new SEXGRP variable using the following statements:

```
SEXGRP = .;
if SEX = 'M' then SEXGRP = 1;
else if SEX = 'F' then SEXGRP = 2;
```

Using the IN Operator

The IN operator makes it easy to determine whether a given value is among a specified list of values. Because of this, a single IF statement including the IN operator can perform comparisons that could otherwise require a large number of IF statements. The general form for using the IN operator is as follows:

```
if  variable  in  value-1,value-2, ...value-n  then  statement;
```

Notice that each value in the preceding list must be separated by a comma.

For example, assume that you have a variable in your dataset called MONTH. The values assumed by this variable are the numbers 1 through 12. With these values, 1 represents January, 2 represents February, 3 represents March, and so forth. Assume that these values for MONTH indicate the month in which a given participant was born, and that you have data for 100 participants.

Imagine that you now wish to create a new variable called SEASON. This variable will indicate the season in which each participant was born. Participants are assigned values for SEASON according to the following guidelines:

- participants are assigned a value of 1 for SEASON if they were born in January, February, or March (months 1, 2, or 3);
- participants are assigned a value of 2 for SEASON if they were born in April, May, or June (months 4, 5, or 6);
- participants are assigned a value of 3 for SEASON if they were born in July, August, or September (months 7, 8, or 9); and
- participants are assigned a value of 4 for SEASON if they were born in October, November, or December (months 10, 11, or 12).

One way to create the new SEASON variable involves using four IF-THEN statements, as shown here:

```
SEASON = .;
if MONTH = 1  or MONTH = 2  or MONTH = 3  then SEASON = 1;
if MONTH = 4  or MONTH = 5  or MONTH = 6  then SEASON = 2;
if MONTH = 7  or MONTH = 8  or MONTH = 9  then SEASON = 3;
if MONTH = 10 or MONTH = 11 or MONTH = 12 then SEASON = 4;
```

However, the same results can be achieved somewhat more easily by using the IN operator within the IF-THEN statements, as shown here:

```
SEASON = .;
if MONTH in (1,2,3)    then SEASON = 1;
if MONTH in (4,5,6)    then SEASON = 2;
if MONTH in (7,8,9)    then SEASON = 3;
if MONTH in (10,11,12) then SEASON = 4;
```

In the preceding example, all variable values are numbers. However, the IN operator may also be used with character variables. As always, it is necessary to enclose all character variable values within apostrophes or single quotation marks. For example, assume that MONTH is actually a character variable that assumes values such as “Jan,” “Feb,” “Mar,” and so forth. Assume further that SEASON assumes the values “Winter,” Spring,” Summer,” and “Fall.” Under these circumstances, the preceding statements would be modified in the following way:

```
SEASON = '.';
if MONTH in ('Jan', 'Feb', 'Mar') then SEASON = 'Winter';
if MONTH in ('Apr', 'May', 'Jun') then SEASON = 'Spring';
if MONTH in ('Jul', 'Aug', 'Sep') then SEASON = 'Summer';
if MONTH in ('Oct', 'Nov', 'Dec') then SEASON = 'Fall';
```

Data Subsetting

Using a Simple Subsetting Statement

Often, it is necessary to perform an analysis on only a subset of the participants who are included in the dataset. For example, you may wish to review survey responses provided by just the female participants. A subsetting IF statement may be used to accomplish this, and the general form is presented here:

```
data   new-dataset-name;
   set   existing-dataset-name;
if   comparison;

proc   name-of-desired-statistical-procedure    data=new-dataset-name;
run;
```

The comparison described in the preceding statements generally includes some existing variable and at least one comparison operator. The following statements allow you to calculate the mean survey responses for only the female participants.

```
.
.

5433224 19 107 10 F
640 590 625
;
run;

data D2;
   set D1;

   if SEX = 'F';

proc means   data=D2;
run;
```

The preceding statements tell SAS to create a new dataset called D2 and to make it identical to D1; however, the program keeps a participant's data only if her SEX has a value of F. Then the program executes the MEANS procedure for the data that are retained.

Using Comparison Operators

All of the comparison operators previously described can be used in a subsetting IF statement. For example, consider the following:

```
data D2;
   set D1;

   if SEX = 'F' and AGE ge 65;

proc means   data=D2;
run;
```

The preceding statements analyze only data from women who are over 65 years of age.

Eliminating Observations with Missing Data for Some Variables

One of the most common difficulties encountered by social scientists is the problem of missing data. Briefly, missing data involves not having scores for all variables for all participants in a dataset. This section discusses the problem of missing data, and shows how a subsetting IF statement may be used to deal with it.

Assume that you administer your volunteerism survey to 100 participants, and you use their scores to calculate a single volunteerism score for each. You also obtain a number of additional variables for participants. The SAS names for the study's variables and their descriptions are as follows:

VOLUNTEER
: participant scores on the volunteerism questionnaire, where higher scores reveal greater likelihood of engaging in unpaid prosocial activities;

SATREAD
: participant scores on the SAT critical reading test;

SATWRITE
: participant scores on the SAT writing test;

SATMATH
: participant scores on the SAT mathematics test;

IQ
: participant intelligence quotient.

Assume further that you obtained scores for VOLUNTEER, SATREAD, SATWRITE, and SATMATH for all 100 participants. However, you were able to obtain IQ scores for only 75 of the participants.

You now wish to analyze your data using a procedure called multiple regression. (This procedure is covered in O'Rourke, Hatcher, and Stepanski [2005]; you do not need to understand multiple regression to understand the points to be made here.) In analysis #1, VOLUNTEER is the criterion or dependent variable, and SATREAD, SATWRITE, and SATMATH are the predictor or independent variables. The multiple regression equation for analysis #1 is represented in the following PROC REG statement:

```
proc reg   data=D1;
   model VOLUNTEER  =  SATREAD   SATWRITE   SATMATH ;
run;
```

When you review the results of the analysis, note that the analysis is based on 100 participants. This makes sense because you had complete data on all of the variables included in this analysis.

In analysis #2, VOLUNTEER is again the criterion variable, but this time the predictor variables will include SATREAD, SATWRITE, and SATMATH as well as IQ. The equation for Analysis #2 is as follows:

```
proc reg   data=D1;
   model VOLUNTEER  =  SATREAD   SATWRITE   SATMATH  IQ;
run;
```

When you review the results of analysis #2, you see that you have encountered a problem. The SAS output indicates that the analysis is based on only 75 participants. At first you may not understand this because you know that there are 100 participants in the dataset. But then you remember that you did not have complete data for one of the variables; you had values for the IQ variable for only 75 participants. The REG procedure (and many other SAS procedures) includes in the analysis only those who have complete data for *all* of the variables analyzed with that procedure. For analysis #2, this means that any participant with missing data for IQ will be eliminated from the analysis. Twenty-five participants had missing data for IQ and were therefore eliminated.

Why were these 25 participants not eliminated from analysis #1? These participants were not eliminated because that analysis did not involve the IQ variable. It only involved VOLUNTEER, SATREAD, and SATMATH, and all 100 participants had complete data for each of those three variables.

In a situation such as this, you have a number of options with respect to how you might perform these analyses and summarize the results. One option is to retain the results described previously. You could report that you performed one analysis on all 100 participants and a second analysis on just the 75 who had complete data for the IQ variable.

This approach might leave you open to criticism, however. The beginning of your research paper probably reported demographic characteristics for all 100 participants (e.g., how many were female, mean age).

However, you may not have a section providing demographics for the subgroup of 75. This might lead readers to wonder if the subgroup differed in some important way from the aggregate group.

There are statistical reasons why this approach might cause problems as well. For example, you might wish to test the significance of the difference between the squared multiple correlation (R^2) value obtained from analysis #1 and the R^2 value obtained from analysis #2. (This test is described in Chapter 14 of O'Rourke, Hatcher, and Stepanski [2005].) When performing this test, it is important that both R^2 values be based on the same participants in both analyses. This is obviously not the case in your study as 25 of the participants used in analysis #1 were not included in analysis #2.

In such situations, it is often better to ensure that all analyses are performed on exactly the same sample. This means that, in general, any participant who has missing data for variables to be included in any (reported) analysis should be deleted before the analyses are performed. In this instance, therefore, it is best to ensure that both analysis #1 and analysis #2 are performed on only those 75 participants who had complete data for all four variables (i.e., VOLUNTEER, SATREAD, SATWRITE, SATMATH, and IQ). Fortunately, this may easily be done using a subsetting IF statement.

Recall that with SAS, a missing value is represented with a period ("."). You can take advantage of this to eliminate any participant with missing data for any analyzed variable. For example, consider the following subsetting IF statement:

```
data D2;
   set D1;
if VOLUNTEER ne . and SATREAD ne . and
   SATWRITE ne . and SATMATH ne . and IQ   ne . ;
```

The preceding statements tell the system to:

1. create a new dataset named D2, and make it an exact copy of D1;
2. retain a participant in this new dataset *only* if (for that participant):
   - VOLUNTEER is not equal to missing;
   - SATREAD is not equal to missing;
   - SATWRITE is not equal to missing;
   - SATMATH is not equal to missing;
   - IQ is not equal to missing.

In other words, the system creates a new dataset named D2; this new dataset contains only the 75 participants who have complete data for all four variables of interest. You may now specify data=D2 in all SAS procedures, such that all analyses will be performed on exactly the same 75 participants.

The following SAS program shows where these statements should be placed:

```
.
.
.
5433224 19 107 10 F
640 590 625
;
run;

data D2;
   set D1;

if VOLUNTEER ne . and SATREAD ne . and
   SATWRITE  ne . and SATMATH ne . and IQ ne . ;
```

```
proc reg   data=D2;
   model VOLUNTEER = SATREAD  SATWRITE  SATMATH ;
run;

proc reg   data=D2;
   model VOLUNTEER = SATREAD  SATWRITE  SATMATH  IQ ;
run;
```

As evident above, the subsetting IF statement must appear in the program before the procedures or PROC statements that request the modified dataset (dataset D2, in this case).

How should I enter missing data? If you are entering data and come to a participant with a missing value for some variable, you do not need to record a "." to represent the missing data. So long as your data are being input using the DATALINES statement and the conventions discussed here, it is acceptable to simply leave that column (or those columns) blank by hitting the space bar on your keyboard. SAS will assign that participant a missing data value (".") for that variable. In some cases, however, it may be useful to enter a "." for variables with missing data, as this may make it easier to keep your place when entering information.

When using a subsetting IF statement to eliminate participants with missing data, exactly *which* variables should be included in that statement? In most cases, it should be those variables, and only those variables, that are ultimately discussed. This means that you may not know exactly which variables to include until you actually begin analyzing the data. For example, imagine that you conduct your study and obtain data for the following number of participants for each of the following variables:

Variable	Number of Participants with Valid Data for This Variable
VOLUNTEER	100
SATREAD	100
SATWRITE	100
SATMATH	100
IQ	75
AGE	10

As before, you obtained complete data for all 100 participants for VOLUNTEER, SATREAD, SATWRITE, and SATMATH, and you obtained data for 75 participants on IQ. But notice the last variable. You obtained information regarding age for only 10 participants. What would happen if you included the variable AGE in the subsetting IF statement, as shown here?

```
if VOLUNTEER ne . and SATREAD ne . and SATWRITE ne .
 and SATMATH ne . and IQ  ne .    and  AGE     ne . ;
```

This IF statement causes the system to eliminate from the sample anyone who does not have complete data for all five variables. Since only 10 participants have values for the AGE variable, you know that the resulting dataset includes just these 10 participants. This sample, however, is too small for virtually all statistical procedures. At this point, you have to decide whether to gather more data or forget about doing analyses with the AGE variable.

In summary, one approach for identifying those variables to be included in the subsetting IF statement is to:

- perform some initial analyses;
- decide which variables will be included in the final analyses (for your study);
- include all of those variables in the subsetting IF statement;
- perform all analyses on this reduced dataset so that all analyses reported are performed on exactly the same sample.

Of course, there will be some circumstances in which it is neither necessary nor desirable that all analyses be performed on exactly the same group of participants. The purpose of the research, along with other considerations, should determine when this is appropriate.

A More Comprehensive Example

Often, a single SAS program will contain a large number of data-manipulation and subsetting statements. Consider the following example which makes use of the INFILE statement rather than the DATALINES statement:

```
data D1;
   infile 'A:/volunteer.dat' ;
   input   #1   @1    Q1-Q7          1.
                @9    AGE            2.
                @12   IQ             3.
                @16   NUMBER         2.
                @19   SEX           $1.
           #2   @1    SATREAD        3.
                @5    SATWRITE       3.
                @9    SATMATH        3. ;
❶   data D2;
❷       set D1;

    Q3 = 6 - Q3;
    Q6 = 6 - Q6;
    RESPONSE = (Q1 + Q2 + Q3) / 3;
❸   TRUST    = (Q4 + Q5 + Q6) / 3;
❹   SHOULD   =  Q7;

❺   proc means   data=D2;
    run;

❻   data D3;
❼   set D2;
      if SEX = 'F';

    proc means   data=D3;
    run;

❽   data D4;
❾        set D2;
    if SEX = 'M';

     proc means   data=D4;
     run;
```

In the preceding program, lines ❶ and ❷ create a new dataset called D2 and set it identical to D1. All data-manipulation commands that appear between those lines and PROC MEANS on line ❺ are performed on dataset D2. Notice that a new variable called TRUST is created on line ❸. TRUST is the average of participants' responses to items 4, 5, and 6. Look over these items on the volunteerism survey to see why the name TRUST makes sense. On line ❹, variable Q7 is duplicated, and the resulting new variable is called SHOULD. Why does this make sense? PROC MEANS appears on line ❺, so the means and other descriptive statistics are calculated for all of the quantitative variables in the most recently created dataset, which is D2. This includes all variables inputted in dataset D1 as well as the new variables that were just created.

In lines ❻ through ❼, a new dataset called D3 is created; only responses from female participants are retained in this dataset. Notice that the SET statement sets D3 equal to D2 rather than D1. This enables the newly created variables such as TRUST and SHOULD to appear in this all-female dataset. In lines ❽ through ❾, a new dataset called D4 is created which is also set equal to D2 (not D3). This new dataset contains data only from males.

After this program is submitted for analysis, the SAS output contains three tables of means. The first table gives the means based on all participants. The second table gives the means based on the responses from females. The third table is based on the responses from males.

Concatenating and Merging Datasets

The techniques described thus far in this appendix are designed to help you transform data within a single dataset (e.g., to recode a variable within a single dataset). However, often you need to perform transformations that involve combining more than one dataset to create a new dataset. For example, **concatenating** involves creating a new dataset by combining two or more previously existing datasets. With concatenation, the same variables typically appear in both of the previously existing datasets, but the two sets contain data from *different participants.* By concatenating the two previously existing sets, you create a new set that contains data from all participants.

In contrast, **merging** combines datasets in a different way. With merging, each of the previously existing datasets typically contains data from the same participants. However, the different, previously existing sets usually contain *different variables*. By merging these sets, you can create a new dataset that contains all variables found in the previously existing datasets. For example, assume that you conduct a study with 100 participants. Dataset A contains each participant's age, while dataset B contains questionnaire responses from the same 100 participants. By merging datasets A and B, you can create a new dataset called C that, again, contains just 100 observations. A given observation in dataset C contains a given participant's age as well as the questionnaire responses made by that participant. Now that the datasets are merged, it is possible to correlate participant age with responses to the questionnaire. These coefficients could not be calculated when AGE was in one dataset and the questionnaire responses were in another.

Concatenating Datasets

Imagine that you are conducting research that involves the Scholastic Assessment Test (SAT). You obtain data from four participants: Karl, Bjarne, Noam, and Vasek. You enter information about these four participants into a SAS dataset called A. This dataset contains three variables:

- NAME, which contains the participant's first name;
- SATREAD, which contains the participant's critical reading test score;
- SATWRITE, which contains the participant's writing test score;
- SATMATH, which contains the participant's mathematics test score.

The contents of dataset A appear below in Table A.3.1. You can see that Karl has a score of 520 for SATREAD, 490 for SATWRITE, and 500 for SATMATH; Bjarne had a score of 610 for SATREAD, 590 for SATWRITE, and 640 for SATMATH, and so forth.

Table A.3.1: Contents of Dataset A

NAME	SATREAD	SATWRITE	SATMATH
Karl	520	490	500
Bjarne	610	590	640
Noam	490	510	470
Vasek	550	575	560

Imagine that later you create a second dataset called B that contains data from four different participants: Jan, Filip, Zarah, and Tibor. Values for these participants for SATREAD, SATWRITE, and SATMATH appear in Table A.3.2.

Table A.3.2: Contents of Dataset B

NAME	SATREAD	SATWRITE	SATMATH
Jan	710	690	650
Filip	450	570	400
Zarah	570	580	600
Tibor	680	675	700

Assume that you would like to perform analyses on a single dataset that contains scores from all eight of these participants. But you encounter a problem; the values in dataset A were entered differently than the values of dataset B, making it impossible to read data from both sets with a single INPUT statement. For example, perhaps you entered SATREAD in columns 10 to 12 in dataset A, but entered it in columns 11 to 13 in dataset B. Because the variable was entered in different columns in the two datasets, it is not possible to write a single INPUT statement that will input this variable (assuming that you use a formatted input approach).

One way to deal with this problem is to input A and B as separate datasets and then concatenate them to create a single dataset that contains all eight observations. You can then perform analyses on the new dataset. The following is the general form for concatenating multiple datasets into a single dataset:

```
data  new-dataset-name;
      set  dataset-1  dataset-2 ... dataset-n;
```

In the present situation, you wish to concatenate two datasets (A and B) to create a new dataset named C. This could be done in the following statements:

```
data C;
   set A B;
```

The entire program follows that places these statements in context. This program:

- inputs dataset A;
- inputs dataset B;
- concatenates A and B to create C;
- uses PROC PRINT to print the contents of dataset C. (PROC PRINT will be discussed in greater detail in Appendix A.4 of this text.)

```
data  A;
     input     #1     @1   NAME       $7.
                      @10  SATREAD     3.
                      @14  SATWRITE    3.
                      @18  SATMATH     3. ;
     datalines;
     Karl       520 490 500
     Bjarne      610 590 640
     Noam    490 510 470
     Vasek       550 575 560
     ;
     run;

     data  B;
        input  #1  @1    NAME       $7.
                   @11   SATREAD     3.
                   @15   SATWRITE    3.
                   @19   SATMATH     3. ;
     datalines;
     Jan      710 690 650
     Filip       450 570 400
     Zarah      570 580 600
     Tibor      680 675 700
     ;
     run;

❶    data  C;
❷      set  A  B;

❸    proc print  data=C;
❹    run;
```

In lines ❶ and ❷ in the preceding program, the two datasets are concatenated to create dataset C. In lines ❸ and ❹, PROC PRINT is used to print the contents of dataset C, and the results of this procedure are reproduced as Output A.3.1. The results of Output A.3.1 show that dataset C contains eight observations: The four observations from dataset A along with the four observations from dataset B. To perform additional statistical analyses on this combined dataset, you would simply specify data=C in the PROC statement of your SAS program.

Output A.3.1: Results of Performing PROC PRINT on Dataset C

Obs	NAME	SATREAD	SATWRITE	SATMATH
1	Karl	520	490	500
2	Bjarne	610	590	640
3	Noam	490	510	470
4	Vasek	550	575	560
5	Jan	710	690	650
6	Filip	450	570	400
7	Zarah	570	580	600
8	Tibor	680	675	700

Merging Datasets

As stated earlier, you would normally merge datasets when:

- you are working with two (or more) datasets;
- both datasets contain information for the same participants, but one dataset contains one set of variables, while the other dataset contains a different set of variables.

Once these two datasets have been merged, you will have a single dataset that contains all variables. Having all variables in one set allows you to assess the associations among variables, should you wish to do so.

As an illustration, assume that your sample consists of just four participants: Karl, Bjarne, Noam, and Vasek. Assume that you have obtained the social security number for each participant, and that these numbers are included as a SAS variable named SOCSEC in both previously existing datasets. In dataset D, you have critical reading, writing, and mathematics test scores for these participants (represented as variables SATREAD, SATWRITE, and SATMATH, respectively). In dataset E, you have college cumulative grade point averages for the same four participants (represented as GPA). Table A.3.3 and Table A.3.4 show the content of these two datasets.

Table A.3.3: Contents of Dataset D

NAME	SOCSEC	SATREAD	SATWRITE	SATMATH
Karl	232882121	520	490	500
Bjarne	222773454	610	590	640
Noam	211447653	490	510	470
Vasek	222671234	550	575	560

Table A.3.4: Contents of Dataset E

NAME	SOCSEC	GPA
Karl	232882121	2.70
Bjarne	222773454	3.25
Noam	211447653	2.20
Vasek	222671234	2.50

Assume that, in conducting your research, you would like to compute the correlation coefficient between SATREAD and GPA. (Let's forget for the moment that you really shouldn't perform a correlation using such a small sample!) Computing this coefficient should be possible because you do have values for these two variables for all four of your participants. You will not, however, be able to compute this coefficient until both variables appear in the same dataset. Therefore, it will be necessary to merge the variables contained in datasets D and E.

There are actually two ways of merging datasets. Perhaps the simplest way is the one-to-one approach. With **one-to-one merging**, observations are simply merged according to their order of appearance in the datasets. For example, imagine that you were to merge datasets D and E using one-to-one merging. In doing this, SAS would take the first observation from dataset D and pair it with the first observation from dataset E, and the result would become the first observation in the new dataset (dataset F). If the observations in datasets D and E were in exactly the same sequence, this method would work fine. Unfortunately, if any of the observations were out of sequence, or if one dataset contained more observations than another, then this approach could result in the incorrect pairing of observations. For this reason, we recommend a different strategy for merging: The match-merging approach next described.

Match-merging seems to be the method that is least likely to produce undesirable results and errors. With **match-merging**, both datasets must contain a common variable, so that values for this common variable can be used to combine observations from the two previously existing datasets into observations for the new dataset (often the participant identification number). For example, consider datasets D and E from Table A.3.3 and Table A.3.4. The variable SOCSEC appears in both of these datasets, thus it is a common variable. When SAS uses match-merging to merge these two datasets according to values on SOCSEC, it will:

- read the social security number for the first participant in dataset D;
- look for a participant in dataset E who has the same social security number;
- merge the information from that participant's observation in dataset D with his or her information from dataset E (if it finds a participant in dataset E with the same social security number);
- combine the information into a single observation in the new dataset, F;
- repeat this process for all participants.

As the preceding description suggests, the variable that you use as your common variable must be chosen carefully. Ideally, each participant should be assigned a *unique value* for this common variable. This means that no two participants should have the same value for the common variable. This objective would be achieved when social security numbers are used as the common variable, because no two people have the same social security number (assuming that the data are entered correctly).

The SAS procedure for match-merging datasets is somewhat more complex than for concatenating datasets. In part, this is because both previously existing datasets must be sorted according to values for the common variable prior to merging. This means that the observations must be rearranged in a consistent order with respect to values for the common variable. Fortunately, this is easy to do with PROC SORT, a SAS procedure that allows you to sort variables. This section shows how PROC SORT can be used to achieve this.

The general form for match-merging two previously existing datasets is presented as follows:

```
proc sort  data=dataset-1;
   by  common-variable;
run;

proc sort  data=dataset-2;
   by  common-variable;
run;

data  new-dataset-name;
   merge  dataset-1  dataset-2;
   by  common-variable;
run;
```

To illustrate, assume that you wish to match-merge datasets D and E from Table A.3.3 and Table A.3.4; to do this, use SOCSEC as the common variable. In the following program, these two datasets are entered, sorted, and then merged using the match-merge approach:

```
data  D;
  input  #1   @1  NAME       $9.
              @10 SOCSEC      9.
              @20 SATREAD     4.
              @23 SATWRITE    4.
              @27 SATMATH     4. ;
datalines;
Karl     232882121 520 490 500
Bjarne    222773454 610 590 640
Noam   211447653 490 510 470
Vasek     222671234 550 575 560
;
run;

data  E;
   input  #1  @1   NAME     $9.
              @10  SOCSEC   9.
              @20  GPA      4.  ;

datalines;
Karl     232882121 2.70
Bjarne    222773454 3.25
Noam   211447653 2.20
Vasek     222671234 2.50
;
run;

proc sort  data=D;
   BY  SOCSEC;
run;

proc sort  data=E;
   by  SOCSEC;
run;

data  F;
   merge  D  E;
   by  SOCSEC;
run;

❶   proc print  data=F;
❷   run;
```

In the preceding program, both datasets were sorted according to values for SOSEC, and the two datasets were merged according to values of SOSEC. Finally, the PROC PRINT on lines ❶ and ❷ requests a printout of the raw data contained in the new dataset.

Output A.3.2 contains the results of PROC PRINT, which printed the raw data now contained in dataset F. You can see that each observation in this new dataset now contains the merged data from the two previous datasets D and E. For example, the line for the participant named Noam now contains his scores on the critical reading, writing, and mathematics test scores (which came from dataset D), as well as his grade point average score (which came from dataset E). The same is true for the remaining participants. It would now be possible to correlate SATREAD with GPA, if that analysis were desired.

Output A.3.2: Results of Performing PROC PRINT on Dataset F

Obs	NAME	SOCSEC	SATREAD	SATWRITE	SATMATH	GPA
1	Noam	211447653	490	510	470	2.20
2	Vasek	222671234	550	575	560	2.50
3	Bjarne	222773454	610	590	640	3.25
4	Karl	232882121	520	490	500	2.70

Notice that the observations in Output A.3.2 are not in the same order in which they appeared in Tables A.3.3 and A.3.4; this is because they have now been sorted according to values for SOCSEC by the PROC SORT statements in the preceding SAS program.

Conclusion

After completing this appendix, you should be prepared to modify datasets, isolate subgroups of participants for analysis, and perform other tasks that are often required when performing quantitative research in the social sciences. At this point, you should be prepared to proceed to the stage of analyzing data to determine what they mean. Some of the most basic statistics for this purpose (descriptive statistics and related procedures) are covered in the following appendix.

Reference

O'Rourke, N., Hatcher, L., & Stepanski, E. J. (2005). *A step-by-step approach to using SAS for univariate and multivariate statistics* (2nd Ed.). Cary, NC: SAS Institute Inc.

Appendix A.4: Exploring Data with PROC MEANS, PROC FREQ, PROC PRINT, and PROC UNIVARIATE

Introduction: Why Perform Simple Descriptive Analyses?

The procedures discussed in this appendix are useful for (at least) three important purposes. The first is data screening. **Data screening** is the process of carefully reviewing data to ensure that they were entered correctly and are being read correctly by the computer. Before conducting the more sophisticated analyses described in this text, you should carefully screen your data to avoid common errors (e.g., numbers that were accidentally entered, out of range values, numbers which were entered in the wrong column). The process of data screening does not guarantee that your data are correct, but it does increase the likelihood by identifying obvious errors.

Second, these procedures allow you to examine the shape or distribution of your data. Among other things, knowing the shape of your data will help you choose an appropriate measure of central tendency (i.e., the mean, mode, or median). Also, many statistical procedures require that sample data are normally distributed, or at least that data do not display a *marked* departure from normality. You can use the procedures discussed herein to graph or produce plots to test the null hypothesis that the data are drawn from a normal population.

Finally, your research question may require use of a procedure such as PROC MEANS or PROC FREQ to obtain a desired statistic. For example, if your research question is "what is the average age at which women married in 2012?" you could obtain data from a representative sample of women who married in that year, analyze their ages with PROC MEANS, and review the results to determine their average age.

In almost any research study, it is desirable to report descriptive information about your sample. For example, if a study is performed that includes participants from a variety of demographic groups, it would be desirable to report the percent of men and women (or boys and girls), the percent of participants by race, their mean age, and so forth. You can also use PROC MEANS and PROC FREQ to obtain this information.

Example: An Abridged Volunteerism Survey

To help illustrate these procedures, assume that you conduct a scaled-down version of the volunteerism study described in the last appendix. You construct a new questionnaire which asks just one question related to helping behavior; this questionnaire also contains an item that asks participants their sex, and another that asks their year in college (e.g., freshman, sophomore, senior). See the questionnaire below:

Please indicate the extent to which you agree or disagree with the following statement:

1. I feel a personal responsibility to help needy people in my community (please select only one response):

(5) _____ Agree Strongly

(4) _____ Agree Somewhat

(3) _____ Neither Agree nor Disagree

(2) _____ Disagree Somewhat

(1) _____ Disagree Strongly

2. Your sex (please check one):

(F) _____ Female

(M) _____ Male

(O) _____ Other

3. Your year in college:

(1) _____ Freshman

(2) _____ Sophomore

(3) _____ Junior

(4) _____ Senior

(5) _____ Other

Notice that this instrument has been printed so that entering the data will be relatively simple. With each variable, the value that will be entered appears to the left of the corresponding response. For example, with question 1 the value "5" appears to the left of "Agree Strongly"; this means that the number "5" will be entered for any participant selecting that response. For participants checking "Disagree Strongly," a "1" will be entered. Similarly, notice that, for question 2, the letter "F" appears to the left of "Female," so an "F" will be entered for participants selecting this response.

The following format is used when entering the data:

Column	Variable Name	Explanation
1	RESNEEDY	Responses to question 1: Participant's perceived responsibility to help the needy
2	blank	
3	SEX	Responses to question 2: Participant's sex
4	blank	
5	CLASS	Responses to question 3: Participant's classification as a college student

You administer the questionnaire to 14 students. The following is the entire SAS program used to analyze their responses, including the raw data:

```
    data D1;
       input   #1   @1   RESNEEDY   1.
                         @3   SEX        $1.
                         @5   CLASS       1.   ;
    datalines;
❶   5 F 1
    4 M 1
    5 F 1
❷     F 1
    4 F 1
    4 F 2
    1 F 2
    4 F 2
    1 F 3
❸   5 M
    4 F 4
    4 M 4
❹   3 F
    4 F 5
    ;
    run;

    proc means   data=D1;
      var RESNEEDY CLASS;
    run;

    proc freq   data=D1;
       tables SEX CLASS RESNEEDY;
    run;

    proc print   data=D1;
       var RESNEEDY SEX CLASS;
    run;
```

The data obtained from the first participant appears on line ❶ of the preceding program. This participant provided a response of "5" to the RESNEEDY statement (indicating that she checked "Agree Strongly"), indicated that her SEX was "F" or female, and responded "1" to the CLASS question (indicating that she is a freshman).

Notice that there are some missing responses in this dataset. On line ❷ in the program, you can see that this participant indicated that she was a female freshman, but did not answer question 1. That is why the corresponding space in column 1 is left blank. There also appears to be missing responses to the CLASS question on lines ❸ and ❹. Unfortunately, missing data are common in questionnaire research.

Computing Descriptive Statistics with PROC MEANS

You can use PROC MEANS to analyze quantitative (numeric) variables. For each variable analyzed, it provides the following information:

- the number of observations on which calculations were performed (abbreviated "N" in the output);
- the mean or average;
- the standard deviation (Std Dev);
- the minimum (smallest) value observed; and
- the maximum (largest) value observed.

These statistics are produced by default, and some additional statistics (to be described later) may also be requested as options.

Here is the general form for PROC MEANS:

```
proc means  data=dataset-name
            option-list
            statistic-keyword-list ;
   var  variable-list  ;
run;
```

The PROC MEANS Statement

The PROC MEANS statement begins with "proc means" and ends with a semicolon. It is recommended that the statement should also specify the name of the dataset to be analyzed with the data = option.

The "option-list" appearing in the preceding program indicates that you can request a number of options with PROC MEANS. Some options especially useful for social science research are:

maxdec=n
: Specifies the maximum number of decimal places (digits to the right of the decimal point) to be reported when printing results; possible range is 0 to 8.

vardef=divisor
: Specifies the devisor to be used when calculating variances and covariances. Two possible divisors are:

    vardef=df
    : Divisor is the degrees of freedom for the analysis: (n–1). This is the default.

    vardef=n
    : Divisor is the number of observations, n.

The "statistic-keyword-list" appearing in the program indicates that you can request a number of statistics in place of the default output. Some statistics that may be of particular value in social science research include the following:

Range
: The range of values in the sample.

Sum
: The sum.

Css
: The corrected sum of squares.

Uss
: The uncorrected sum of squares.

Var
: The variance.

Stderr
: The standard error of the mean.

Skewness
: The skewness displayed by sample data. **Skewness** refers to the extent to which the distribution of sample data departs from the normal or bell-shaped curve because of a long "tail" on either side of the distribution. If the long tail appears on the right side of the sample distribution (where the higher values appear), it is described as being **positively skewed**. If the long tail appears on the left side of the distribution (where the lower values appear), it is described as being **negatively skewed.**

Kurtosis
: The kurtosis displayed by the sample. Kurtosis refers to the extent to which the sample distribution departs from the normal curve because it is either peaked or flat. If the sample distribution is relatively peaked (tall and skinny), it is described as being **leptokurtic**. If the distribution is relatively flat, it is described as being **platykurtic**.

T
: The obtained value for student's t test calculated to test the null hypothesis that the population mean is zero.

Prt
: The p value for the preceding t test; that is, the probability of obtaining a t value this large or larger if the population mean were zero.

To illustrate the use of these options and statistic keywords, assume that you wish to use the maxdec option to limit the printing of results to two decimal places, use the VAR keyword to request that the variances of all quantitative variables be printed, and use the kurtosis keyword to request that the kurtosis of all quantitative variables be printed. You could do this with the following PROC MEANS statement:

```
proc means    data=D1    maxdec=2    var    kurtosis ;
```

The VAR Statement

Here again is the general form of the statements requesting the MEANS procedure, including the VAR statement:

```
proc means  data=dataset-name
        option-list
        statistic-keyword-list ;
    var variable-list  ;
run;
```

In the place of "variable-list" in the preceding VAR statement, you may list the quantitative variables to be analyzed. Each variable name should be separated by at least one blank space. If no VAR statement is specified, SAS will perform PROC MEANS on all quantitative variables in the dataset. This is true for many SAS procedures as explained in the following note:

> **What happens if I do not include a VAR statement?** For many SAS procedures, failure to include a VAR statement causes the system to perform the requested analyses on *all* variables in the dataset. For datasets with a large number of variables, leaving off the VAR statement may unintentionally result in a very long output file.

The program used to analyze your dataset included the following statements: RESNEEDY and CLASS were specified in the VAR statement so that descriptive statistics would be calculated for both variables:

```
proc means   data=D1;
   var RESNEEDY CLASS;
run;
```

Output A.4.1: Results of the MEANS Procedure

The MEANS Procedure

Variable	N	Mean	Std Dev	Minimum	Maximum
RESNEEDY	13	3.6923077	1.3155870	1.0000000	5.0000000
CLASS	12	2.2500000	1.4222262	1.0000000	5.0000000

Reviewing the Output

Output A.4.1 presents the results generated by the preceding program. Before undertaking more sophisticated analyses, you should perform PROC MEANS for each quantitative variable and review the output to ensure that the results appear correct. Under the heading "Variable" is the name of each variable analyzed; descriptive statistics appear to the right of the variable name. Below the heading "N" is the number of valid cases, or observations, on which calculations were performed. Notice that in this instance, calculations were performed on only 13 observations for RESNEEDY even though the dataset contains 14 cases. This is because one participant did not respond to this statement on the survey. It is for this reason that N is equal to 13 rather than 14 for RESNEEDY in this output.

You should next examine the mean for the variable to verify that it appears correct. Remember that for statement 1, responses could range from 1 "Disagree Strongly" to 5 "Agree Strongly." Therefore, the mean for this RESNEEDY variable should be somewhere between 1.00 and 5.00. If the average is outside of this range, you will know that some type of error has been made. In this instance the mean for RESNEEDY is 3.69, which is within acceptable range, so everything appears correct so far.

Using the same reasoning, it is prudent to next check the column headed "Minimum." Here you will find the lowest value on RESNEEDY that appeared in the dataset. If this is less than 1.00, you will again know than an error was made because 1 was the lowest possible value. On the printout, the minimum is 1.00, which indicates no problems. The largest value provided for that variable appears under the "Maximum" heading, which for RESNEEDY should not exceed 5.00. This is because 5 is the largest possible score for that variable. The reported maximum value in this case is 5.00, so again it appears that there were no obvious errors in entering the data or the program syntax.

Once you have examined the results for RESNEEDY, you should also review the results for the CLASS variable. If any of the observed values are out of range, you should carefully review the program for data entry or programming errors. In some cases, you may use PROC PRINT to reproduce the raw data to make it easier to review. PROC PRINT is described later in this appendix.

Creating Frequency Tables with PROC FREQ

The FREQ procedure produces frequency distributions for quantitative variables as well as classification variables. For example, you may use PROC FREQ to determine the percent of participants who "agreed strongly" with a statement on a questionnaire, the percent who "agreed somewhat," and so forth.

The PROC FREQ and TABLES Statements

The general form for the procedure is as follows:

```
proc freq    DATA=dataset-name;
   tables  variable-list  /   options;
run;
```

In the TABLES statement, you list the names of the variables to be analyzed, with each name separated by at least one blank space. Below are the PROC FREQ and TABLES statements from the program presented earlier in this appendix (analyzing data from the volunteerism survey):

```
proc freq   DATA=D1;
   tables SEX CLASS RESNEEDY;
run;
```

Reviewing the Output

These statements will cause SAS to create three frequency distributions: one for the SEX variable; one for CLASS; and one for RESNEEDY. This output appears in Output A.4.2.

Output A.4.2: Results of the Freq Procedure

The FREQ Procedure

SEX	Frequency	Percent	Cumulative Frequency	Cumulative Percent
F	11	78.57	11	78.57
M	3	21.43	14	100.00

CLASS	Frequency	Percent	Cumulative Frequency	Cumulative Percent
1	5	41.67	5	41.67
2	3	25.00	8	66.67
3	1	8.33	9	75.00
4	2	16.67	11	91.67
5	1	8.33	12	100.00

Frequency Missing = 2

RESNEEDY	Frequency	Percent	Cumulative Frequency	Cumulative Percent
1	2	15.38	2	15.38
3	1	7.69	3	23.08
4	7	53.85	10	76.92
5	3	23.08	13	100.00

Frequency Missing = 1

Output A.4.2 shows that the name for the variable being analyzed appears on the far left side of the frequency distribution. Possible values for this variable appear below its name. The first table provides information for the SEX variable, and below the word "SEX" appear the values "F" and "M." Information about female participants appears to the right of "F," and information about males appears to the right of "M." (None of these participants indicated that their sex was "other.") When reviewing a frequency distribution, it is useful to think of these different values as representing categories to which participants may belong.

Under the heading "Frequency," the output indicates the number of individuals in a given category. Here, you can see that 11 participants were female while 3 were male. Below "Percent," the percent of participants in each category appears. The table shows that 78.57% of the participants were female while 21.43% were male.

Under "Cumulative Frequency" is the number of observations that appear in the current category plus all of the preceding categories. For example, the first (top) category for SEX was "female." There were 11 participants so the cumulative frequency was 11. The next category was "male," and there were 3 participants. The cumulative frequency for the "male" category was therefore 14 (because 11 + 3 = 14). In the same way, "Cumulative Percent" provides the percent of observations in the current category plus all of the preceding categories.

The next table provides results for the CLASS variable. Unlike the previous table, two participants did not respond to this question. Under this table as a result we see "Frequency Missing = 2." Note that PROC FREQ does not include missing cases when calculating "Percent," "Cumulative Frequency," or "Cumulative Percent." In other words, these calculations are based only on available participant data.

Notice that below the RESNEEDY table, we see "Frequency Missing = 1" indicating that the one participant skipped this question. Also in this table, note that under the "RESNEEDY" heading, you may find only values "1," "3," "4," and "5." The value "2" does not appear because none of the participants checked "Disagree Somewhat" for this statement. If none of the participants select a response alternative, the value representing that response alternative will not appear in the frequency table as in this case.

Printing Raw Data with PROC PRINT

PROC PRINT can be used to reproduce your raw data. PROC PRINT output shows each participant's value for each requested variable. This procedure can be used with both quantitative and classification variables. The general form is:

```
proc print   data=dataset-name;
   var  variable-list  ;
run;
```

In the variable list, you may request any variable that has been specified in the INPUT statement, as well as any new variable that has been created from existing variables. If you do not include the VAR statement, then all existing variables will be printed. The program presented earlier in this appendix included the following PROC PRINT statements:

```
proc print   data=D1;
   var RESNEEDY SEX CLASS;
run;
```

These statements produce Output A.4.3.

Output A.4.3: Results of the PRINT Procedure

Obs	RESNEEDY	SEX	CLASS
1	5	F	1
2	4	M	1
3	5	F	1
4	.	F	1
5	4	F	1
6	4	F	2
7	1	F	2
8	4	F	2
9	1	F	3
10	5	M	.
11	4	F	4
12	4	M	4
13	3	F	.
14	4	F	5

The first column of output is headed "Obs" for "observation." This variable is created by SAS to assign an observation number to each participant. The second column provides the raw data for the RESNEEDY variable, the third column displays the SEX variable, and the last displays responses for the CLASS variable. The output shows that participant 1(observation 1) provided a response of 5 on RESNEEDY, was a female, and a freshman (i.e., value of 1 for the CLASS variable). Notice that SAS prints periods in place of missing values.

PROC PRINT is helpful to verify that your data are entered correctly and that SAS is reading these data correctly. It is particularly useful for studies with a large number of variables, for instance, when you use a questionnaire with a large number of questions. In these situations, it is often difficult to visually inspect the data as they exist in the SAS program file.

You should compare the results of PROC PRINT with several of the questionnaires as completed by participants. Verify that participant responses as they appear in the PROC PRINT output correspond to the original questionnaires. If not, it is likely that mistakes were made in either entering the data or in writing the SAS program.

Testing for Normality with PROC UNIVARIATE

A normal distribution is a symmetrical, bell-shaped distribution of values. The shape of the normal distribution is shown in Figure A.4.1.

Figure A.4.1: The Normal Distribution

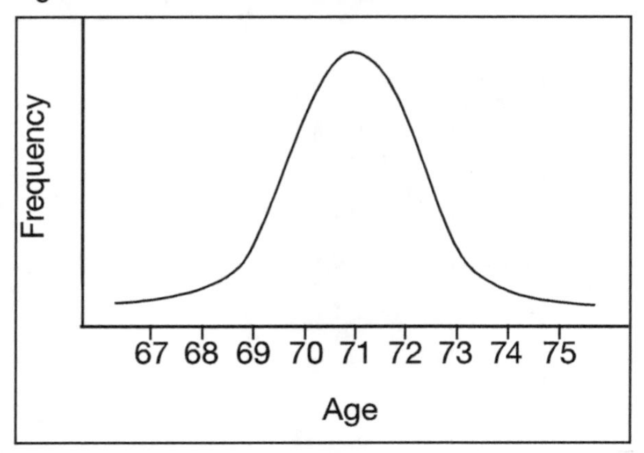

To understand the distribution in Figure A.4.1, assume that you are interested in conducting research on people who live in retirement communities. Imagine it is possible to assess the age of every person in this population. To summarize this distribution, you prepare a figure similar to Figure A.4.1 with the variable AGE plotted on the horizontal axis, and the frequency of persons at each age plotted on the vertical axis. Figure A.4.1 suggests that many of your participants are around 71 years of age since the distribution "peaks" near the age of 71. This suggests that the mean of this distribution will likely be somewhere around 71. Notice also that most of your participants are between 67 (near the lower end of the distribution) and 75 years of age (near the upper end of the distribution).

Why Test for Normality?

Normality is an important concept in quantitative analyses because there are at least two problems that may result when data are not normally distributed. The first is that markedly non-normal data may lead to incorrect conclusions in inferential statistical analyses. Many inferential procedures are based on the assumption that the sample of observations is normally distributed. If this assumption is violated, the statistic may give misleading results. For example, the independent group's t test assumes that both samples in the study were drawn from normally distributed populations. If this assumption is violated, then performing the analysis may cause you to incorrectly reject the null hypothesis (or incorrectly accept the null hypothesis). Under these circumstances, you should instead analyze the data using a procedure that does not assume normality (e.g., a nonparametric procedure).

The second problem is that markedly non-normal data may have a biasing effect on correlation coefficients, as well as more sophisticated procedures that are based on correlation coefficients. For example, assume that you compute the Pearson product moment correlation coefficient between two variables. If the distribution(s) of values for one or both of these variables are markedly non-normal, this may cause your coefficient to be much larger (or much smaller) than the true correlation between these variables in the population; your obtained coefficient is essentially misleading. To make matters worse, many more sophisticated statistical procedures such as principal component analysis are performed on an array of correlation coefficients. If some or all of these correlations are distorted due to departures from normality, then the results of the analyses may again be misleading. For this reason, you should routinely check your data for major departures from normality prior to performing more sophisticated analyses (Cozby & Bates, 2012).

Departures from Normality

Assume that you draw a random sample of 18 participants from the population of persons living in retirement communities. There are a wide variety of ways that your data may depart from normality.

Figure A.4.2 shows the age distribution for two samples of participants drawn from the population of retirees. This figure is somewhat different from Figure A.4.1 because the distributions have been turned on their sides so that age is now plotted on the vertical axis rather than on the horizontal axis; this is so that these figures will be more similar to the Stem-Leaf plots produced by PROC UNIVARIATE discussed later in this appendix. Each small circle in Figure A.4.2 represents one participant in a given distribution. For example, in the distribution for Sample A, you can see that there is one participant at age 75, one at age 74, two at age 73, three at age 72, and so forth. The ages of the 18 participants in Sample A range from a low of 67 to a high of 75.

Figure A.4.2: Sample with an Approximately Normal Distribution and a Sample with an Outlier

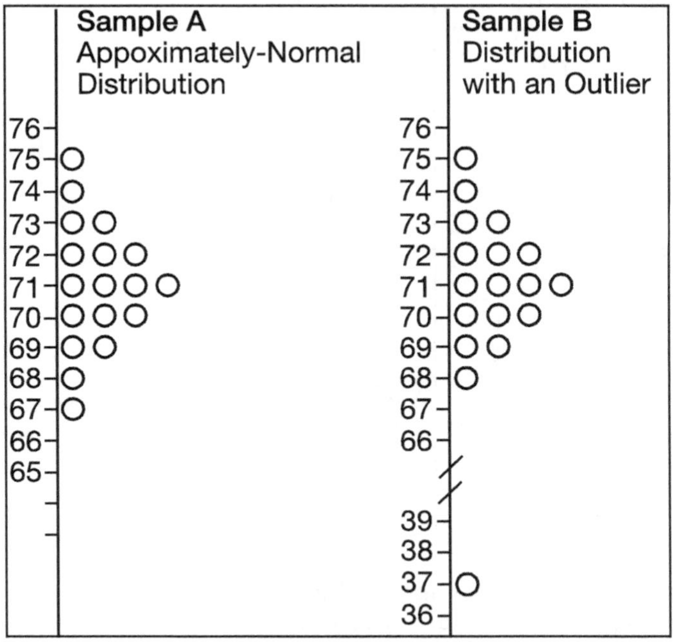

```
   | Sample A                  | Sample B
   | Appoximately-Normal       | Distribution
   | Distribution              | with an Outlier
76-|                       76-|
75-|O                      75-|O
74-|O                      74-|O
73-|O O                    73-|O O
72-|O O O                  72-|O O O
71-|O O O O                71-|O O O O
70-|O O O                  70-|O O O
69-|O O                    69-|O O
68-|O                      68-|O
67-|O                      67-|
66-|                       66-|
65-|                          //
   -                       39-|
   -                       38-|
                           37-|O
                           36-|
```

The data in Sample A form an approximately normal distribution (called approximately normal because it is difficult to form a perfectly normal distribution using a small sample of just 18 cases). An inferential test (discussed later) will show that Sample A does not significantly differ from normal. Therefore, it would be appropriate to analyze Sample A data.

In contrast, there is a problem with Sample B. These data are very similar to Sample A except that there is an outlier at the lower end of the distribution. An **outlier** is an extreme value that differs substantially from the other values in the distribution. In this case, the outlier is the participant who is only 37 years of age. Later, you will see that this outlier causes the dataset to differ significantly from normal, making the data inappropriate for many statistical procedures. When you detect an outlier, it is important to determine whether it should be corrected or deleted from the dataset. Obviously, if the outlier exists because an error was made in entering the data, it should be corrected (e.g., should have been entered as 73, not 37).

Data may also depart from normality because it is kurtotic. **Kurtosis** refers to the peakedness of the distribution (from the Greek word *kurtos* meaning bulging). The two samples displayed in Figure A.4.3 depict the two different types of kurtosis:

Figure A.4.3: Samples Displaying Positive versus Negative Kurtosis

	Sample C Positive Kurtosis (Leptokurtic)		Sample D Negative Kurtosis (Platykurtic)
78		78	
77		77	O
76		76	O
75		75	O
74		74	O
73	O	73	O O
72	O O O O	72	O O
71	O O O O O O O O O	71	O O
70	O O O O	70	O O
69	O	69	O O
68		68	O
67		67	O
66		66	O
65		65	O
64		64	

Sample C in Figure A.4.3 depicts **positive kurtosis**, which means that the distribution is relatively peaked (tall and skinny) rather than flat. Notice that, with Sample C, there are a relatively large number of participants who cluster around the center of the distribution (around age 71). This is what makes the distribution peaked (relative to Sample A, for example). Distributions with positive kurtosis are also called **leptokurtic**. A mnemonic device to remember the meaning of this word is to think of the distribution *leaping* upward (i.e., a leptokurtic distribution has leapt up).

In contrast, Sample D in the same figure displays **negative kurtosis**, which means that the distribution is relatively flat. Flat distributions are described as being **platykurtic**. A mnemonic device to remember this word is to think of a platykurtic distribution as *flat as a plate*.

In addition to kurtosis, distributions may also demonstrate varying degrees of **skewness**, or sidedness. A distribution is skewed if the tail on one side of the distribution is extended making it longer than the tail on the other side. The distributions in Figure A.4.4 show the two different types of skewness:

Figure A.4.4: Samples Displaying Positive versus Negative Skewness

	Sample E Positive Skewness		Sample F Negative Skewness
78-		78-	
77-	O	77-	O
76-		76-	O O
75-	O	75-	O O O
74-		74-	O O
73-	O	73-	O O
72-	O	72-	O O
71-	O	71-	O
70-	O	70-	O
69-	O O	69-	O
68-	O O	68-	O
67-	O O	67-	
66-	O O O	66-	O
65-	O O	65-	
64-	O	64-	O
63-		63-	

Consider Sample E in Figure A.4.4. Notice that the largest number of participants in this distribution tends to cluster around age 66. The tail of the distribution that stretches above 66 (from 67 to 77) is relatively long, while the tail of the distribution that stretches below 66 (from 65 to 64) is comparatively short; clearly, this distribution is skewed. A distribution is said to be **positively skewed** if the longer tail of a distribution extends in the direction of *higher* values. You can see that Sample E displays positive skewness, because its longer tail points toward larger numbers such as 75, 77, and so forth.

On the other hand, if the longer tail of a distribution points in the direction of lower values, the distribution is said to be **negatively skewed**. You can see that Sample F of Figure A.4.4 displays negative skewness because in that sample the longer tail points downward, in the direction of lower values (such as 66 and 64).

General Form for PROC UNIVARIATE

Like the MEANS procedure, PROC UNIVARIATE provides a number of descriptive statistics for quantitative variables, including the mean, standard deviation, kurtosis, and skewness. PROC UNIVARIATE, however, has the added advantage of printing a significance test for the null hypothesis that data come from a normally distributed population. The procedure also provides plots that will help you understand the shape of your sample's distribution, along with additional information that will help understand *why* your data depart from normality (if, indeed, they do). This text describes just a few of the features of PROC UNIVARIATE.

Here is the general form for the PROC UNIVARIATE statements that produce the output discussed in this appendix:

```
proc univariate   data=dataset-name   normal   plot;
   var variable-list;
   id identification-variable;
run;
```

In the preceding program, the normal option requests a significance test for the null hypothesis that sample data are from a normally distributed population. The Shapiro-Wilk statistic is applied for samples of 2,000 or fewer whereas the Kolmogorov-Smirnov statistic is applied for larger samples.

The plot option of the preceding program produces a Stem-Leaf plot, a Boxplot, and a Normal Probability Plot, each of which is useful for understanding the shape of a sample's distribution. This appendix describes how to interpret the Stem-Leaf plot.

The names of the variables to be analyzed should be listed in the VAR statement. The ID statement is optional but is useful (and recommended) for identifying outliers. PROC UNIVARIATE prints an "Extreme Observations" table that lists the five largest and five smallest values in the dataset; these values are identified by the identification variable listed in the ID statement. For example, assume that AGE (participant age) is listed in the VAR statement, and SOCSEC (for social security number) is listed in the ID statement. PROC UNIVARIATE will print social security numbers for participants with the five largest and five smallest AGE values. This should make it easier to identify which participant is the outlier in your dataset. (This use of the extreme observations table is illustrated here.)

Results for an Approximately Normal Distribution

For purposes of illustration, assume that you wish to analyze the data that are illustrated as Sample A of Figure A.4.2 (the approximately normal distribution). You prepare a SAS program in which participant age is entered as a variable called AGE, and participant identification numbers are entered as a variable called PARTICIPANT. Here is the entire program that will input these data and analyze them using PROC UNIVARIATE:

```
data D1;
   input   #1   @1   PARTICIPANT  2.
                @4   AGE          2.   ;
datalines;
 1 72
 2 69
 3 75
 4 71
 5 71
 6 73
 7 70
 8 67
 9 71
10 72
11 73
12 68
13 69
14 70
15 70
16 71
17 74
18 72
;
run;
proc univariate   data=D1   normal   plot;
   var AGE;
   id PARTICIPANT;
run;
```

The preceding program requests that PROC UNIVARIATE be performed on the variable AGE. Values of the variable PARTICIPANT will be used to identify outlying values of AGE in the extremes table.

This output would contain:

- a **moments table** that includes the mean, standard deviation, variance, skewness, kurtosis, along with other statistics;
- a table of **basic statistical measures** that provides indices of central tendency and variability estimates;
- tests for location;
- tests for normality such as the Shapiro-Wilk statistic;

- a **quartiles table** that provides the median, 25th percentile, 75th percentile, and related information;
- **Extreme Observations table** that provides the five highest values and five lowest values on the variable being analyzed;
- a Stem-Leaf plot, Boxplot, and Normal Probability Plot.

Output A.4.4 includes the moments table, basic statistical measures, tests of normality, quartiles table, Extreme Observations table, and a Stem-Leaf plot for Sample A.

Output A.4.4: Tables from PROC UNIVARIATE for Sample A

The UNIVARIATE Procedure
Variable: AGE

Moments			
N	18	Sum Weights	18
Mean	71	Sum Observations	1278
Std Deviation	2.05798302	Variance	4.23529412
Skewness	0	Kurtosis	-0.1357639
Uncorrected SS	90810	Corrected SS	72
Coeff Variation	2.89856764	Std Error Mean	0.48507125

Basic Statistical Measures			
Location		Variability	
Mean	71.00000	Std Deviation	2.05798
Median	71.00000	Variance	4.23529
Mode	71.00000	Range	8.00000
		Interquartile Range	2.00000

Tests for Location: Mu0=0				
Test	Statistic		p Value	
Student's t	t	146.3702	Pr > |t|	<.0001
Sign	M	9	Pr >= |M|	<.0001
Signed Rank	S	85.5	Pr >= |S|	<.0001

Tests for Normality				
Test	Statistic		p Value	
Shapiro-Wilk	W	0.983895	Pr < W	0.9812
Kolmogorov-Smirnov	D	0.111111	Pr > D	>0.1500
Cramer-von Mises	W-Sq	0.036122	Pr > W-Sq	>0.2500
Anderson-Darling	A-Sq	0.196144	Pr > A-Sq	>0.2500

Quantiles (Definition 5)	
Quantile	**Estimate**
100% Max	75
99%	75
95%	75
90%	74
75% Q3	72
50% Median	71
25% Q1	70
10%	68
5%	67
1%	67
0% Min	67

Extreme Observations					
Lowest			Highest		
Value	PARTICIPANT	Obs	Value	PARTICIPANT	Obs
67	8	8	72	18	18
68	12	12	73	6	6
69	13	13	73	11	11
69	2	2	74	17	17
70	15	15	75	3	3

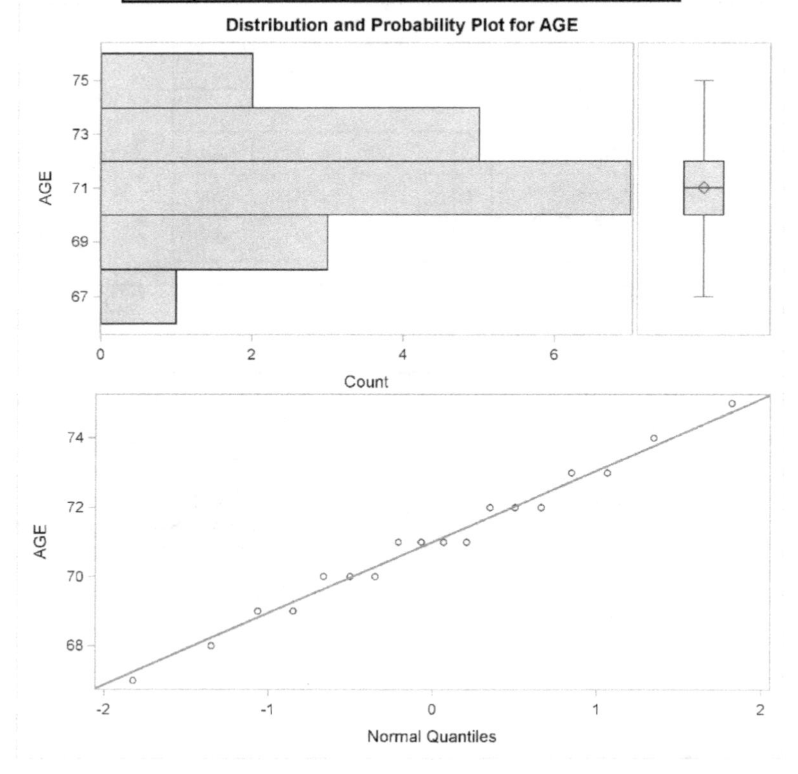

At the top of Output A.4.4, the note "Variable: AGE" indicates that AGE is the name of the variable being analyzed by PROC UNIVARIATE. The moments table is the first table reproduced in Output A.4.4. On the upper-left side of the moments table is the heading "N," and to the right of this you can see that the analysis was based on 18 observations. Below "N" are the headings "Mean" and "Std Deviation." To the right of these you can see that the mean and standard deviation for AGE were 71 and 2.06 (rounded to two decimal places), respectively.

To the right of "Skewness" you can see that the skewness statistic for AGE is zero. In interpreting the skewness statistic, keep in mind the following:

- a skewness value of zero means that the distribution is not skewed. In other words, this means that the distribution is symmetrical, that neither tail is longer than the other;
- a positive skewness value means that the distribution is positively skewed, that the longer tail points toward higher values in the distribution (as with Sample E of Figure A.4.4);
- a negative skewness value means that the distribution is negatively skewed, that the longer tail points toward lower values in the distribution (as with Sample F of Figure A.4.4).

Since the AGE variable of Sample A displays a skewness value of zero, we know that neither tail is longer than the other in this sample.

A closer look at the moments table of Output A.4.4 shows that it actually consists of two columns of statistics. The column on the left provides statistics such as the sample size, the mean, the standard deviation, and so forth. The column on the right contains headings such as "Sum Weights," "Sum Observations," and "Variance." Notice that in this right-hand column, the fourth entry down has the heading "Kurtosis" (just below "Variance"). To the right of "Kurtosis," you can see that the kurtosis statistic for AGE is approximately –.14. When interpreting this kurtosis statistic, keep in mind the following:

- a kurtosis value of zero means that the distribution displays no kurtosis. In other words, the distribution is neither relatively peaked nor is it relatively flat compared to the normal distribution;
- a positive kurtosis value means that the distribution is relatively peaked, or leptokurtic;
- a negative kurtosis value means that the distribution is relatively flat, or platykurtic.

The small negative kurtosis value of –.14 in Output A.4.4 indicates that Sample A is slightly flat, or platykurtic.

In the fourth table (Tests for Normality), the Shapiro-Wilk statistic appears at the top of the left-hand column. As stated previously in this appendix, this statistic tests the null hypothesis that sample data are normally distributed. To the right of the "W," you can see that the value for the Shapiro-Wilk statistic is 0.98. To the immediate right of this statistic is its corresponding p value; this value appears as the first value in the right-hand column, to the right of the heading "Pr < W." In this instance, $p = 0.98$. Remember that this statistic tests the null hypothesis that the data are normally distributed. This value is very large at .98, meaning that there are approximately 98 chances in 100 that you would obtain the present results if the data were drawn from a normal population. Because this statistic gives no evidence to reject the null hypothesis, you can tentatively accept it. This makes sense when you review the shape of the distribution of Sample A in Figure A.4.2 as the sample data appear to be normally distributed. In general, you should reject the null hypothesis of normality when p values are less than .05.

Results for a Distribution with an Outlier

The data of Sample A in Figure A.4.2 displayed an approximately normal distribution. For purposes of contrast, assume that you now use PROC UNIVARIATE to analyze the data of Sample B from Figure A.4.2. You will remember that Sample B was similar in shape to Sample A except that Sample B contained an outlier. The lowest value in Sample B was 37, which was extremely low compared to other values in the sample. (If necessary, turn back to Figure A.4.2 at this time to verify this.)

The raw data from Sample B follow. Columns 1 to 2 contain values of PARTICIPANT, the participant identification number, and columns 4 to 5 contain AGE values. Notice that these data are identical to those of

Sample A, except for participant 8. In Sample A, this participant's age was listed as 67; in Sample B, it is listed as 37.

```
 1 72
 2 69
 3 75
 4 71
 5 71
 6 73
 7 70
 8 37
 9 71
10 72
11 73
12 68
13 69
14 70
15 70
16 71
17 74
18 72
```

When analyzed with PROC UNIVARIATE, the preceding data would again produce the following output. Some of the results of this analysis are presented in Output A.4.5.

Output A.4.5: Selected Tables from PROC UNIVARIATE for Sample B

The UNIVARIATE Procedure
Variable: AGE

Moments			
N	18	Sum Weights	18
Mean	69.3333333	Sum Observations	1248
Std Deviation	8.26758376	Variance	68.3529412
Skewness	-3.9049926	Kurtosis	16.0332475
Uncorrected SS	87690	Corrected SS	1162
Coeff Variation	11.9243996	Std Error Mean	1.94868818

Basic Statistical Measures			
Location		Variability	
Mean	69.33333	Std Deviation	8.26758
Median	71.00000	Variance	68.35294
Mode	71.00000	Range	38.00000
		Interquartile Range	2.00000

Tests for Location: Mu0=0				
Test	Statistic		p Value	
Student's t	t	35.57949	Pr > |t|	<.0001
Sign	M	9	Pr >= |M|	<.0001
Signed Rank	S	85.5	Pr >= |S|	<.0001

Tests for Normality				
Test	Statistic		p Value	
Shapiro-Wilk	W	0.458117	Pr < W	<0.0001
Kolmogorov-Smirnov	D	0.380384	Pr > D	<0.0100
Cramer-von Mises	W-Sq	0.696822	Pr > W-Sq	<0.0050
Anderson-Darling	A-Sq	3.681039	Pr > A-Sq	<0.0050

Quantiles (Definition 5)	
Quantile	Estimate
100% Max	75
99%	75
95%	75
90%	74
75% Q3	72
50% Median	71
25% Q1	70
10%	68
5%	37
1%	37
0% Min	37

Extreme Observations					
Lowest			Highest		
Value	PARTICIPANT	Obs	Value	PARTICIPANT	Obs
37	8	8	72	18	18
68	12	12	73	6	6
69	13	13	73	11	11
69	2	2	74	17	17
70	15	15	75	3	3

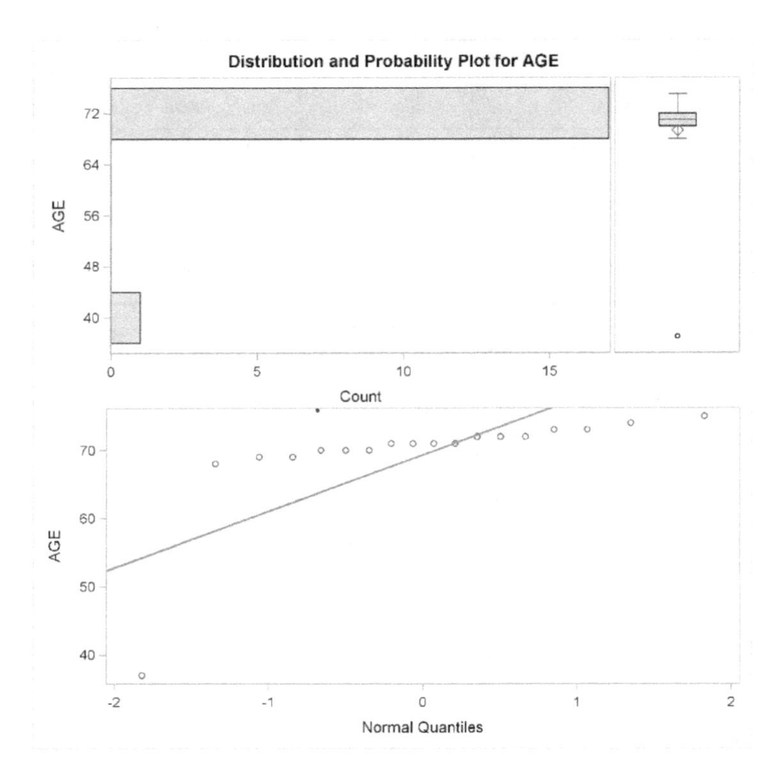

By comparing the moments table of Output A.4.5 (for Sample B) to that of Output A.4.4 (for Sample A) you can see that inclusion of the outlier has had a considerable effect on some of the descriptive statistics for AGE. The mean of Sample B is now 69.33, down from the mean of 71 found for Sample A. More dramatic is the effect that the outlier has had on the standard deviation. With the approximately normal distribution, the standard deviation was only 2.06. With the outlier included, the standard deviation was much larger at 8.27.

Output A.4.5 shows that the skewness index for Sample B is –3.90. Negative skewness is just what you would expect; the outlier has, in essence, created a long tail that points toward the lower values in the AGE distribution. You will remember that this generally results in negative skewness.

Output A.4.5 shows that the test for normality for Sample B results in a Shapiro-Wilk statistic of .46 (to the right of "W") with a corresponding p value less than .01 (to the right of "Pr < W"). Because this p value is well below .05, you reject the null hypothesis and conclude that Sample B data are not normally distributed. In other words, you can conclude that Sample B displays a statistically significant departure from normality.

The Extreme Observations table for Sample B appears just below the quartiles table in Output A.4.5. On the left side of the extremes table, below the heading "Lowest," PROC UNIVARIATE prints the lowest values observed for the variable specified in the VAR statement (AGE, in this case). Here, you can see that the lowest five values were 37, 68, 69, 69, and 70. To the immediate right of each value is the identification number for the participant who contributed that value to the dataset. The participant identification variable is specified in the ID statement (PARTICIPANT, in this case). Reviewing these values shows you that participant 8 contributed the AGE value of 37, participant 12 contributed the AGE value of 68, and so forth. Compare these results in the extremes table with the actual raw data (reproduced earlier) to verify that these are in fact the specific participants who provided these values on AGE.

On the right side of the extremes table, similar information is provided, though in this case, it is provided for the five *highest* values observed in the dataset. Under the heading "Highest" (and reading from the bottom up), you can see that the highest value on age was 75, and it was provided by participant 3, the next highest value was 74, provided by participant 17, and so forth.

This Extreme Observations table is useful for quickly identifying those participants who may have contributed outliers to a dataset. For example, in the present case you were able to determine that it was participant 8 who contributed the outlier on AGE. Using the Extreme Observations table may not be necessary when working with a very small dataset as in the present situation, but it can be invaluable when dealing with a large number of responses. For example, if you know that you have an outlier in a dataset with 1,000 observations, the Extreme Observations table can help quickly identify outliers. This will save you from the tedious chore of individually examining each of the 1,000 data lines.

Understanding the Stem-Leaf Plot

A Stem-Leaf plot provides a visual depiction of your data with conventions somewhat similar to Figures A.4.2, A.4.3, and A.4.4. Output A.4.6 provides the Stem-Leaf plot for Sample A (the approximately-normal distribution).

Output A.4.6: Stem-Leaf Plot from PROC UNIVARIATE for Sample A (Approximately Normal Distribution)

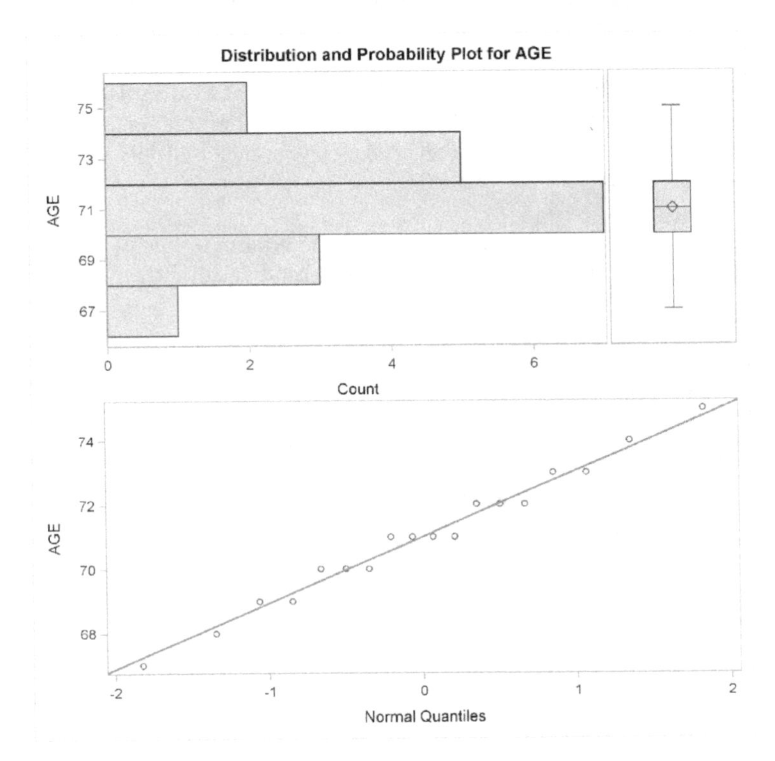

To understand a Stem-Leaf plot, think of each participant's score on AGE as consisting of a "stem" and a "leaf." The **stem** is that part of the value that appears to the left of the decimal point, and the **leaf** consists of that part that appears to the right of the decimal point. For example, participant 8 in Sample A had a value on AGE of 67. For this participant, the stem is 67 (because it appears to the left of the decimal point) and the leaf is 0 (because it appears to the right). Participant 12 had a value on age of 68, so the stem for this value is 68, and the leaf is again 0.

In the Stem-Leaf plot of Output A.4.6, the vertical axis (running up and down) plots the various stems that could be encountered in the dataset (these appear under the heading "Stem"). Reading from the top down, these stems are 75, 74, 73, and so forth. Notice that at the very bottom of the plot is the stem 67. To the right of this stem appears a single leaf (a single 0). This means that there was only one participant in Sample A with a stem-leaf of 67 (i.e., a value on AGE of 67). Move up one line, and you see the stem 68. To the right of this, again one leaf appears (i.e., one zero appears), meaning that only one participant had a score on AGE of 68. Move up an additional line, and you see the stem 69. To the right of this, two leaves appear (i.e., two zeros appear). This means that there were two participants with a stem-leaf of 69 (two participants with values on AGE of 69). Continuing up the plot in this manner, you can see that there were three participants at age 70, four participants at age 71, three at age 72, two at age 73, one at 74, and one at 75.

On the right side of the Stem-Leaf plot appears a column headed "#." This column prints the number of observations that appear at each stem. Reading from the bottom up, this column again confirms that there was one participant with a score on age of 67, one with a score of 68, two with a score of 69, and so forth.

Reviewing Output A.4.6 shows that the shape of this Stem-Leaf plot is very similar to Sample A in Figure A.4.2. This is to be expected, since both figures apply similar conventions and both describe the data of Sample A. In Output A.4.6, notice that the shape of the distribution is symmetrical (i.e., neither tail is longer than the other). This, too, is to be expected since Sample A data has a skewness value of zero

In most cases, the Stem-Leaf plot produced by UNIVARIATE will be more complex than the one depicted in Output A.5.6. For example, Output A.4.7 shows the Stem-Leaf plot for Sample B data from Figure A.4.2 (the distribution with an outlier). Consider the Stem-Leaf at the bottom of this plot. The stem for this entry is 3, and the leaf is 7, meaning that the stem-leaf is 3.7. Does this mean that some participant had a score on AGE of 3.7?

Notice the note at the bottom of this plot, which says "Multiply Stem.Leaf by 10**+1." This means "Multiply the stem-leaf by 10 raised to the first power." Ten raised to the first power (or 10^1), of course, is merely 10. This means that to find a participant's *actual* value on AGE, you must multiply a stem-leaf for that participant by 10.

For example, consider what this means for the stem-leaf at the bottom of this plot. This stem-leaf was 3.7. To find the actual score that corresponds to this stem-leaf, you would perform the following multiplication:

```
3.7 X 10 = 37
```

This means that for the participant who had a stem-leaf of 3.7, the actual value of AGE was 37.

Output A.4.7: Stem-Leaf Plot from PROC UNIVARIATE for Sample B (Distribution with Outlier)

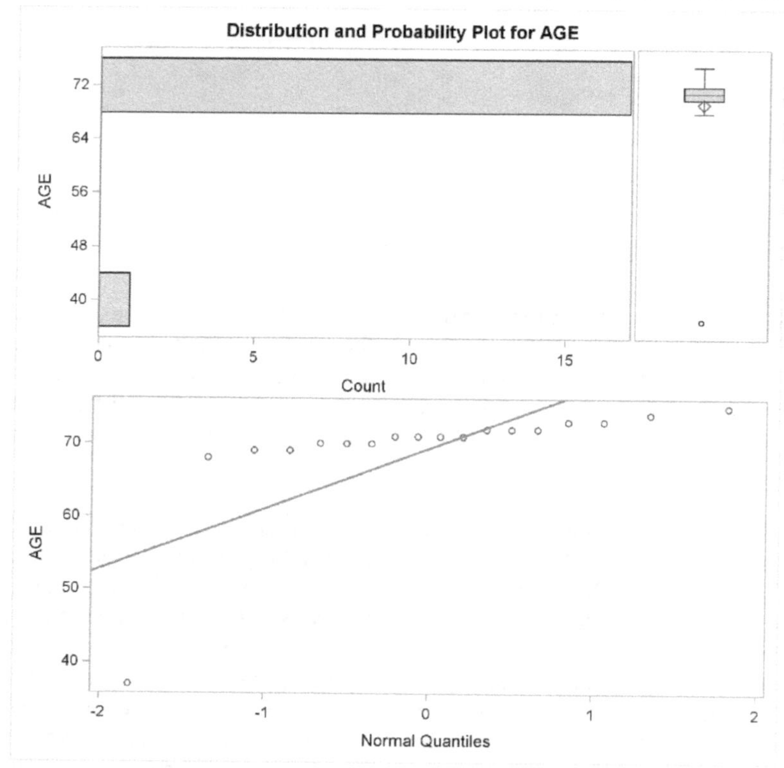

Move up one line in the plot, and you come to the stem "4." Note, however, that there are no leaves for this stem which means that there were no participants with a stem of 4.0 (i.e., no participant age 40). Reading up the plot, note that no leaves appear until you reach the stem "6." The leaves on this line suggest that there was one participant with a stem-leaf of 6.8, and two participants with a stem-leaf of 6.9. Multiply these values by 10 to determine their actual values on AGE:

```
6.8 X 10 = 68
6.9 X 10 = 69
```

Move up an additional line, and note that there are actually two stems for the value 7. The first stem (moving up the plot) includes stem-leaf values from 7.0 through 7.4, while the next stem includes stem-leaf values from 7.5 through 7.9. Reviewing values in these rows, you can see that there are three participants with a stem-leaf of 7.0, four with a stem-leaf of 7.1, and so forth.

The note at the bottom of the plot told you to multiply each stem-leaf by 10 raised to the first power. However, sometimes this note will tell you to multiply by 10 raised to a different power. For example, consider the following note:

```
Multiply Stem.Leaf by 10**+2
```

This note tells you to multiply by 10 raised to the second power (i.e., 10^2) or 100. Notice what some of the actual values on AGE would have been if this note had appeared (needless to say, such large values would not have made sense for the AGE variable):

```
6.8 X 100 = 680
6.9 X 100 = 690
```

All of this multiplication probably seems somewhat tedious at this point, but there is a simple rule to interpret the note that sometimes appears at the bottom of a Stem-Leaf plot. Remember that the power to which 10 is raised indicates the *number of decimal places* you should move the decimal point in the stem-leaf. Once you have moved the decimal point this number of spaces, your stem-leaf will represent the actual value of interested. For example, consider the following note:

```
Multiply Stem.Leaf by 10**+1
```

This note tells you to multiply the stem-leaf by 10 raised to the power of one; in other words, move the decimal point *one space to the right.* Imagine that you start with a stem-leaf of 3.7. Moving the decimal point one space to the right results in an AGE value of 37. If you begin with a stem-leaf of 6.8, this becomes 68.

On the other hand, consider if the plot had included this note:

```
Multiply Stem.Leaf by 10**+2
```

It would have been necessary to move the decimal point *two* decimal spaces to the right. In this case, a stem-leaf of 3.7 would become 370; 6.8 would become 680. (Again, these values would not make sense for AGE; they are used only for purposes of demonstration.) Finally, remember that, if no note appears at the bottom of the plot, it is not necessary to move the decimal points for the stem-leaf values at all.

Results for Distributions Demonstrating Skewness

Output A.4.8 provides some results from the PROC UNIVARIATE analysis of Sample E from Figure A.4.4. You will recall that this sample demonstrated a positive skew.

Output A.4.8: Tables and the Stem-Leaf Plot from PROC UNIVARIATE for Sample E (Positive Skewness)

The UNIVARIATE Procedure
Variable: AGE

Moments			
N	18	Sum Weights	18
Mean	68.7777778	Sum Observations	1238
Std Deviation	3.62273143	Variance	13.124183
Skewness	0.86982584	Kurtosis	0.11009602
Uncorrected SS	85370	Corrected SS	223.111111
Coeff Variation	5.26729933	Std Error Mean	0.85388599

Basic Statistical Measures			
Location		Variability	
Mean	68.77778	Std Deviation	3.62273
Median	68.00000	Variance	13.12418
Mode	66.00000	Range	13.00000
		Interquartile Range	5.00000

Tests for Location: Mu0=0				
Test	Statistic		p Value	
Student's t	t	80.54679	Pr > |t|	<.0001
Sign	M	9	Pr >= |M|	<.0001
Signed Rank	S	85.5	Pr >= |S|	<.0001

Tests for Normality				
Test	Statistic		p Value	
Shapiro-Wilk	W	0.929575	Pr < W	0.1909
Kolmogorov-Smirnov	D	0.14221	Pr > D	>0.1500
Cramer-von Mises	W-Sq	0.074395	Pr > W-Sq	0.2355
Anderson-Darling	A-Sq	0.465209	Pr > A-Sq	0.2304

Quantiles (Definition 5)	
Quantile	**Estimate**
100% Max	77
99%	77
95%	77
90%	75
75% Q3	71
50% Median	68
25% Q1	66
10%	65
5%	64
1%	64
0% Min	64

Extreme Observations					
Lowest			**Highest**		
Value	**PARTICIPANT**	**Obs**	**Value**	**PARTICIPANT**	**Obs**
64	18	18	71	5	5
65	17	17	72	4	4
65	16	16	73	3	3
66	15	15	75	2	2
66	14	14	77	1	1

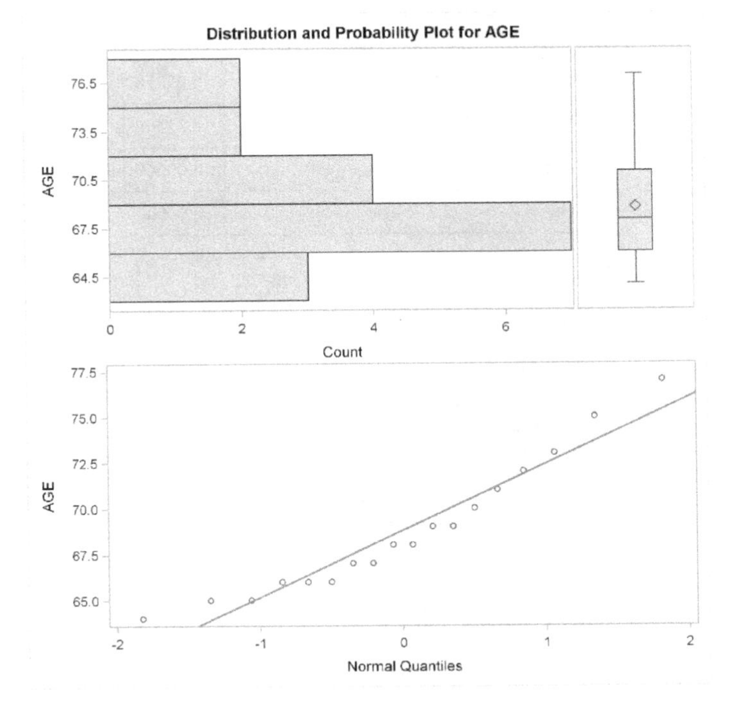

Remember that when the approximately normal distribution was analyzed, it displayed a skewness index of zero. In contrast, note that the skewness index for Sample E in Output A.4.8 is close to .87. This positive skewness index is what you would expect, given the positive skew of the data. The skew is also reflected in the Stem-Leaf plot that appears in Output A.4.8. Notice the relatively long tail that points in the direction of higher values for age (such as 74 and 76).

Although this sample displays positive skewness, it does not display a significant departure from normality. In the Tests for Normality table of Output A.4.8, you can see that the Shapiro-Wilk statistic (to the right of "W") is .93; its corresponding p value (to the right of "Pr < W") is .19. Because this p value is greater than .05, you need not reject the null hypothesis. With small samples such as the one examined here, this test is not very powerful (i.e., is not very sensitive). This is why the sample was not found to display a significant departure from normality, even though it was clearly skewed.

For purposes of contrast, Output A.4.9 presents the results of an analysis of Sample F from Figure A.4.4. Sample F displayed negative skewness, and this is reflected in the skewness index of –.87 that appears in Output A.4.9. Once again, the Shapiro-Wilk test shows that the sample does not demonstrate a significant departure from normality.

Output A.4.9: Tables and Stem-Leaf Plot from PROC UNIVARIATE for Sample F (Negative Skewness)

The UNIVARIATE Procedure
Variable: AGE

Moments			
N	18	Sum Weights	18
Mean	72.2222222	Sum Observations	1300
Std Deviation	3.62273143	Variance	13.124183
Skewness	-0.8698258	Kurtosis	0.11009602
Uncorrected SS	94112	Corrected SS	223.111111
Coeff Variation	5.01608967	Std Error Mean	0.85388599

Basic Statistical Measures			
Location		Variability	
Mean	72.22222	Std Deviation	3.62273
Median	73.00000	Variance	13.12418
Mode	75.00000	Range	13.00000
		Interquartile Range	5.00000

Tests for Location: Mu0=0				
Test	Statistic		p Value	
Student's t	t	84.58064	Pr > |t|	<.0001
Sign	M	9	Pr >= |M|	<.0001
Signed Rank	S	85.5	Pr >= |S|	<.0001

Tests for Normality				
Test	Statistic		p Value	
Shapiro-Wilk	W	0.929575	Pr < W	0.1909
Kolmogorov-Smirnov	D	0.14221	Pr > D	>0.1500
Cramer-von Mises	W-Sq	0.074395	Pr > W-Sq	0.2355
Anderson-Darling	A-Sq	0.465209	Pr > A-Sq	0.2304

Quantiles (Definition 5)	
Quantile	**Estimate**
100% Max	77
99%	77
95%	77
90%	76
75% Q3	75
50% Median	73
25% Q1	70
10%	66
5%	64
1%	64
0% Min	64

Extreme Observations					
Lowest			**Highest**		
Value	**PARTICIPANT**	**Obs**	**Value**	**PARTICIPANT**	**Obs**
64	18	18	75	5	5
66	17	17	75	6	6
68	16	16	76	2	2
69	15	15	76	3	3
70	14	14	77	1	1

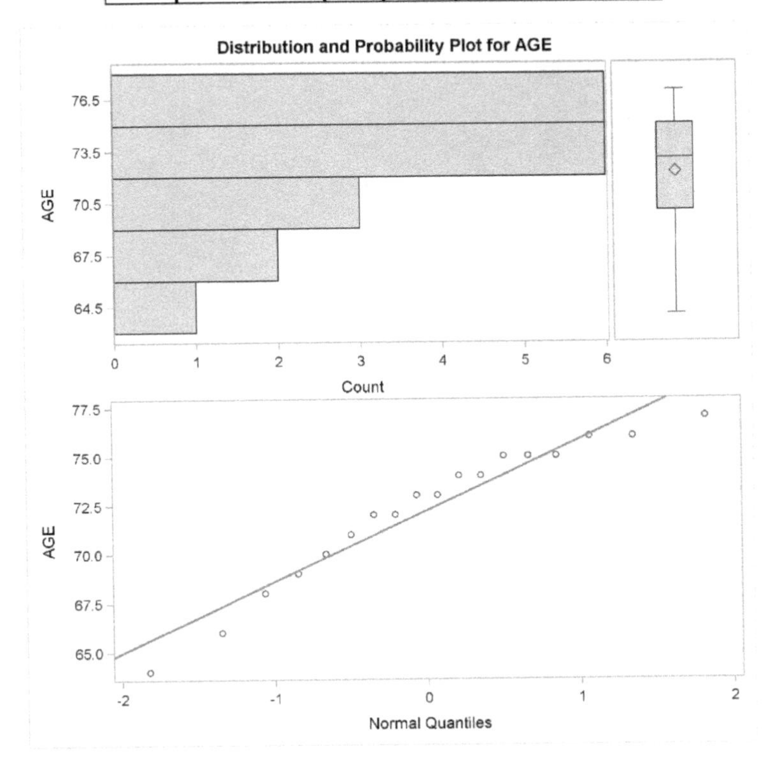

The Stem-Leaf plot of Output A.4.9 reveals a long tail that points in the direction of lower values for AGE (such as 64 and 66). This, of course, is the type of plot that you would expect for a negatively skewed distribution.

Conclusion

Regardless of what other statistical procedures you use, always begin the data analysis process by performing the simple analyses described herein. This will help to ensure that the data and program do not contain obvious errors that, if left unidentified, could lead to incorrect conclusions. Once the data have undergone this initial screening, you may move forward to the more sophisticated procedures described in this text.

Reference

Cozby, P. C., & Bates, S. C. (2012). *Methods in Behavioral Research* (11th Ed.). Toronto, ON: McGraw-Hill.

Appendix A.5: Preparing Scattergrams and Computing Correlations

Introduction: When Are Pearson Correlations Appropriate?

The Pearson product-moment correlation coefficient (symbolized with the lower-case letter r) can be used to assess the nature of the relationship between variables when both are measured on either an interval- or ratio-levels of measurement (e.g., IQ scores). It is further assumed that both variables should include a relatively large number of values. For example, you would not use this statistic if one of the variables could assume only three values.

It would be appropriate to compute a Pearson correlation coefficient to identify the nature of the relationship between SAT critical reading test scores and grade point average (GPA). Critical reading scores are measured on an interval-level and may assume a wide variety of values (i.e., possible scores range from 200 through 800). Grade point ratio is also assessed on an interval level and may also assume a wide variety of values from 0.00 through 4.00.

There are a number of additional assumptions that should be met before computing Pearson correlations between sets of variables (e.g., bivariate normal distribution). These assumptions are listed at the end of this appendix.

Interpreting the Coefficient

To more fully understand the nature of the relationship between the two variables, it is necessary to interpret two characteristics of a Pearson correlation coefficient. First, the **sign of the coefficient** tells you whether there is a positive or negative relationship between variables. A positive correlation indicates that as values for one variable increase, values for the second variable also increase. A **positive correlation** is illustrated in Figure A.5.1 which shows the relationship between SAT critical reading test scores and GPA in a fictitious sample.

Figure A.5.1: A Positive Correlation

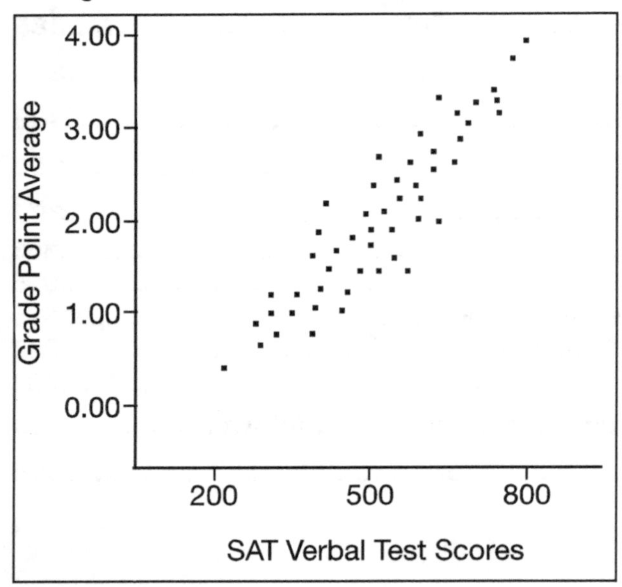

You can see that participants who received low scores on the predictor variable (critical reading) also received low scores on the criterion variable (GPA). At the same time, participants who received high critical reading test scores also received high GPA scores. The two variables are therefore positively correlated.

With a **negative correlation**, as values for one variable increase, values for the second variable decrease. For example, you might expect to see a negative correlation between critical reading test scores and the number of errors that participants make on a vocabulary test (i.e., the students with high critical reading scores tend to make few mistakes, and the students with low test scores tend to make many mistakes). This relationship is illustrated with fictitious data in Figure A.5.2.

Figure A.5.2: A Negative Correlation

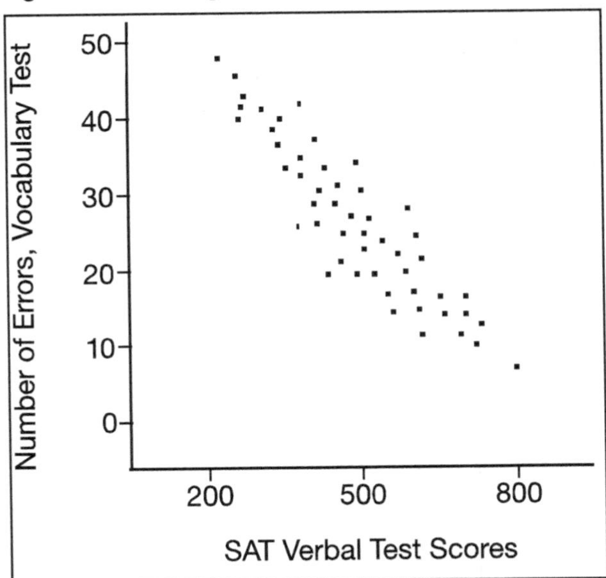

The second characteristic of a correlation coefficient is its **magnitude**: The greater the absolute value of a correlation coefficient, the stronger the relationship between variables. Pearson correlation coefficients range from –1.00 through 0.00 to +1.00; coefficients of 0.00 indicate no relationship between variables. For example, if there were a coefficient of zero between critical reading test scores and GPA, then knowing a person's critical reading test score would tell you nothing about his or her GPA. In contrast, coefficients of –1.00 or +1.00 indicate perfect correlation. If the coefficient between critical reading test scores and GPA were 1.00, it would mean that knowing someone's critical reading test score would allow you to predict his or her GPA with complete accuracy. In the real world, however, critical reading test scores are not that strongly related to GPA, so you would expect the correlation to be considerably less than 1.00.

The following is an approximate guide for interpreting the strength of the association between two variables, based on the absolute value of the coefficient:

```
±1.00 = Perfect correlation
 ±.80 = Strong correlation
 ±.50 = Moderate correlation
 ±.20 = Weak correlation
 ±.00 = No correlation
```

We recommend that you consider the magnitude of correlation coefficients as opposed to whether or not coefficients are statistically significant. This is because significance estimates (i.e., p values) are strongly influenced by sample sizes. For instance, an r value of .15 (weak correlation) would be statistically significant with samples 700+ whereas a coefficient of .50 (moderate correlation) would not be statistically significant with a sample of only 15.

Remember that one considers the *absolute value* when interpreting the size of correlation coefficients. This is to say that a correlation of –.50 is just as strong as a correlation of +.50, a correlation of –.75 is just as strong as a correlation of +.75, and so forth.

Linear versus Nonlinear Relationships

Computing the Pearson correlation coefficient is appropriate only if there is a linear relationship between variables. There is a **linear relationship** between two variables when their scattergram follows the form of a straight line. For example, it is possible to draw a straight line through the center of the scattergram presented in Figure A.5.3, and this straight line fits the pattern of the data fairly well. This means that there is a linear relationship between critical reading test scores and GPA.

Figure A.5.3: A Linear Relationship

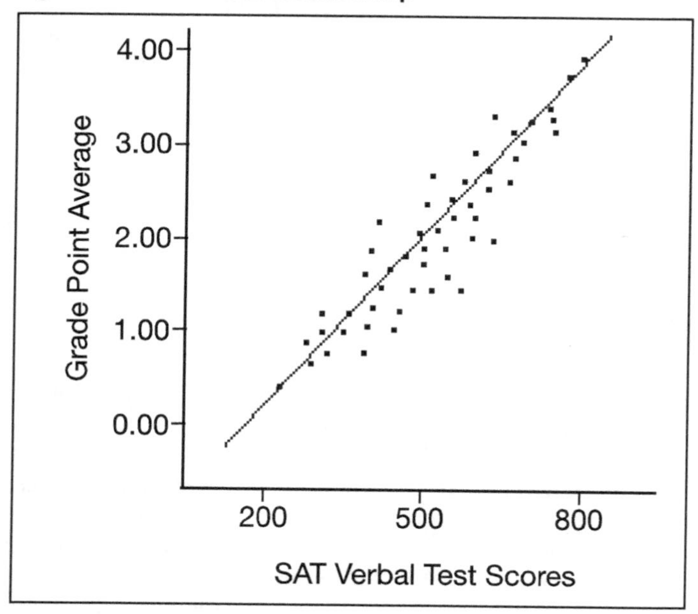

In contrast, there is a **nonlinear relationship** between two variables if their scattergram does not follow the general form of a straight line. For example, imagine that you have constructed a test of creativity and have administered it to a large sample of college students. With this test, higher scores reflect higher levels of creativity. Imagine further that you obtain the critical reading test scores for these students, plot their SAT scores against their creativity scores, and obtain the scattergram presented in Figure A.5.4.

Figure A.5.4: A Nonlinear Relationship

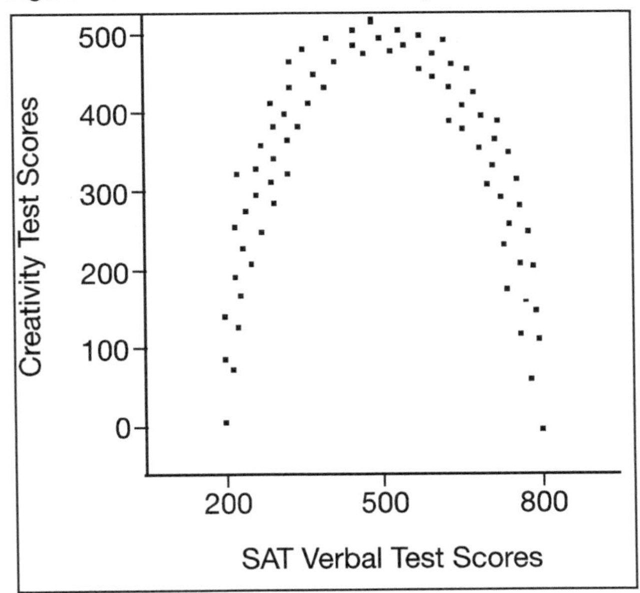

The scattergram in Figure A.5.4 reveals a nonlinear relationship between critical reading test scores and creativity. It shows that:

- students with low critical reading test scores tend to have low creativity scores;
- students with moderate critical reading test scores tend to have high creativity scores;
- students with high critical reading test scores tend to have low creativity scores.

It is not possible to draw a good-fitting straight line through the data points in Figure A.5.4. This is why we say that there is a *nonlinear* (here, *curvilinear*) relationship between critical reading test scores and creativity scores.

When one computes the Pearson correlation coefficient to assess the relationship between variables reflecting a nonlinear relationship, the resulting coefficient *underestimates* the actual strength of association between variables. For example, computing the Pearson correlation between critical reading and creativity scores presented in Figure A.5.4 might result in a coefficient of .10 which would suggest a very weak relationship between variables. From the diagram, however, there is clearly a strong relationship between critical reading test scores and creativity. The figure shows that if you know someone's SAT score, you can accurately predict his or her creativity score.

The conclusion here is that you should verify that there is a linear relationship between variables before computing a Pearson correlation coefficient between those variables. One of the easiest ways of verifying that the relationship is linear is to prepare a scattergram similar to those presented in the preceding figures. Fortunately, this is easily done with the SAS SGPLOT procedure.

Producing Scattergrams with PROC SGPLOT

Here is the general form for requesting a scattergram with the SGPLOT procedure:

```
proc sgplot   data=dataset-name;
   scatter   y=criterion-variable x=predictor-variable ;
run;
```

The variable listed as the "criterion-variable" in the preceding program will be plotted on the vertical axis, and the "predictor-variable" will be plotted on the horizontal axis.

To illustrate this procedure, imagine that you have conducted a study dealing with the *investment model*, a theory of commitment in romantic associations (Rusbult, 1980;). The investment model identifies a number of variables that are believed to influence one's commitment to a romantic association (Le and Agnew, 2003). **Commitment** refers to the person's intention to remain in the relationship. These are some of the variables that are predicted to influence commitment:

Satisfaction
The person's affective response to the relationship.

Investment size
The amount of time and personal resources that the person has put into the relationship.

Alternative value
The attractiveness of alternatives to one's current relationship (e.g., the attractiveness of other prospective romantic partners).

Assume that you have developed a 16-item questionnaire to measure these four variables. The questionnaire is administered to 20 participants who are currently involved in a romantic relationship; participants are asked to complete the instrument while thinking about their current relationship. When they have completed the questionnaire, it is possible to compute four scores for each participant. First, each receives a score on the *commitment scale* based on his or her responses. Higher values on the commitment scale reflect greater commitment to the relationship. Each participant also receives a score on the *satisfaction scale*, where higher scores reflect greater satisfaction with the relationship. Higher scores on the *investment scale* mean that the participant believes that he or she has invested a great deal of time and effort in the relationship. Finally, with the *alternative value scale,* higher scores mean that it would be attractive to the respondent to find a different romantic partner.

Once data have been entered, you can use the SGPLOT procedure to prepare scattergrams for various combinations of variables. The following SAS program inputs some fictitious data and requests that a scattergram be prepared in which commitment scores are plotted against satisfaction scores:

```
 data D1;
   input   #1   @1   COMMITMENT    2.
                @4   SATISFACTION 2.
                @7   INVESTMENT    2.
                @10  ALTERNATIVES 2.    ;
datalines;
20 20 28 21
10 12  5 31
30 33 24 11
 8 10 15 36
22 18 33 16
31 29 33 12
 6 10 12 29
11 12  6 30
25 23 34 12
10  7 14 32
31 36 25  5
 5  4 18 30
31 28 23  6
 4  6 14 29
36 33 29  6
22 21 14 17
15 17 10 25
19 16 16 22
12 14 18 27
24 21 33 16
;
run;

❶     proc sgplot   data=D1;
        scatter y=COMMITMENT x=SATISFACTION
/markerattrs=(symbol=circlefilled);
      run;
```

In the preceding program, scores on the commitment scale are entered in columns 1 to 2, and are given the SAS variable name COMMITMENT. Similarly, scores on the satisfaction scale are entered in columns 4 to 5, and are given the name SATISFACTION; scores on the investment scale appear in columns 7 to 8 and are given the name INVESTMENT; and scores on the alternative value scale appear as the last column of data, and are given the name ALTERNATIVES.

The data for the 20 participants appear in this program; there is one line of data for each participant.

Line ❶ of the program requests the SGPLOT procedure, specifying that the dataset to be analyzed is dataset D1. The SCATTER statement specifies COMMITMENT as the criterion variable and SATISFACTION as the predictor variable for this analysis. The results of this analysis appear in Output A.5.1.

Output A.5.1: Scattergram of Commitment Scores Plotted against Satisfaction Scores

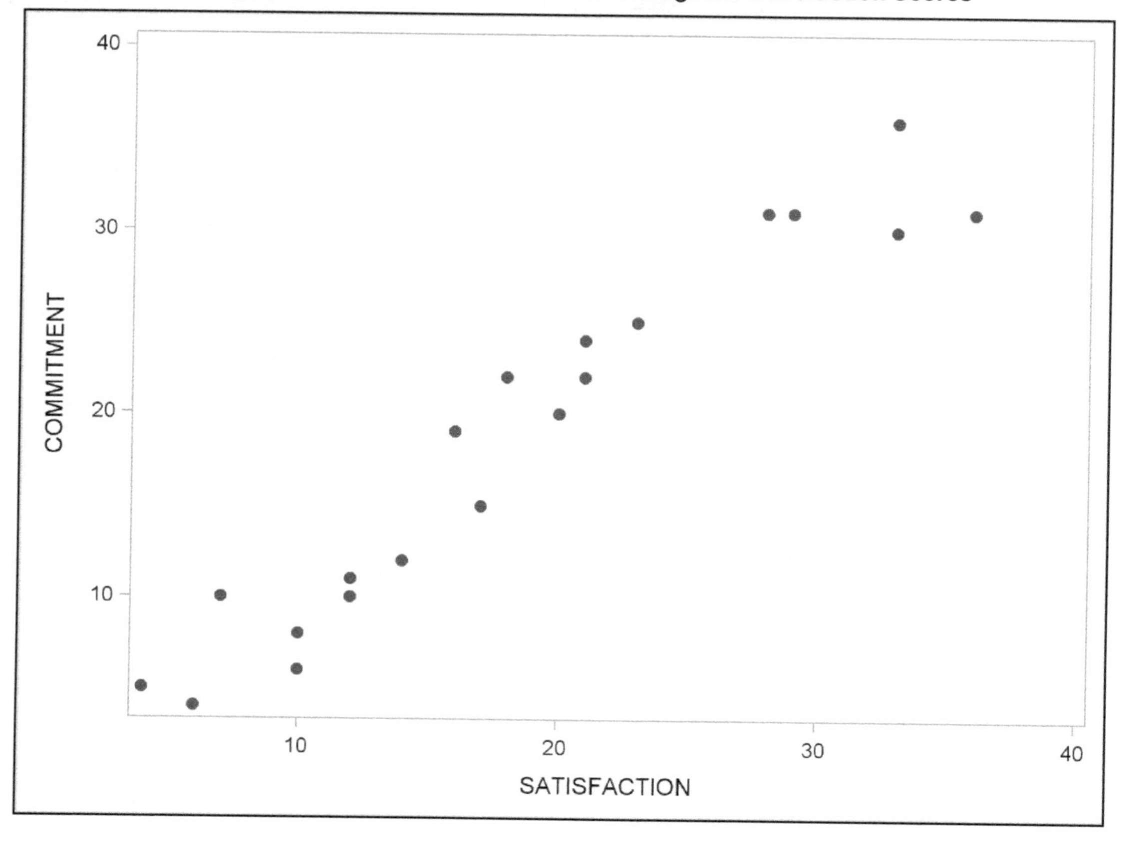

Notice that in this output, the criterion variable (COMMITMENT) is plotted on the vertical axis and the predictor variable (SATISFACTION) is plotted on the horizontal axis. The shape of the scattergram indicates that there is a linear relationship between SATISFACTION and COMMITMENT. This is evident by the fact that it would be possible to draw a relatively good-fitting straight line through the center of the scattergram. Given that the relationship is linear, it seems appropriate to compute the Pearson correlation coefficient for this pair of variables.

The general shape of the scattergram also suggests that there is a fairly strong relationship between variables: Knowing where a participant stands on the SATISFACTION variable allows you to predict with some accuracy where that participant will stand on the COMMITMENT variable. Later, you will compute the correlation coefficient to determine just how strong the relationship is between these two variables.

Output A.5.1 also indicates that the relationship between SATISFACTION and COMMITMENT is *positive* (i.e., large values on SATISFACTION are associated with large values on COMMITMENT and small values on SATISFACTION are associated with small values on COMMITMENT). This makes intuitive sense; you would expect that participants who are highly satisfied with their relationships would also be highly committed to those relationships. To illustrate a negative relationship, you can plot COMMITMENT against ALTERNATIVES. To do this, include the following statements in the preceding program:

```
proc sgplot   data=D1;
         scatter y=COMMITMENT x=ALTERNATIVES/
markerattrs=(symbol=circlefilled);
run;
```

These statements are identical to the earlier statements except that ALTERNATIVES has now been specified as the predictor variable. These statements produce the scattergram presented in Output A.5.2.

Output A.5.2: Scattergram of Commitment Scores Plotted against Alternative Value Scores

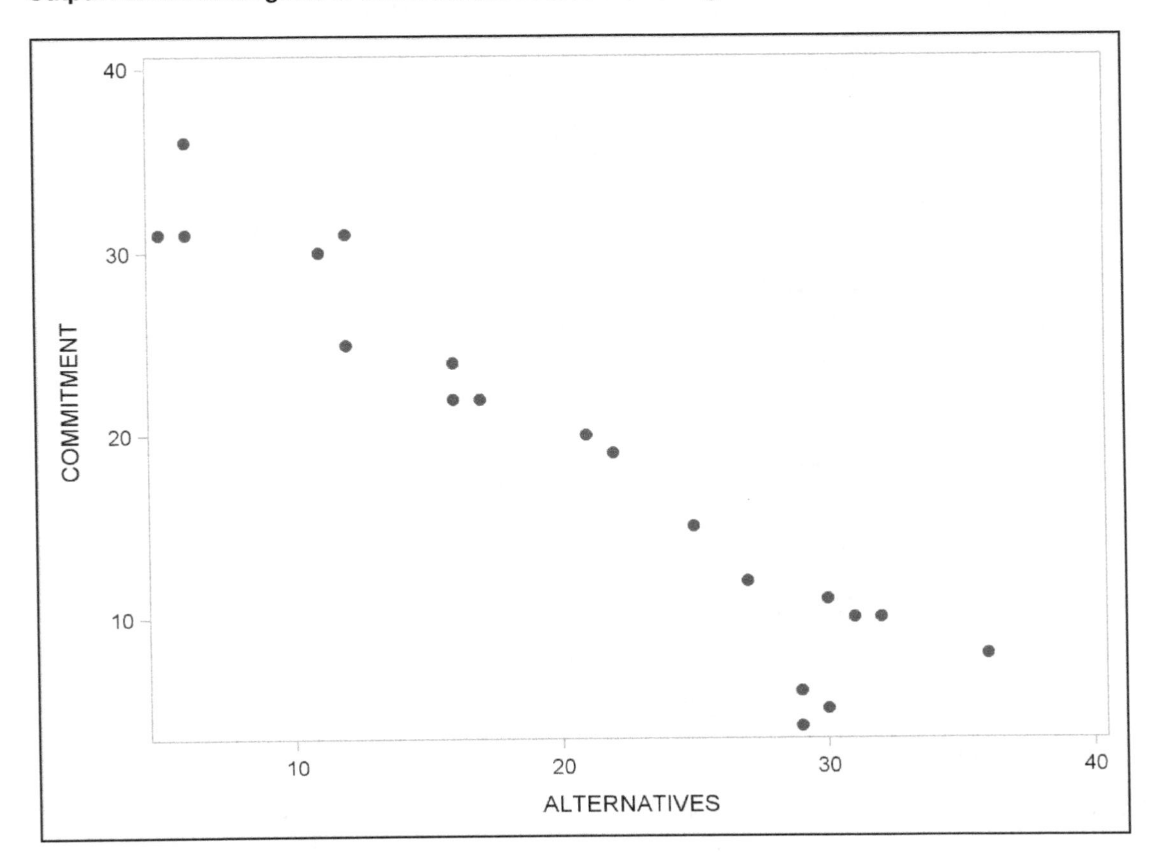

Notice that the relationship between these variables is *negative.* This is what you would expect as it makes intuitive sense that participants who indicate that alternatives to their current romantic partners are attractive would not be overly committed to their current partners. The relationship between ALTERNATIVES and COMMITMENT also appears to be linear. It is therefore appropriate to assess the strength of the relationship between these variables using the Pearson correlation coefficient.

Computing Pearson Correlations with PROC CORR

The CORR procedure offers a number of options regarding what *type* of coefficient will be computed as well as a number of options regarding the *way* they will appear. Some of these options are discussed here.

Computing a Single Correlation Coefficient

In some instances, you may wish to compute the correlation between just two variables. Here is the general form for the statements that will accomplish this:

```
proc corr   data=dataset-name   options;
   var   variable1   variable2;
run;
```

The choice of which variable is "variable1" and which is "variable2" is arbitrary. For a specific example, assume that you want to compute the correlation between commitment and satisfaction. These are the required statements:

```
proc corr   data=D1;
   var COMMITMENT SATISFACTION;
run;
```

This program command results in a single page of output, reproduced here as Output A.5.3:

Output A.5.3: Computing the Pearson Correlation between Commitment and Satisfaction

The CORR Procedure

2 Variables:	COMMITMENT SATISFACTION

Simple Statistics						
Variable	N	Mean	Std Dev	Sum	Minimum	Maximum
COMMITMENT	20	18.60000	10.05459	372.00000	4.00000	36.00000
SATISFACTION	20	18.50000	9.51177	370.00000	4.00000	36.00000

Pearson Correlation Coefficients, N = 20
Prob > |r| under H0: Rho=0

	COMMITMENT	SATISFACTION
COMMITMENT	1.00000	0.96252 <.0001
SATISFACTION	0.96252 <.0001	1.00000

The first part of Output A.5.3 presents simple descriptive statistics (Simple Statistics) for the variables being analyzed. This allows you to verify that everything looks appropriate (e.g., the correct numbers of cases were analyzed, no out-of-range variables). The names of the variables appear below the "Variable" heading, and statistics for the variables appear to the right of the variable names. These descriptive statistics show that 20 participants provided usable data for the COMMITMENT variable, that the mean for COMMITMENT is 18.6 and the standard deviation is 10.05. It is important to review the "Minimum" and "Maximum" columns to verify that no impossible scores appear. With COMMITMENT, the lowest possible score was 4 and the highest possible score was 36. The "Minimum" and "Maximum" columns of Output A.5.3 show that no observed values were out of range (i.e., no evidence of incorrectly entered data). Again, these procedures do not *guarantee* no data entry errors were made, but they are useful for identifying some types of errors. Since the descriptive statistics provide no obvious evidence of entering or programming mistakes, you are now free to review the correlations themselves.

The bottom table of Output A.5.3 reports the correlation coefficients requested in the VAR statement. There are actually four coefficients in the output because your statement requested that SAS compute every possible correlation between COMMITMENT and SATISFACTION. This caused SAS to compute the correlation between COMMITMENT and SATISFACTION, between SATISFACTION and COMMITMENT, between COMMITMENT and COMMITMENT, and between SATISFACTION and SATISFACTION.

The correlation between COMMITMENT and COMMITMENT appears in the upper-left corner of the matrix of coefficients in Output A.5.3. You can see that the correlation between these variables is 1.00; this makes sense, because the correlation of any variable with itself is always equal to 1.00. Similarly, in the lower-right corner, you see that the correlation between SATISFACTION and SATISFACTION is also 1.00.

The coefficient you are actually interested in appears where the column headed COMMITMENT intersects with the row headed SATISFACTION. The top number in the "cell" where this column and row intersect is .96, which is the correlation between COMMITMENT and SATISFACTION (rounded to two decimal places).

Just below the correlation is the p value associated with this coefficient. This is the significance estimate obtained from a test of the null hypothesis that the correlation between COMMITMENT and SATISFACTION is zero in the population. More technically, the p value gives us the probability that you would obtain a coefficient this large (or larger) if the correlation between COMMITMENT and SATISFACTION was zero in the population. For this coefficient, r = .96 and the corresponding p value is less than .01. This means that, given your sample size, there is less than 1 chance in 100 of obtaining a correlation of .96 or larger from this population by chance alone. You may therefore reject the null hypothesis and tentatively conclude that COMMITMENT is significantly associated with SATISFACTION in the population. The alternative hypothesis for this statistical test is that the correlation is not equal to zero in the population. This alternative hypothesis is 2-sided (or 2-tailed) which means that it does not predict whether the correlation coefficient is positive or negative, only that it is not equal to zero.

Determining Sample Size

The size of the sample used in computing the correlation coefficient may appear in one of two places in the output. If all correlations in the analysis were based on the same number of participants, the sample size appears only once in the line above the matrix of correlations. This line appears just below the descriptive statistics. In Output A.5.3 the line says:

Pearson Correlation Coefficients, N = 20

The "N =" portion of this output indicates the sample size. In Output A.5.3, the sample size is 20.

However, if one is requesting correlations between several different pairs of variables, it is possible that certain coefficients will be based on more participants than others due to missing data. In this case, the sample size will be printed for each correlation coefficient. Specifically, the sample size will appear immediately just below the correlation coefficient and its associated significance level (i.e., p value), following this format:

Correlation

P value

N

Computing All Possible Correlations for a Set of Variables

Here is the general form for computing all possible Pearson correlation coefficients for a set of variables:

```
proc corr   data=dataset-name   options;
   var   variable-list;
run;
```

Each variable name in the preceding "variable-list" should be separated by at least one space. For example, assume that you now wish to compute all possible correlations for the variables COMMITMENT, SATISFACTION, INVESTMENT, and ALTERNATIVES. The statements that request these correlations are as follows:

```
proc corr   DATA=D1;
   var COMMITMENT SATISFACTION INVESTMENT ALTERNATIVES;
run;
```

The preceding program produced the output reproduced here as Output A.5.4:

Output A.5.4: Computing All Possible Pearson Correlation Coefficients

The CORR Procedure

4 Variables:	COMMITMENT SATISFACTION INVESTMENT ALTERNATIVES

Simple Statistics						
Variable	N	Mean	Std Dev	Sum	Minimum	Maximum
COMMITMENT	20	18.60000	10.05459	372.00000	4.00000	36.00000
SATISFACTION	20	18.50000	9.51177	370.00000	4.00000	36.00000
INVESTMENT	20	20.20000	9.28836	404.00000	5.00000	34.00000
ALTERNATIVES	20	20.65000	9.78869	413.00000	5.00000	36.00000

Pearson Correlation Coefficients, N = 20 Prob > |r| under H0: Rho=0				
	COMMITMENT	SATISFACTION	INVESTMENT	ALTERNATIVES
COMMITMENT	1.00000	0.96252 <.0001	0.71043 0.0004	-0.95604 <.0001
SATISFACTION	0.96252 <.0001	1.00000	0.61538 0.0039	-0.93355 <.0001
INVESTMENT	0.71043 0.0004	0.61538 0.0039	1.00000	-0.72394 0.0003
ALTERNATIVES	-0.95604 <.0001	-0.93355 <.0001	-0.72394 0.0003	1.00000

You can interpret the correlations and significance values in this output in exactly the same way as with the preceding output. For example, to find the correlation between INVESTMENT and COMMITMENT, you find the cell where the row for INVESTMENT intersects with the column for COMMITMENT. The top number in this cell is .71, which is the Pearson correlation coefficient for these two variables. Just below this coefficient is the p value which is less than .01, meaning that there is less than 1 chance in 100 of obtaining a sample correlation this large if the population correlation is really zero. The observed coefficient is statistically significant.

Notice that the pattern of the correlations supports some of the predictions of the investment model: Commitment is positively related to satisfaction and investment size and is negatively related to alternative value. With respect to magnitude, the correlations range from being moderately strong to very strong. (Remember, however, that these data are fictitious.)

What happens if I omit the VAR statement? It is possible to run PROC CORR without the VAR statement. This causes every possible correlation to be computed between all quantitative variables in the dataset. Use caution in doing this, however. With large datasets, leaving off the VAR statement may result in a very long printout.

Computing Correlations between Subsets of Variables

Using the WITH statement in the SAS program, it is possible to compute correlations between one subset of variables and a second subset of variables. The general form is as follows:

```
proc corr   DATA=dataset-name   options;
   var   variables-that-will-appear-as-columns;
   with  variables-that-will-appear-as-rows;
run;
```

Any number of variables may appear in the VAR statement and any number of variables may also appear in the WITH statement. To illustrate, assume that you want to prepare a matrix of correlation coefficients in which there is one column of coefficients, representing the COMMIT variable and there are three rows of coefficients representing the SATISFACTION, INVESTMENT, and ALTERNATIVES variables. The following statements would create this matrix:

```
proc corr   DATA=D1;
   var  COMMITMENT;
   with SATISFACTION INVESTMENT ALTERNATIVES;
run;
```

Output A.5.5 presents the results generated by this program. Note, surprisingly, that the correlations in this output are identical to those obtained in Output A.5.4, though the Output A.5.5 is more succinct. This is why it is often a good idea to use the WITH statement in conjunction with the VAR statement as this can produce smaller and more manageable printouts than obtained if you use only the VAR statement.

Output A.5.5: Computing Pearson Correlation Coefficients for Subsets of Variables

The CORR Procedure

3 With Variables:	SATISFACTION INVESTMENT ALTERNATIVES
1 Variables:	COMMITMENT

Simple Statistics						
Variable	N	Mean	Std Dev	Sum	Minimum	Maximum
SATISFACTION	20	18.50000	9.51177	370.00000	4.00000	36.00000
INVESTMENT	20	20.20000	9.28836	404.00000	5.00000	34.00000
ALTERNATIVES	20	20.65000	9.78869	413.00000	5.00000	36.00000
COMMITMENT	20	18.60000	10.05459	372.00000	4.00000	36.00000

Pearson Correlation Coefficients, N = 20 Prob > |r| under H0: Rho=0	
	COMMITMENT
SATISFACTION	0.96252 <.0001
INVESTMENT	0.71043 0.0004
ALTERNATIVES	-0.95604 <.0001

Options Used with PROC CORR

The following items are some of the PROC CORR options that you might find especially useful when conducting social science research. Remember that the option names should appear before the semicolon that ends the PROC CORR statement:

ALPHA
: prints Cronbach's alpha which is an estimate of internal consistency of responses for the variables listed in the VAR statement; this is an index of scale reliability. Chapter 3 of this text deals with coefficient alpha in detail.

COV
: prints the Covariance Matrix for variables. This is useful when you need a variance-covariance table instead of a table of correlation coefficients.

KENDALL
: prints Kendall Tau-b Correlation coefficients, a measure of bivariate association between variables assessed at the ordinal level.

NOMISS
: drops from the analysis any observation with missing data on any of the variables listed in the VAR statement. Using this option ensures that all correlations will be based on exactly the same observations and, therefore, on the same *number* of observations.

NOPROB
: prevents printing the p values associated with the correlation coefficients.

RANK
for each variable, reorders the correlations from highest to lowest (in absolute value) and displays them in this order.

SPEARMAN
prints Spearman Correlation Coefficients, which are appropriate for variables measured on an ordinal level.

Appendix: Assumptions Underlying the Pearson Correlation Coefficient

- **Interval or ratio-level measurement.** Both predictor and criterion variables should be measured on the interval- or ratio-levels of measurement.
- **Linearity.** The relationship between the criterion and predictor variables should be linear. This means that the mean criterion scores at each value of the predictor variable should fall on a straight line. The Pearson correlation coefficient is not appropriate for assessing the strength of the relationship between two variables with a curvilinear relationship.
- **Bivariate normal distribution.** The pairs of scores should follow a bivariate normal distribution (i.e., criterion variable scores should form a normal distribution at each value of the predictor variable). Similarly, scores of the predictor variable should form a normal distribution at each value of the criterion variable. When scores represent a bivariate normal distribution, they form an *elliptical scattergram* when plotted (i.e., their scattergram is shaped like a rugby ball [i.e., fat in the middle and tapered at both ends]).

References

Le, B., & Agnew, C. R. (2003). Commitment and its theorized determinants: A meta-analysis of the investment model. *Personal Relationships, 10,* 37–57.

Rusbult, C. E. (1980). Commitment and satisfaction in romantic associations: A test of the investment model. *Journal of Experimental Social Psychology, 16,* 172–186.

Appendix A.6: Simplifying PROC CALIS Programs

Introduction

In this book, we have presented a series of rules to perform path analysis in a detailed way so that you understand also how to perform confirmatory factor analysis and structural equation modeling. This is done to introduce the elements of covariance structure analyses for the PROC CALIS novice.

For those familiar with PROC CALIS, these rules can be reduced. This simplified approach is presented in this appendix using the example presented in Chapter 4 (Path Analysis), More information about PROC CALIS can be found at http://support.sas.com/documentation.

Reviewing the Rules for Performing Path Analysis

Summarized below are the 14 rules first presented in Chapter 4 to produce the figures and develop the SAS program.

RULE 1: Only exogenous variables are supposed to have covariance parameters in the model.

RULE 2: A residual term is identified for each endogenous variable in the model.

RULE 3: Exogenous variables do not require residual terms.

RULE 4: Variance should be estimated for every exogenous variable in the model, including residual terms.

RULE 5: In most cases, covariance should be estimated for every possible pair of manifest exogenous variables; covariance is not estimated for endogenous variables.

RULE 6: For simple recursive models, covariance is generally not estimated for residual terms.

RULE 7: One equation should be created for each endogenous variable, with that variable's name to the left of the equals sign.

RULE 8: Variables that have a direct effect on that endogenous variable are listed to the right of the equals sign.

RULE 9: Exogenous variables, including residual terms, are never listed to the left of the equals sign.

RULE 10: To estimate a path coefficient for a given independent variable, a unique path coefficient name should be created for the path coefficient associated with that independent variable

RULE 11: The last term in each equation should be the residual (disturbance) term for that endogenous variable; this term will have no name for its path coefficient.

RULE 12: To *estimate* a path coefficient parameter, create a name for that parameter.

RULE 13: To *fix* a parameter at a given numerical value, insert that value in place of the parameter's name.

RULE 14: To constrain two or more parameters to be equal, use the same name for those parameters.

PROC CALIS simplifies these steps for you so that many of them are handled automatically as listed here:

RULE 1: PROC CALIS automatically ensures that only exogenous variables will have covariance parameters in the model.

RULE 4: PROC CALIS automatically estimates variance for every exogenous variable in the model, including residual terms.

RULE 6: By default, PROC CALIS does not estimate covariance for residual terms for simple recursive models.

RULE 10: PROC CALIS can automatically name coefficients.

RULE 12: To estimate a path coefficient parameter, specify the variable in a LINEQS equation. Most commonly needed parameters are automatically estimated in PROC CALIS unless you specify otherwise.

Simplifying the SAS Program

Below is the entire PROC CALIS program, including the DATA step used to analyze the model presented in Figure 4.11:

```
data D1(type=corr) ;
   input _type_ $ _name_ $ V1-V6 ;
      label
         V1 = COMMITMENT
         V2 = SATISFACTION
         V3 = REWARDS
         V4 = COSTS
         V5 = INVESTMENTS
         V6 = ALTERNATIVES   ;
datalines;
n      .    240      240      240      240      240      240
std    .   2.3192   1.7744   1.2525   1.4086   1.5575   1.8701
corr  V1   1.0000    .        .        .        .        .
corr  V2    .6742   1.0000    .        .        .        .
corr  V3    .5501    .6721   1.0000    .        .        .
corr  V4   -.3499   -.5717   -.4405   1.0000    .        .
corr  V5    .6444    .5234    .5346   -.1854   1.0000    .
corr  V6   -.6929   -.4952   -.4061    .3525   -.3934   1.0000
;
run;
```

```
❶  proc calis  modification ;
❷     lineqs
        V1 = PV1V2 V2 + PV1V5 V5 + PV1V6 V6 + E1,
        V2 = PV2V3 V3 + PV2V4 V4            + E2;
❸     variance
        E1 = VARE1,
        E2 = VARE2,
        V3 = VARV3,
        V4 = VARV4,
        V5 = VARV5,
        V6 = VARV6;
❹     cov
        V3 V4 = CV3V4,
        V3 V5 = CV3V5,
```

```
          V3 V6 = CV3V6,
          V4 V5 = CV4V5,
          V4 V6 = CV4V6,
          V5 V6 = CV5V6;
❺    var  V1 V2 V3 V4 V5 V6 ;
     run;
```

All lines appearing before ❶ of the preceding program constitute the DATA step, here entered as a correlation matrix. Line ❶ includes the PROC CALIS statement that initiates this procedure will be performed. Equations included in the LINEQS statement (beginning line ❷) identify the predicted relationships among the model's variables (i.e., which variables are assumed to predict the model's endogenous variables).

The VARIANCE statement, beginning at line ❸, identifies the variances to be estimated, while the subsequent COV statement (beginning line ❹) indicates which covariances are to be estimated. Finally, the VAR statement on line ❺ indicates which variables are to be analyzed. Each of these sections is discussed in detail below.

Because PROC CALIS handles many of the requirements automatically, the code can be greatly simplified to this:

```
     data D1(type=corr) ;
        input _type_ $ _name_ $ V1-V6 ;
           label
              V1 = COMMITMENT
              V2 = SATISFACTION
              V3 = REWARDS
              V4 = COSTS
              V5 = INVESTMENTS
              V6 = ALTERNATIVES   ;
     datalines;
     n      .     240      240      240      240      240      240
     std    .   2.3192   1.7744   1.2525   1.4086   1.5575   1.8701
     corr  V1   1.0000    .        .        .        .        .
     corr  V2    .6742   1.0000    .        .        .        .
     corr  V3    .5501    .6721   1.0000    .        .        .
     corr  V4   -.3499   -.5717   -.4405   1.0000    .        .
     corr  V5    .6444    .5234    .5346   -.1854   1.0000    .
     corr  V6   -.6929   -.4952   -.4061    .3525   -.3934   1.0000
     ;
     run;

❶    proc calis  modification ;
❷        lineqs
           V1 = PV1V2 V2 + PV1V5 V5 + PV1V6 V6 + E1,
           V2 = PV2V3 V3 + PV2V4 V4            + E2;
❸     var  V1 V2 V3 V4 V5 V6 ;
     run;
```

All lines appearing before ❶ of the preceding program constitute the DATA step, here entered as a correlation matrix. Line ❶ includes the PROC CALIS statement that initiates this procedure will be performed. Equations included in the LINEQS statement (beginning line ❷) identify the predicted relationships between the model's variables (i.e., which variables are assumed to predict the model's endogenous variables). Finally, the VAR statement on line ❸ indicates which variables are to be analyzed.

Appendix B: Datasets

Dataset from Chapter 1: Principal Component Analysis

Fictitious data from the Prosocial Orientation Inventory:

```
data D1;
   input   #1  @1  (V1-V6)   (1.) ;
datalines;
556754
567343
777222
665243
666665
353324
767153
666656
334333
567232
445332
555232
546264
436663
265454
757774
635171
667777
657375
545554
557231
666222
656111
464555
465771
142441
675334
665131
666443
244342
464452
654665
775221
657333
666664
545333
353434
666676
```

```
667461
544444
666443
676556
676444
676222
545111
777443
566443
767151
455323
455544
;
```

Datasets from Chapter 2: Exploratory Factor Analysis

Fictitious data from the investment model study:

```
data D1;
   input   #1  @1  (V1-V6)    (1.)
               @8  (COMMITMENT)   (2.) ;
datalines;
776122 24
776111 28
111425  4
222633 24
551666  4
666524  4
633112 24
766212 23
454444 17
111332  8
444343 21
556212 20
543332 11
677222 27
666234 15
557322  6
555221 18
544111 17
424232 28
445435 13
767232 13
444422  8
653211 15
555323 16
655123 17
221121 11
666421 20
454332  9
655321 25
444332 12
433222 27
777314 16
555212  7
443221  4
243334 11
666111 15
423412 11
555222 18
223332  8
333335 13
445433 22
```

```
444323 12
455556 22
444112 17
334445 12
444321 16
655222 15
433344 15
557332 20
655222 13
;
```

Intercorrelations between 25 scales from the Job Search Skills Questionnaire, decimals omitted; standard deviations appear in parentheses along the diagonal, N = 220:

```
VALUES      2.32
ABILITY      .50 1.77
ASSESS       .37  .40 1.25
STRATEGY     .32  .45  .62 1.41
EXPERIENCE   .39  .41  .46  .49 1.56
ORGCHAR      .58  .51  .31  .40  .48 1.87
RESOCCUP     .37  .40  .53  .57  .62  .47 2.14
RESEMPLOY    .39  .38  .54  .58  .57  .46  .69 1.76
GOALS        .53  .45  .32  .34  .44  .56  .40  .41 1.86
BARRIER      .40  .40  .41  .48  .50  .50  .49  .54  .53 2.04
MOTIVATED    .49  .46  .39  .38  .41  .53  .44  .43  .58  .51 1.73
RESUMES      .32  .40  .38  .53  .50  .43  .50  .53  .40  .44  .46 1.59
RECOMMEND    .42  .39  .48  .52  .52  .45  .51  .62  .42  .53  .45  .64 1.37
DIRECT       .30  .34  .43  .55  .48  .35  .50  .60  .33  .45  .37  .70  .65 1.99
APPLICAT     .37  .35  .41  .43  .46  .34  .44  .52  .33  .47  .42  .49  .59  .49 2.11
IDEMPLOY     .35  .36  .49  .62  .59  .42  .75  .72  .40  .53  .42  .58  .60  .67  .48 2.05
CARDEVEL     .27  .24  .51  .51  .59  .31  .50  .49  .30  .44  .33  .49  .51  .55  .42  .59 1.61
AGENCY       .28  .32  .45  .53  .42  .34  .46  .61  .28  .36  .32  .53  .49  .59  .44  .56  .55 1.84
FAIRS        .20  .27  .42  .53  .47  .26  .51  .48  .27  .43  .30  .50  .53  .55  .44  .59  .63  .52 2.02
ADVERT       .37  .42  .46  .62  .53  .43  .58  .67  .39  .54  .41  .60  .61  .65  .54  .70  .52  .62  .58 1.77
COUNSEL      .24  .30  .53  .56  .49  .33  .56  .59  .33  .48  .34  .51  .55  .61  .41  .64  .70  .62  .65  .67 1.48
UNADVERT     .27  .32  .45  .54  .45  .33  .52  .62  .37  .49  .44  .45  .54  .52  .52  .61  .51  .58  .49  .68  .65 1.68
NETWORK      .34  .36  .46  .57  .53  .42  .53  .55  .42  .48  .47  .49  .50  .53  .39  .59  .52  .48  .46  .66  .56  .62 1.39
INTERVIEW    .31  .38  .33  .52  .43  .42  .50  .59  .38  .48  .39  .51  .63  .54  .59  .55  .42  .48  .46  .62  .52  .62  .55 2.17
SALARY       .28  .35  .43  .52  .45  .37  .64  .67  .36  .50  .39  .52  .56  .52  .44  .68  .49  .56  .52  .64  .62  .63  .54  .64 1.73
```

Dataset from Chapter 3: Assessing Scale Reliability with Coefficient Alpha

The dataset described in Chapter 3 is identical to the dataset from Chapter 1, which appears earlier.

Appendix C: Critical Values for the Chi-Square Distribution

Chi Square Distribution Table

d.f.	$\chi^2_{.25}$	$\chi^2_{.10}$	$\chi^2_{.05}$	$\chi^2_{.025}$	$\chi^2_{.010}$	$\chi^2_{.005}$	$\chi^2_{.001}$
1	1.32	2.71	3.84	5.02	6.63	7.88	10.8
2	2.77	4.61	5.99	7.38	9.21	10.6	13.8
3	4.11	6.25	7.81	9.35	11.3	12.8	16.3
4	5.39	7.78	9.49	11.1	13.3	14.9	18.5
5	6.63	9.24	11.1	12.8	15.1	16.7	20.5
6	7.84	10.6	12.6	14.4	16.8	18.5	22.5
7	9.04	12.0	14.1	16.0	18.5	20.3	24.3
8	10.2	13.4	15.5	17.5	20.1	22.0	26.1
9	11.4	14.7	16.9	19.0	21.7	23.6	27.9
10	12.5	16.0	18.3	20.5	23.2	25.2	29.6
11	13.7	17.3	19.7	21.9	24.7	26.8	31.3
12	14.8	18.5	21.0	23.3	26.2	28.3	32.9
13	16.0	19.8	22.4	24.7	27.7	29.8	34.5
14	17.1	21.1	23.7	26.1	29.1	31.3	36.1
15	18.2	22.3	25.0	27.5	30.6	32.8	37.7
16	19.4	23.5	26.3	28.8	32.0	34.3	39.3
17	20.5	24.8	27.6	30.2	33.4	35.7	40.8
18	21.6	26.0	28.9	31.5	34.8	37.2	42.3
19	22.7	27.2	30.1	32.9	36.2	38.6	32.8
20	23.8	28.4	31.4	34.2	37.6	40.0	45.3
21	24.9	29.6	32.7	35.5	38.9	41.4	46.8
22	26.0	30.8	33.9	36.8	40.3	42.8	48.3
23	27.1	32.0	35.2	38.1	41.6	44.2	49.7
24	28.2	33.2	36.4	39.4	32.0	45.6	51.2
25	29.3	34.4	37.7	40.6	44.3	46.9	52.6
26	30.4	35.6	38.9	41.9	45.6	48.3	54.1
27	31.5	36.7	40.1	43.2	47.0	49.6	55.5
28	32.6	37.9	41.3	44.5	48.3	51.0	56.9
29	33.7	39.1	42.6	45.7	49.6	52.3	58.3
30	34.8	40.3	43.8	47.0	50.9	53.7	59.7
40	45.6	51.8	55.8	59.3	63.7	66.8	73.4
50	56.3	63.2	67.5	71.4	76.2	79.5	86.7
60	67.0	74.4	79.1	83.3	88.4	92.0	99.6
70	77.6	85.5	90.5	95.0	100	104	112
80	88.1	96.6	102	107	112	116	125
90	98.6	108	113	118	124	128	137
100	109	118	124	130	136	140	149

Table from Ronald J. Wonnacott and Thomas H. Wonnacott, *Statistics: Discovering Its Power*, New York: John Wiley and Sons, 1982, p.352.

Index

A

B

C

N

O

P

T

U

V

W

CPSIA information can be obtained
at www.ICGtesting.com
Printed in the USA
BVHW01s1328170818
524177BV00055B/61/P